AF356156

Integrated Computer Technologies in Mechanical Engineering – Synergetic Engineering II

Integrated Computer Technologies in Mechanical Engineering – Synergetic Engineering II

Editors

Mykola Nechyporuk
Volodymyr Pavlikov
Dmytro Krytskyi

Basel • Beijing • Wuhan • Barcelona • Belgrade • Novi Sad • Cluj • Manchester

Editors

Mykola Nechyporuk	Volodymyr Pavlikov	Dmytro Krytskyi
Cars and Transport	Aerospace Radioelectronic	Information Technology
Infrastructure	Systems	Design
National Aerospace	National Aerospace	National Aerospace
University Kharkiv Aviation	University Kharkiv Aviation	University Kharkiv Aviation
Institute	Institute	Institute
Kharkiv	Kharkiv	Kharkiv
Ukraine	Ukraine	Ukraine

Editorial Office
MDPI
St. Alban-Anlage 66
4052 Basel, Switzerland

This is a reprint of articles from the Special Issue published online in the open access journal *Computation* (ISSN 2079-3197) (available at: www.mdpi.com/journal/computation/special_issues/JNH07929E2).

For citation purposes, cite each article independently as indicated on the article page online and as indicated below:

Lastname, A.A.; Lastname, B.B. Article Title. *Journal Name* **Year**, *Volume Number*, Page Range.

ISBN 978-3-7258-0832-8 (Hbk)
ISBN 978-3-7258-0831-1 (PDF)
doi.org/10.3390/books978-3-7258-0831-1

Contents

About the Editors

Mykola Nechyporuk

Mykola Nechyporuk was born on November 10, 1952, in a village in the Malevo Rivne region. In 1973, he came to the Kharkov Aviation Institute at the Radio Engineering Faculty to study the design and manufacture of radio equipment. After graduation in 1979, he worked as an engineer at the Kharkov Aviation Institute. In the same year, he became the director of the KhAI campus.

Since 1998, M. Nechyporuk has been vice-rector for administrative and financial activities at the KhAI. From 1998 to 2010, he worked as a deputy of the Kiev district council of Kharkiv. From 2010 to 2015, he was a deputy of the Kharkiv City Council. From 2004 to 2018, he served as vice-rector for scientific and pedagogical work at the National Aerospace University, Kharkiv Aviation Institute. In his practice, he has conducted a number of scientific studies: problems of integrated technologies for the assembly and utilization of aircraft and other vehicles; technologies for the implementation of protective coatings for aircraft and automotive vehicles; ontology modeling; and ontological decision support systems for the choice of impulse devices. He has developed a system for supporting decision-making in the activities of a university, on the basis of which, for the first time, a computer-based decision support system was introduced among universities in Ukraine.

He is the rector of the National Aerospace University, Kharkiv Aviation Institute, 2018–2023. He has more than a hundred scientific papers, including monographs, textbooks, publications in foreign journals, and publications in world scientometric databases. His achievements include the following: Doctor of Technical Sciences (2012), Professor (2011), Mark of Excellence in Education of Ukraine (2002), Honored Educator of Ukraine (2005), and being awarded with the Diploma of the Verkhovna Rada of Ukraine for merits to the Ukrainian people (2013).

Volodymyr Pavlikov

Vladimir Pavlikov was the Dean of the Radio Electronics, Computer Systems, and Info-Communications Faculty (April 2017–September 2018). And he was head of the department of aircraft radio-electronic systems design (April 2015–March 2017). He has received an academic degree, a Doctor of Technical Sciences (since January 2014). He has a scientific rank of senior researcher (since April 2015).

In 2014, he was awarded the Prize of the Verkhovna Rada of Ukraine for young scientists for a series of scientific works that developed the theory of the synthesis of wideband radiometric systems. In 2015, he won the First Prize (European Microwave Association). In the period 2013–2018, he was a scholar of the Cabinet of Ministers of Ukraine for young scientists. He is a reviewer of the following scientific journals: *IEEE Access*, *Physical Bases of Instrumentation*, and *Radioelectronic and Computer Systems*.

In the period 2016–2018, he was the head of the scientific study titled "New principles of processing signals of own radio-thermal radiation of objects of different physical nature and technologies for their implementation". This project is one of 79 that were selected from more than 400 works recognized as important by the Ministry of Education and Science of Ukraine and assessed by Ukrainian and European scholars for relevance.

From 2019 to 2021, he was the head of the scientific study titled "The development of the theory of ultra-wideband active aperture synthesis systems for high-precision remote sensing from high-speed aerospace platforms".

Dmytro Krytskyi

Dmitriy Kritskiy has a Ph.D. in project and program management. In 2002–2008, he received his specialist degree in information technology from the National Aerospace University, Kharkiv Aviation Institute. He is the Dean of the Faculty of Aircraft Engineering. He was Associate Professor at the National Aerospace University Kharkiv Aviation Institute (2018–2021) and Senior Lecturer at the National Aerospace University Kharkiv Aviation Institute (2015–2017). He was an assistant at the National Aerospace University Kharkiv Aviation Institute in 2014 and a Ph.D. student at the National Aerospace University Kharkiv Aviation Institute (2010–2013). He was a junior researcher at the National Aerospace University Kharkiv Aviation Institute (2008–2009). His current position is as Dean of the Aircraft Building Faculty and lecturer on decision-making theory, information system design, digital information protection, and virtual reality technology. His research interests include developing software for unmanned aerial vehicles (UAVs) and automatic control systems for multi-copter group fights. He has 100 peer-reviewed papers and book chapters, participated in nine research projects, four of which he headed, and has 15 patents.

Preface

The best works were selected after the conference titled "Integrated Computer Technologies in Mechanical Engineering"–Synergetic Engineering (ICTM).

The conference provided technical exchange between the scientific community in the form of keynote addresses, panel discussions, and a special session. In addition, participants were treated to a series of techniques that facilitated collaboration among fellow researchers. ICTM'2022 received 137 applications from different countries.

All this gives us a lot of valuable information and will greatly benefit the exchange of experience between scientists in the field of modeling and simulation. The organizers of ICTM 2022 have made great efforts to ensure the success of this conference. We would hereby like to thank all members of the ICTM 2022 Advisory Committee for their guidance and advice; members of the program committee and organizing committee; reviewers for their efforts in reviewing and collecting articles; and all authors for their contributions to creating a unified intellectual environment for solving current scientific problems.

We express our gratitude to the MDPI publishing house and the editors of the *Computation* journal for the opportunity to publish the best research.

Mykola Nechyporuk, Volodymyr Pavlikov, and Dmytro Krytskyi
Editors

computation

Article

Agile Software Development Lifecycle and Containerization Technology for CubeSat Command and Data Handling Module Implementation

Oleksandr Liubimov [1,2,*], Ihor Turkin [2], Vladimir Pavlikov [3] and Lina Volobuyeva [2]

1   Ektos-Ukraine LLC, 1 Academika Proskury Str., 61070 Kharkiv, Ukraine
2   Department of Software Engineering, National Aerospace University "Kharkiv Aviation Institute", 17 Chkalova Str., 61070 Kharkiv, Ukraine; i.turkin@khai.edu (I.T.); l.volobuieva@khai.edu (L.V.)
3   Department of Aerospace Radio-Electronic Systems, National Aerospace University "Kharkiv Aviation Institute", 17 Chkalova Str., 61070 Kharkiv, Ukraine; v.pavlikov@khai.edu
*   Correspondence: oleksandr.liubimov@gmail.com

Abstract: As a subclass of nanosatellites, CubeSats have changed the game's rules in the scientific research industry and the development of new space technologies. The main success factors are their cost effectiveness, relative ease of production, and predictable life cycle. CubeSats are very important for training future engineers: bachelor's and master's students of universities. At the same time, using CubeSats is a cost-effective method of nearest space exploration and scientific work. However, many issues are related to efficient time-limited development, software and system-level quality assurance, maintenance, and software reuse. In order to increase the flexibility and reduce the complexity of CubeSat projects, this article proposes a "hybrid" development model that combines the strengths of two approaches: the agile-a-like model for software and the waterfall model for hardware. The paper proposes a new computing platform solution, "Falco SBC/CDHM", based on Microchip (Atmel) ATSAMV71Q21 with improved performance. This type of platform emphasizes low-cost space hardware that can compete with space-grade platforms. The paper substantiates the architecture of onboard software based on microservices and containerization to break down complex software into relatively simple components that undergraduates and graduates can handle within their Master's studies, and postgraduates can use for scientific space projects. The checking of the effectiveness of the microservice architecture and the new proposed platform was carried out experimentally, involving the time spent on executing three typical algorithms of different algorithmic complexities. Algorithms were implemented using native C (Bare-metal) and WASM3 on FreeRTOS containers on two platforms, and performance was measured on both "Falco" and "Pi Pico" by Raspberry. The experiment confirmed the feasibility of the complex application of the "hybrid" development model and microservices and container-based architecture. The proposed approach makes it possible to develop complex software in teams of inexperienced students, minimize risks, provide reusability, and thus increase the attractiveness of CubeSat student projects.

Keywords: CubeSat; nanosatellite; COTS; CDHM; OBC; containerization; microservices architecture; agile; on-board interpreter; WASM3; software; WebAssembly; FreeRTOS; Falco; Raspberry Pi

Citation: Liubimov, O.; Turkin, I.; Pavlikov, V.; Volobuyeva, L. Agile Software Development Lifecycle and Containerization Technology for CubeSat Command and Data Handling Module Implementation. *Computation* **2023**, *11*, 182. https://doi.org/10.3390/computation11090182

Academic Editor: Demos T. Tsahalis

Received: 29 June 2023
Revised: 27 August 2023
Accepted: 9 September 2023
Published: 14 September 2023

1. Introduction

CubeSat is a type of nanosatellite that began its commercial and research in 1999 when the CubeSat Design Specification was started in California [1]. From the beginning, CubeSats were invented as technology for research and education. Since 1999, this form factor and the paradigm of nanosatellites have become popular in both commercial and military industries [2], as well as in academia and in particular astronomy [3]. Academic availability and access to such technology allowed student and science teams to gain access to research on space and the Earth from low earth orbit (hereafter LEO). A typical CubeSat is a nanosatellite ranging in size from 1U–10 × 10 × 10 cm up to 12U when several 1U units can be assembled together (stacked or placed next to each other). Usually, CubeSats are delivered to orbit as "parasitic load", i.e., a secondary load. This is what makes its delivery to LEO very cheap, in comparison to the dedicated satellite launch. The typical programs for these parasitic launches of CubeSats are NASA's CubeSat Launch Initiative (CLI) [4] and the European Space Agency (ESA) program called Fly Your Satellite (FYS) [5]. The method for the local "from the rocket" launch of satellites to orbit is a mechanically armed spring-based ejection, using particular load "dispensers", for example, P-POD [6]. The key reasons for the high popularity [7], as well as the prosperous future [8], of the CubeSats standard and approach are the low cost of such satellites [9,10]; the relatively short time of their construction and testing (which made it possible to construct, program, test, and launch a satellite during master's degree students studies); and standardization. Standardization in its turn allowed one to reuse both individual parts of satellites and ground stations to receive telemetry and control satellites at LEO. The typical tasks of CubeSats include three types: the remote sensing of the Earth or other space objects, communication infrastructure (especially for CubeSats constellations, examples of OneWeb and StarLink), and the research of problems and tasks of re-entry into the Earth's atmosphere. Looking at the statistics of launches and orbital deliveries for the past few years [7,11,12], we can see rapid growth in popularity and further development of this area of space technology.

It is easy to see that almost 500 CubeSat satellites were launched in 2021 and 2022. Unfortunately, even with 20 years of launch experience and platform flight heritage, a relatively large number of launches remain unsuccessful. Looking at Figure 1, we can see that approximately 18.5% of satellites in 2022 were completely lost, and only 2.9% of satellites fully completed their task (full launch mission). Numerous scientific and commercial teams worldwide strive to minimize the probability of either a full or partial satellite failure and maximize the percentage of satellites that accomplish their programmed mission. Such a need utilizes the following areas: general reliability, hardware and software design and development processes and techniques, hardware and software reuse, and verification and validation methods. As CubeSat design and development is a popular aerospace-related academic scientific project, many teams worldwide are trying to develop an entire satellite from scratch. This, in its turn, leads to the long duration, comprehensive planning, and execution of the project, and it establishes a very high bar for software and hardware development skills, which are not typical for academic people. In other words, teams are starting from scratch, trying to plan from the very beginning to the very end of the project, and running rigid but not yet mature software and hardware development processes. Looking into statistical data, it is clear that software development takes up roughly 2/3 of the time needed to complete a whole CubeSat project. The software development part of a typical CubeSat project always ends up being very complex and time-consuming. It involves defining, designing, developing, and testing (verification and validation, commissioning, and support after launch) the software. Even with a lot of earlier developed software reuse and open-source software use, it is still a significant challenge for the development teams to use the software properly and to obtain solid flight-ready software promptly. This article will provide an alternative approach to using a modern software development lifecycle, i.e., agile-like, and proper technology for the implementation, deployment, and use of the software. After the definition of a new development approach and its technical background is given, the authors make a top-level design of a CubeSat software structure in a newly introduced paradigm. To ensure that the proposed method is sound, the authors carried out a performance check on the selected variety of classical computer science algorithms.

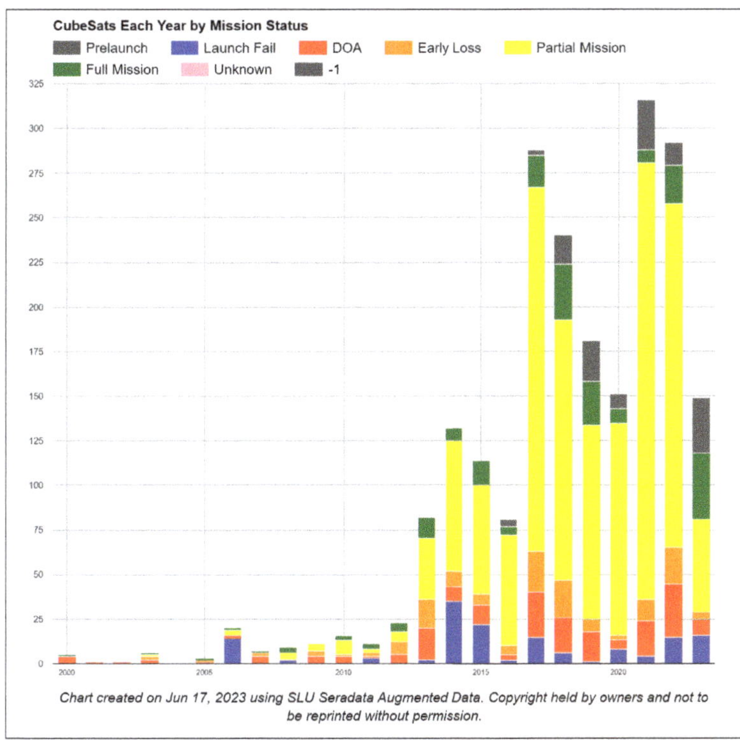

Figure 1. CubeSats launch statistics (annual by mission status).

1.1. How Is the Software Development Complexity and Processes Addressed by the Industry?

If we look into the typical ways of addressing the design and development of complex software, it is clear that the majority of techniques are divided into two major parts:

- Technical design and development methods;
- Improvements in the areas of communication, planning, and processes around the development process and development team.

The technical methods include but are not limited to techniques like decomposition, minimalism adoption, code from data separation, proper abstraction identifications, code reuse, etc., and the process and communication part is normally represented by modern software development life cycles, people and project management principles, and the so-called conscious collaboration paradigm. Looking into NASA's own recommendations [13], it is very clearly stated that the first-time development teams shall: "keep it simple" and "use familiar components", and they "do not design to the limits".

1.2. CubeSat Software State-of-the-Art

The whole idea of making a CubeSat cheap and easy to launch lies in the domain of using commercial-off-the-shelf (COTS) or modifiable-off-the-shelf (MOTS) components in both software and hardware parts of a typical development project. The use of COTS components opens up a wide variety of open-source and proprietary software to be used [14,15]. However, the use of the COTS software creates a huge number of troubles by way of the developing and proper testing of the CubeSat on-board software. Why? Well, many of the software parts are rather provided as the sample or *"take it on your own risk"* and thus simply do not fulfill any reasonable metrics for production-ready software. This means that an increasing number of development teams are leaning towards the very comprehensive testing of the final assembled flight software, simply to ensure its quality and readiness to go to space. The majority of the research is focused on a very thorough verification & validation (V&V) approach [16], using complex data validation models and techniques [17]. Other research papers propose software in the loop (SIL) and/or hardware in the loop (HIL) simulation [18]. Various efforts are also made in the field of failure emulation mechanisms (FEM) [19], as well as the introduction of different fault injection platforms (FIP) [20]. All these methods are very typical for the "waterfall' software development lifecycle and overall project management, as well as for rigidly and thoroughly planned projects. The rigidity and thoroughness are truly nontypical for academia and research projects where the mindset of "greenfield exploration" is commonly used.

1.3. Why the CubeSat Software Is Complex to Develop?

When we look into what actually makes CubeSats development a complex and challenging task, we can see a few major factors. The first and widely underestimated factor is the cyclomatic complexity of making CubeSat software. It is rather easy to see that even on the first order of a functional breakdown, CubeSat software demonstrates a complex set of sub-systems and their functions. The second factor is the skillset of the undergraduate and/or postgraduate students who did not have the option to obtain the industry practices and did not make their own so-called rule of 10,000 h of experience. It leads to a bunch of issues related to both architectural and implementation errors, and low awareness of the more effective tools and approaches for making software solid and robust. Last but not least, a major factor lies in the attempt of "blind" reuse of the open-source software (OSS). The push towards this reuse is normally given by the following factors: other team successes, strict and tight project deadlines, and focusing on the payload rather than on a whole CubeSat system. Such reuse leads to the low quality and maturity of the source code, and a lack of time to properly overview the entire code being developed. At the same time, CubeSat systems, including ground stations, communication signals, and CubeSat spacecraft, are subject to various cyberattacks, the classification of which is given by the standard ISO/IEC 15408 [21]. There are various papers that analyze [22] and propose methods for analyzing CubeSat security threats and solving those, for example, based on the analysis of attack trees [23].

1.4. What Programming Language and Operation System Are Used?

Without a doubt, the "C" programming language is the number one choice and the de-facto market standard for embedded systems and thus CubeSat's software development [24]. The following factors contribute to the popularity of "C":

- It has minimalistic overhead, which is a must for the low power systems;
- It allows one to use both COTS and a truly proprietary hardware (processor) platform;
- Typical CubeSat FSW requires a lot of near-hardware programming, where "C" is very powerful and effective;
- It requires less time to start development (in comparison to object-oriented languages such as C++ or Rust).

Therefore, in this article, the authors concentrate on the use of the "C" language for the experiment. With the experimental work, the authors will later compare the performance of the pure "C" code and "C"-written containers for the WASM3 containers engine. The term "Native C", is introduced later in the article, means the C-language written software that is compiled by a dedicated and specialized to the specific hardware platform compiler. In other words, it is the most effectively compiled source code for a given hardware platform. Looking into the operation systems (OS) used in the CubeSat industry, it is pretty clear that it is an OS called FreeRTOS that is leading the development team's choice [25]. Therefore, FreeRTOS is selected to be used for all the following experiments described in this article.

1.5. How Does It Look from the Process Side?

So we learned the technical challenges of the software source code development. The other major side of the problem lies in the way software development is planned and carried out—this involves addressing the software development lifecycle (SDLC) and general project management model. The common project and thus software development lifecycle for the space projects is still the "waterfall" one. The "waterfall model" is a linear-sequential life cycle model and was the first process model to be introduced and shown in Figure 2. It is very simple to understand and use. It is a breakdown of project activities into linear sequential phases, where each phase depends on the deliverables of the previous one and corresponds to a specialization of tasks. Typical major stages of the waterfall model are:

- "Requirements analysis and specification phase": the phase when a project team gathers together and elicits and writes down all applicable project requirements. This phase usually ends when all project team members and a customer or a client agree that all requirements are final and fully defined. After this phase, it is assumed that requirements never change and are ready for design and implementation;
- "Design phase": the phase where the engineering team designs parts of the project having the requirements compiled in a previous stage. Normally, such a phase delivers two documents: a software design document (SDD) and a hardware design document (HDD). If some of the requirements cannot be fulfilled by the design, the design phase is considered as "non-feasible" and the project returns to the previous phase: "requirements";
- "Implementation (and unit-testing) phase": the main phase where the development team associated with the project gets the software/hardware done. If something from the design cannot be implemented in unit-tested, the phase is considered as unfinished and the project shall return to the previous phase for either design or requirements elaboration;
- "Integration and system testing phase": the phase where all parts of the project are put together for the system-level testing. This is where the majority of problems appear because of either design and/or implementation errors or non-foreseen conflicts. Normally, this phase is the most heavy one and, in fact, is the most eye-opening for all project members due to integration "surprises";
- "Operation and maintenance phase": the phase where the developed software and/or overall system delivers its designed value to the customer. This phase normally consists of adding further new functions to the software, fixing found during the operation bugs, and adopting the developed software to the changing product environment.

The waterfall model is a heritage of 1950s engineering and is bounded to standards like MIL-STD-499 [26], MIL-STD-1521 [27], and IEEE-15288 [28]. It is rather clear from NASA's "CubeSat 101 Handbook" [13] that the recommended approach to the development of a CubeSat software is purely sequential and thus is of a "waterfall" life-cycle too. It is mainly driven by two common facts: industry practice and heritage and the fact of having and actually constructing the hardware in a CubeSat project. These are the factors that lead the entire project to be within the "waterfall" development life-cycle. What does it consist of?

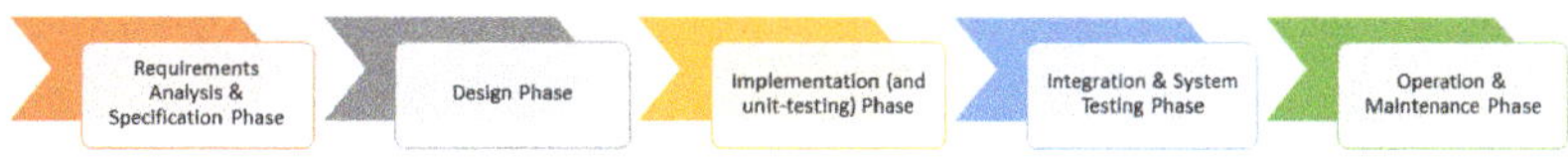

Figure 2. Typical outline of the waterfall development model.

The problem that is created by the waterfall development process is not only in its method of structuring the pipeline of the tasks but also in the the way the source code design, development, testing, and structuring are carried out. After conducting an analysis of the existing CubeSat projects available on GitHub, it became clear that the structure of the source code is weak and often primitive. Even regarding the projects where the code structure is executed in a modular manner, it is still made as a monolithic, tightly bounded, data/code coupled piece of software. Often such a monolithic approach is inherited from the framework being used as part of the re-use strategy. Knowledge of the development time and technical skill constraints in a typical CubeSat software development team, such a monolithic and tightly bounded structure, has never been challenged or changed. So how can the problem of a waterfall development model that is "reinforced" by a tightly coupled data/code and monolithic structured source code be solved? The answer is simple and clearly stipulated by the NASA "CubeSat 101 Handbook" [13]—it involves carrying out proper testing! Testing and quality control and assurance are by themselves very skill-demanding areas, and thus, in the majority of CubeSat projects, software, and hardware testing are pushed to the very last moment and undergo only integration testing. If we take the more mature development teams (mainly those who are part of commercial companies), they are using the so-called V-model that comes from the functional safety world where every step of the SDLC shall be properly verified after being accomplished. The V-model is widely used in applications driven by the following standards: IEC 61508 [29]—"electronic functional safety package", IEC 62304 [30]/ISO 14971 [31]—"medical device software", ISO 26262 [32]—"automotive functional safety", etc. It is represented below in Figure 3.

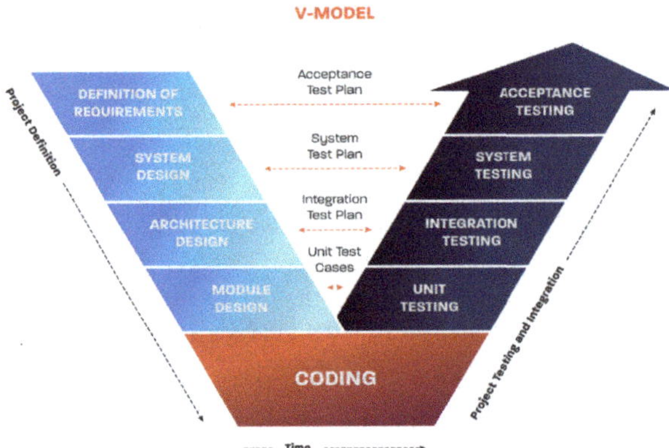

Figure 3. "Waterfall" SDLC reinforced by the V-model.

With the use of the V-model SDLC, the development team can better decompose the "design phase" and "integration and system testing phase" mentioned above and have more time to design/test the system at different abstraction levels. This in return helps one to obtain a better overview of its functions and addresses the functions' complexity better. However, it is important to state that the use of the V-model does not change the fact that the integration is pushed to the very end of the development process and still requires a lot of development and testing efforts in the very end of the development. Using such an approach gives a closer look into the verification and later validation of the software but does not change either the design thinking or design approach of building the CubeSat software. Generally, both "waterfall" and "V-model" are widely criticized by the software industry due to their rigidity, simplicity, inflexibility, and linearity. It is important to emphasize that the use of the V-model does not improve the speed of the integration simplicity of the source code, it just helps to use more time for the design considerations at the different abstraction levels.

1.6. Era of Agile

However, nowadays, the majority of the so-called "big IT" software development teams and specialists are eager to use iterative and/or incremental software development models, i.e., based on agile processes, for example, SCRUM (see Figure 4 or Kanban).

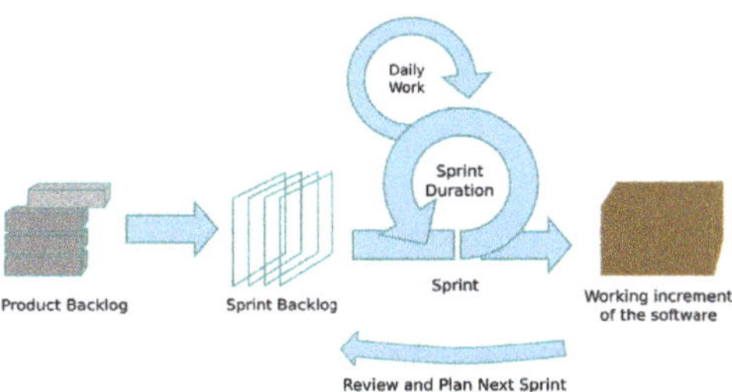

Figure 4. Typical agile-like SCRUM SDLC.

Such an approach means that the whole CubeSat software is broken down into chunks of "features" (functions) that are ready to be implemented, tested, and demonstrated as a separate stand-alone function. Looking into the research "Using the Event-B Formal Method for Disciplined agile Delivery of Safety-critical Systems" [33] available, it became clear that there are many alternations of this classical SCRUM process that fit better into mission-critical software development that might be more applicable for the CubeSat projects and shall be further investigated. One such example is a development framework called disciplined agile (DA) and presented by the Figure 5, which represents the process parts shown below.

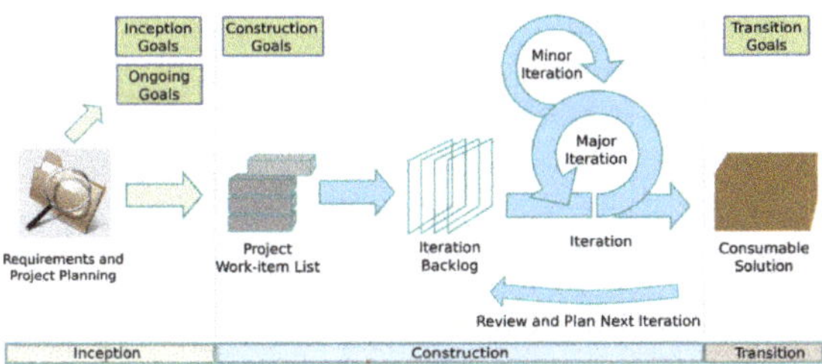

Figure 5. Disciplined agile as the mission-critical systems development response to the classical SCRUM process.

Based on the statements and found problems of the waterfall and the advantages of the agile (SCRUM, DA) development models, the authors suggest moving the CubeSat software development and overall integration of the satellite to the agile model while the hardware parts mature via the classical waterfall model. Accordingly, the authors suggest using the hybrid model, which is becoming commonly used in embedded electronics development. A simple visualization of the benefits is shown below in Figure 6.

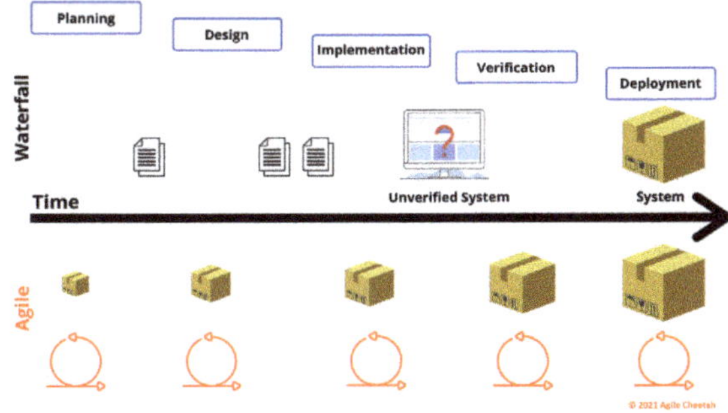

Figure 6. Comparison of the agile and waterfall development models and their value delivery.

As it could be seen, the main difference is shown by the question mark symbol. It emphasizes the problem of too late verification of the entirely developed software, rather than testing smaller but completed deliveries (as shown in the Agile part of the figure).

1.7. Accompanying Development Model and Software Life-Cycle with Proper Software Structure

So, what are the problems that are identified by the authors? Two central problems: the waterfall development process, and monolithic and tightly coupled onboard software. How can these two be cracked? We have already addressed the process side by introducing the "hybrid" development model that combines the strengths of the agile and waterfall processes.

The next problem that was identified and addressed by the authors was the problem of a software design approach that was biased towards a monolithic piece of software being developed. The authors propose to address relatively newly introduced embedded software principles, such as microservices architecture and further containerization. These are good candidates for providing a counter-solution to the monolithic, tightly coupled, and "waterfall"-based developed software.

1.7.1. Microservices

A microservice is a single service built to accommodate an application feature. It handles discrete tasks within a microservices architecture. Each microservice communicates with other services through simple interfaces to solve business problems. The key benefits of the microservices are that they are:

- Independently deployable;
- Loosely coupled;
- Organized around business capabilities;
- Owned by a small team.

As can be easily seen, all of the problems identified in Section 1.3 of a typical CubeSat software are addressed by the nature of the microservice. The idea of using microservices architecture opens up the migration options for the initially monolithic software via the use of the "strangler application" design pattern. The process of such migration is called "Strangling the monolith". The industry of software development is rapidly booming in using such an approach, and there are numerous metrics and tools related to this topic [34].

1.7.2. Microservices Architecture

What is the microservices architecture? It is a development concept in which the entire software to be built is broken down into several small, independent, and loosely coupled services that communicate with each other. The communication between those microservices can be carried out using HTTP, WebSockets, AMQP, or even MQTT. The simplest way to explain the difference between microservice-based and monolithic architectures is to demonstrate the following Figure 7:

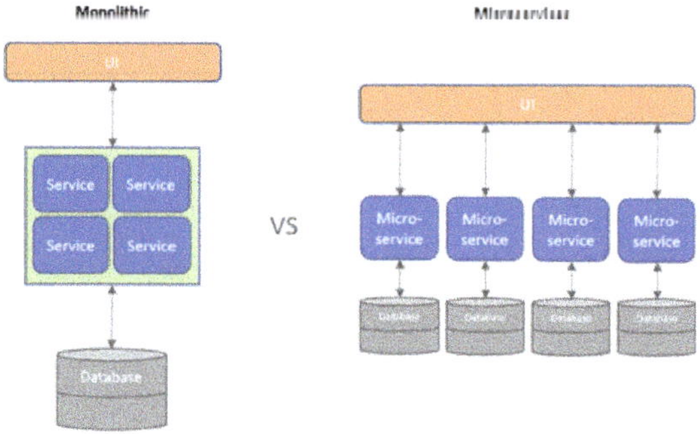

Figure 7. Monolithic vs. microservice-based software structure.

Obviously, connecting such a decomposition approach to the idea of a SCRUM-like decomposed process when developing smaller fractions/modules of the CubeSat software makes perfect sense. Practically, this will allow the CubeSat teams to develop different software modules simultaneously, isolate quality issues and errors in a particular module of the software, reuse someone's else modules in a much simpler manner, and use a bigger development team to shorten the time of the development. Having said that, the one big limitation still remains unsolved is the non-continuous and low-to-medium-skilled development teams. Microservices architecture is based on a strong and well-designed message and event exchange structure, where the typical undergraduate and postgraduate students are simply not skilled enough to reach the interfacing agreement.

1.8. Introducing Containerization

What containerization in a broad sense is? According to IBM [35], one of the frontiers of developing the concept and bringing it to the industry, containerization is the packaging of software code with just the operating system (OS) libraries and dependencies required to run the code to create a single lightweight executable—called a container—that runs consistently on any infrastructure [36]. More portable and resource-efficient than virtual machines (VMs), containers have become the "de facto" compute units of modern cloud-native applications and successfully moving to smaller hardware platforms. Containers are called "lightweight" because they share the OS kernel of the machine and do not have to load an OS for each application. It makes containers smaller and faster than virtual machines and allows more containers to run on the same computing capacity. For the embedded microcontrollers (smaller than a CPU), the classical virtualization would not fly anyway as the resources (Flash, RAM) are still too low for the completely separate VM(s). The main benefit of containerization is that it enables applications to be "written once (on one platform) and run anywhere". This means that developers can create and deploy applications faster and more securely across different platforms and clouds without worrying about bugs or vendor lock-in. Containerization also offers other advantages such as fault isolation and tolerance [37], easy management, simplified security, and more. Perhaps the most essential thing is that containerization allows applications to be truly portable and platform-independent, i.e., "written once and run anywhere". This portability accelerates development; prevents cloud vendor lock-in; and offers other notable benefits such as fault isolation, ease of management, and simplified security. As the importance of embedded applications is rising, hardware capacity is increasing—the development of the microservice architecture and packing of it into containers on embedded systems is booming. There are many examples of home-baked frameworks that are increasingly changing developers' mindsets into microservices-based architectures [38]. Having analyzed the mentioned approaches, the authors suggest using the containerization approach as shown in Figure 8 and implementing the typical onboard software tasks as microservices and placing those among the containers. To be able to do so, the first step would be the selection of a proper container engine (or so-called framework).

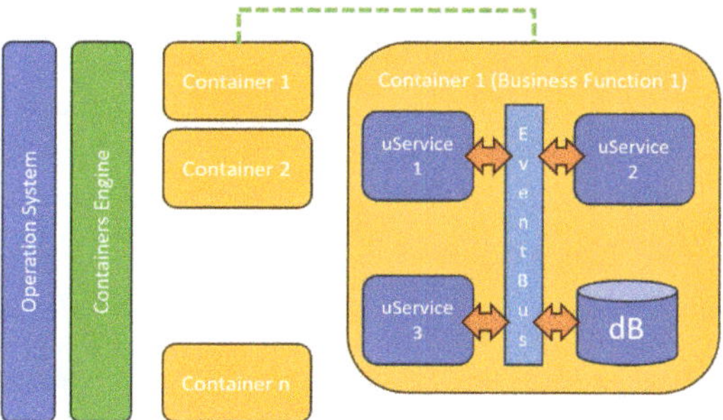

Figure 8. Microservices, implementing a dedicated onboard function, that is distributed to containers.

1.9. Available Containerization Frameworks

During the analysis of the IEEE Xplore, Scope, and Google Scholar papers on the matter of embedded software containerization, the following container engines were found:

- MicroPython;
- Jerry Script;
- Singh;
- Velox VM;
- Toit;
- Femto containers;
- Wasm3;
- Golioth;
- Others.

After conducting a top-level analysis and reviewing it, it became clear that each container engine is more or less based on the WebAssembly principles of real-time code translation. WebAssembly was initially developed by the W3C organization as the translator for web-technologies and applications. Later on, because of its popularity, it became a good choice for the cross-platform engine for non-web applications too. Partially its popularity is driven by the support of the native to the typical embedded software programmings languages such as C, C++, and Rust. As of today, there are more than 35 implementations of high-performance WebAssembly machines, roughly 50% of which are actively supported and continue their lifecycle. During the desktop analysis of the above-mentioned WebAssembly implementations of the container engines, the main focus was on three things: ease of migration and adoption of the most commonly used hardware platforms; the existence of already completed ports for easy-to-get COTS hardware platforms; and, last but not least, the royalty-free nature of the engine, so the concept of COTS components of CubeSats is still kept. As the basis of the solution of the container's engine implementation on the typical CubeSat hardware, the WASM3 [39] interpreter engine was chosen. WASM3 was initially created to deliver outstanding performance for the low-performance targets and thus fully supports the energy-efficient and low-performance hardware of CubeSats. The other neither qualitative nor quantitative selection factor was that WASM3 was developed and actively supported in Ukraine. This will allow the authors to establish a direct connection with the WASM3 development team and by this ensure further work on the use of containers for the CubeSat applications. To be able to better understand how containerization could help CubeSat development teams, let us look into a typical CubeSat system structure and identify the conceptual way of implementing containers for CubeSat projects.

1.10. Combining Benefits of Microservices and Containeriztion

So far, we have solved a monolithic "one-does-it-all" code problem and have moved to a set of microservices, that represent a standalone function of the onboard software. The next step that is proposed to be solved is the proper "isolation" of the microservices so they are not influencing each other's stability, ensuring reliability and allowing for faster integration. The authors suggest solving this via the distribution of the microservices to a set of containers. Each container will consist of a much smaller sub-set of the microservices related to the high-level business function of a CubeSat, i.e., "communication" or "altitude determination and control module".

1.11. Hidden but Yet Important Advantages of Containerization

In a typical CubeSat project, the hardware and the test stand where the software and overall integration testing are carried out are very scarce resources. Therefore, there are a few hidden yet important benefits of the use of containerization:

- The ability to develop and test the container on a regular PC, rather than CubeSat hardware. It minimizes the need for the test stand availability and minimizes the risk of breaking the working hardware;
- Full isolation from the other developers with different (lower) qualifications in a frame of the particular container scope. This means that software errors and malfunctions in other containers will not destroy your own work;
- Easier profiling of a container performance as the developer may just stop other containers at any given point in time.

2. Materials and Methods

2.1. Finding a Unique and Proper Combination of Development Approach and Containerization with Micro-Services Use

In order to evaluate the possibility of using the WASM3 container and microservices-based software, it is essential to evaluate the technology's ability to meet the requirements of real-time systems, that is, to guarantee the execution of a specific task at a predetermined time. Compliance with real-time requirements is a mandatory component of every dependability system. It is especially true of the nanosatellite software when the loss of communication or power supply leads to the loss of the entire satellite. In this part, it is necessary to consider the system's compliance with two requirements: soft and hard real-time. Soft real-time is when, on average, the execution time of the algorithm does not exceed predetermined limits; hard is when each violation is a possible reason for the disruption of the system as a whole. The work aims to estimate the overhead and additional costs of meeting the performance requirements of software algorithms, both from the point of view of soft and hard real-time requirements, if the software architecture includes microservices based on the WASM3 interpreter. Microservices provide more opportunities for the independent implementation of individual student projects and their subsequent integration into a single whole, but the question of how much such an opportunity will cost remains unanswered. It is already clear that the transition from a program compiled and tailored for a specific processor to a program interpreted by a shell requires additional processor time due to additional costs and limited available onboard energy. In order to meet the evaluations required, the next part of the article will provide readers with the essence of the CubeSat structure. Such an overview will show us the typical approaches to a CubeSat software decomposition and implementation, and the author's proposal on the new method of implementing the onboard software based on WASM3 containers with the microservices-based software modules.

2.2. Typical CubeSat Software Structure

To better understand and break down a problem, let's look into a typical CubeSat build shown in Figures 9 and 10, and how the declaration of the author's techniques can be used for its system and software design. A CubeSat normally consists of a few main electronics systems:

- OBC(D): on-board computer (and data);
- ADCS: an altitude determination and control system;
- EPS: electronic power system (could include batteries);
- Comm (or COM): communication system;
- Payload: the "mission" of the CubeSat that brings business value to its creators;
- Propulsion: a propulsion system that is typical for a more advanced and bigger (6U+) CubeSats.

Additionally, there are a few non-electronic systems (which are out of the scope of this document): solar panels, batteries, antenna(s), and the mechanical structures around which all the modules are assembled.

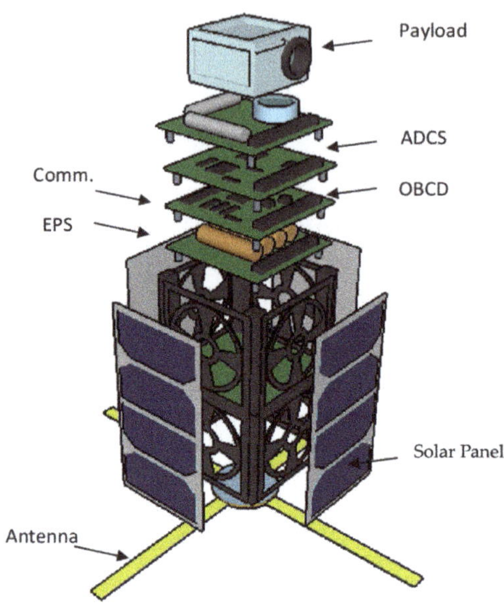

Figure 9. Typical CubeSat structure and components.

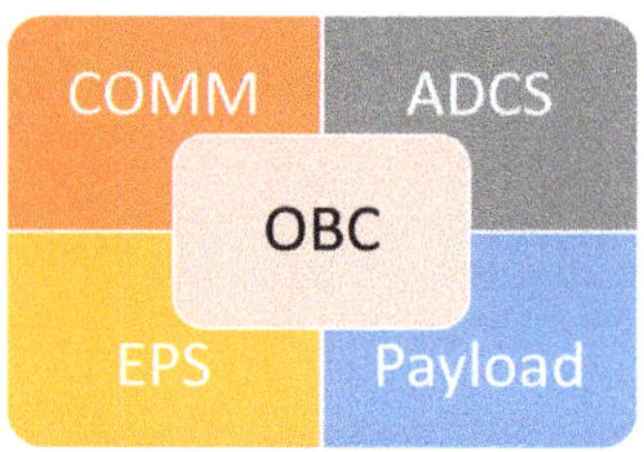

Figure 10. Top-Level typical CubeSat component structure.

In some pretty complex CubeSat build-ups, OBC(D) and ADCS could be combined and called CDHM, which stands for the command and data handling module. This article will address this module in particular as the most software-heavy and complex of any CubeSat project. If we look into a typical method of implementing the software for such a CubeSat platform, there are two main approaches:

- Approach A: An OBC/CDHM only coordinates the data exchange between the rest of the system components, where each component has its firmware and program. See Figure 11

- Approach B: An OBC/CDHM does it all. All other components are as "dummy" as possible. See Figure 12;

To be able to perform the mission, both approaches A and B shall basically implement the same software. However, the number of CubeSat components and sub-systems will impose different complexities on the software, its structure, and its data footprints. The typical OBC/CDHM software (both approaches A and B) could be represented as the following main software building blocks, where the OBC/CDHM handles the main control block and other modules represent different system and payload functions.

The CDHM normally represents either a server in a client-server architecture of CubeSat software, or a central data and task scheduler in a more classic OS-like CubeSat system architecture. Further breakdown of the data flows and the software components of a typical CubeSat can be represented as follows:

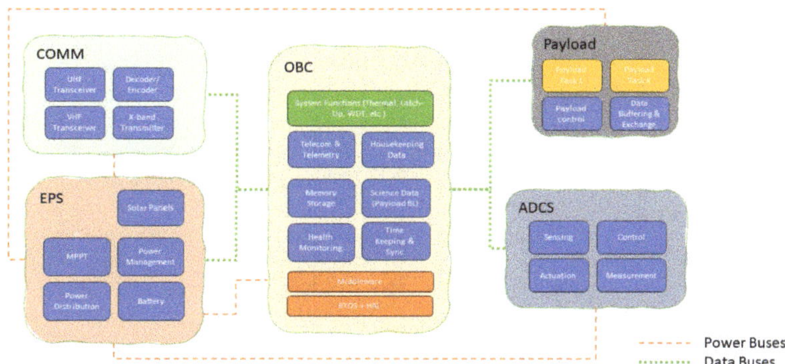

Figure 11. Approach A: CubeSat on-board software and module structure.

It is reasonable to assume that it is a typical Approach A system architecture that requires several comparable calculation power processors/microcontrollers to be embedded in each system board. Therefore, the energy consumption, complexity, cost, and, last but not least, the probability of a hidden error are at their maximum. This is also a very clear example where each separate component of a CubeSat will be designed in a monolithic manner. Making a separate system component (especially one that runs on separate hardware) in a monolithic manner is rather typical for the industry. The big problem with such distributed yet monolithic systems comes with the integration tests complexity, where the bug-finding efforts and time are very costly and the overall project is very much pressed for time.

Instead of such a monolithic system architecture, the software can be designed differently by the use of containerization and microservices. Instead of having many smaller different monolithic components in the system that are compressed into a big one, the business and control logic of those modules is spread towards the containers and corresponding microservices. Each container can represent the particular system component, and in that way, strictly breakdown functions across containers. The suggested approach is represented below in a Figure 12 (Approach B).

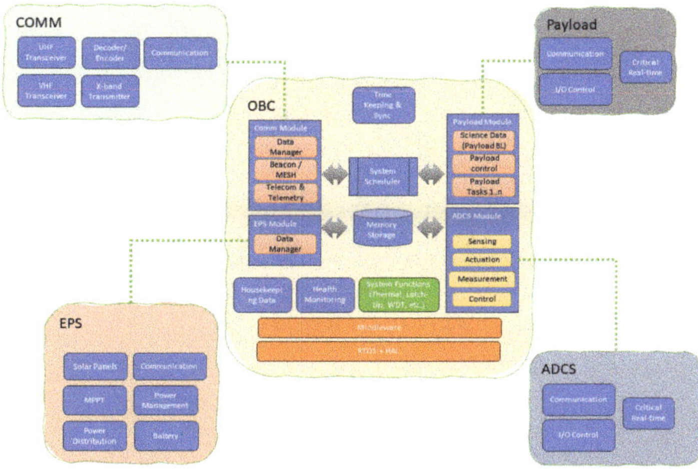

Figure 12. Approach B: The proposed container-based software structure of a CubeSat on-board software.

In this approach, the utilization of the centralized EventHub and EventBus, which is used for the data and commands exchange, can be enhanced by the introduction of the "exposed" EventBus to the communication modules of the other CubeSat components. In this way, the whole system control can be synchronized and orchestrated by the "Saga" pattern or similar. At the same time, such an approach solves the problem of complex and non-uniformed multiple firmware files over the different modules and thus reduces overall project complexity. The majority of the development work moves to the OBC/CDHM side and requires a system architecture response to make it easier to handle, i.e., by using design patterns like "Saga" and adopting a proper database. Surely, the modern approach of using RTOSes, aka FreeRTOS, embOS, and Salvo, allows for a proper abstraction level and the modularity of the complex OBC/CDHM to be designed properly too. However, the main problem of the monolithic firmware, which is still the industry practice, remains active. The monolithic implementation approach of such logically broken down and modularized software brings challenges that are typical for both waterfall development models and overall software quality complexity and cost. Normally, these challenges are integration complexity and time, the many iterations of the re-design of the data exchange and APIs, complex bug-finding and fixing, etc.

2.3. Implementing a Concept CDHM Software on WASM3 Container

So, what are the proposed new methods of developing CubeSat software with the intent of using SCRUM-like development life-cycle and containerization? They are:

- Each container represents either one functional block of the CubeSat (ADCS, EPS, etc.) and/or the separate OBC or payload function, see Figure 13;
- During the development and V&V process of each separate container, the separate team or the team member develops each separate container;
- Due to the fact that the containers are cross-platform, the verification and development of each separate container is carried out at the PC and not at the CubeSat hardware.

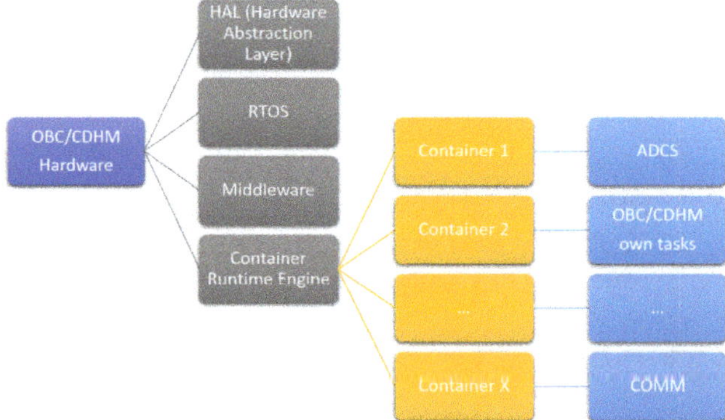

Figure 13. A CubeSat CDHM containerization concept.

As the basis of the solution of the container's engine implementation on the typical CubeSat hardware, the WASM3 interpreter engine was chosen. To be able to run WASM3 on the embedded target of any type, the following minimal infrastructure is to be ported:

- The file system: this is required to store containers and be able to upload those for execution;
- CLI (command line interface): This allows manipulation with the containers (run, stop, load, etc.) and allows the user to see the system parameters in real-time;
- The HTTP server: This will allow for the simple and easy-to-implement OTA (over-the-air) transfer of the container images from a PC to the embedded target (for the communication with the ground stations, AX.25 over HTTP can be used).

As the typical RTOS for the CubeSats is the FreeRTOS by Amazon, the overall architecture of the solution looks as follows (Figure 14):

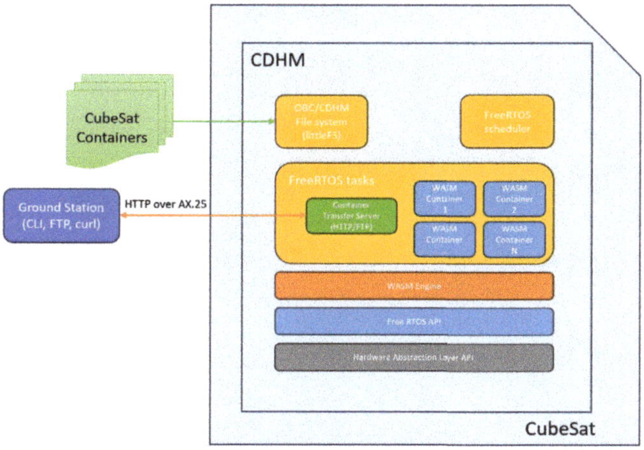

Figure 14. A CubeSat CDHM software architecture based on WASM3 and FreeRTOS.

2.4. Porting WASM3 to the FreeRTOS-Based Environment

The so-called "porting" process is a process of developing and adapting a selected piece of software to the selected hardware platform and later-on to the selected RTOS. In our case, there were already embedded ports provided by the community via the GitHub repository. The most relevant existing port for our tasks was the ESP32-IDF one, and it was chosen to be used as a basis for further experimental work in this article. During the porting, the following assumptions were made:

1. Each container is running as a separate FreeRTOS task;
2. Containers are running with the priority and scheduling by Round Robin principles as for the FreeRTOS;
3. The File System is used for the container storage and each container is uploaded to RAM before its use;
4. The overall porting is carried out by the use of MCU-specific API and FreeRTOS-specific API.

For the file-system implementation, the littleFS embedded low-footprint filesystem was selected. The littleFS is widely used across the industry and is the main implementation candidate for CubeSat use as well. All of the container images are permanently stored on the file system and thus are ready for operation right after the system boot. For the sake of the remote (from the Earth) container upload (Figure 15), a simplified HTTP/FTP server could be considered and thus used further in this article. Schematically, the container upload and corresponding RTOS task start/stop sequence can be shown as follows:

The sequence diagram above assumes that only one container is running at one single moment of time. If the container already exists on the file system, only the "START" command is required to be altered (See Figure 16).

In the case of a CubeSat implementation, such a command can arrive via any of the telemetry communication channels or from the system scheduler rather than the container uploading interface only (GroundStation via HTTP/FTP server here).

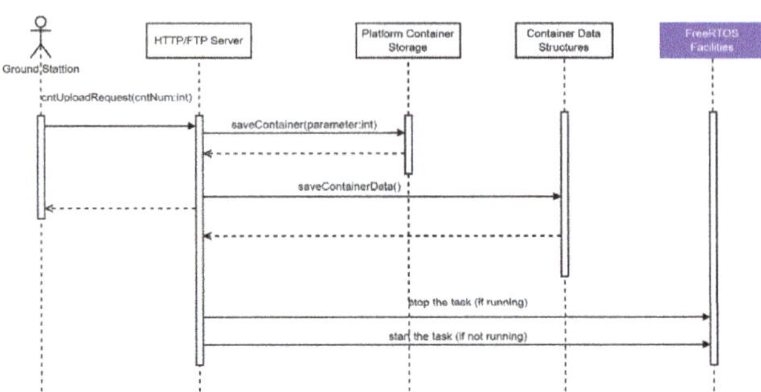

Figure 15. A WASM3 container upload to the CDHM.

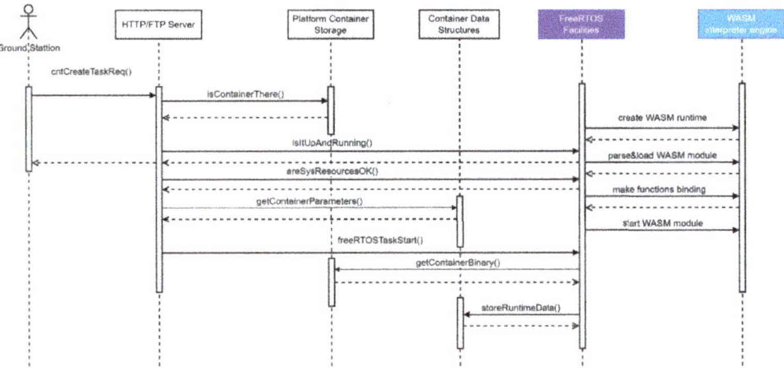

Figure 16. A WASM3 remote container control at the CDHM (START/STOP commands).

2.5. Selection of the Hardware Platforms and Algorithms for the Performance Tests

For the performance testing, two platforms were chosen.

1. The most popular in the open STEM-like hardware projects is Pi Pico by Raspberry (Figure 17), based on an RP2040 processor [40], see Figure 18. Core type: ARM Cortex-M0+, whose calculation performance is roughly 130 DMIPS.
2. The CDHM platform "Falco SBC 1.0" (see Figure 19), which was developed by Oleksandr Liubimov [41] for the Ph.D. thesis, will be used for the upcoming "KhAI-1 spacecraft" 3U CubeSat. The platform is based on the Microchip (Atmel) AT-SAMV71Q21 [42,43] and shown in Figure 20. The core type is ARM Cortex-M7 and it's expected calculation performance is roughly 600 DMIPS.

Figure 17. The COTS Raspberry Pi Pico Dev. kit.

Figure 18. The Pi Pico processor RP2040.

Figure 19. The Falco SBC/CDHM component.

Figure 20. The Falco SBC/CDHM microprocessor ATSAMV71Q21.

At the same time, typical algorithms for the embedded software were used for the benchmarking, namely:

- Fast Fourier transform (FFT);
- bubble sort;
- The CRC-16 checksum calculation algorithm.

For the selected algorithms, the following algorithm's complexity in "O-notation" is expected:

The main reason for measuring performance here is to see how much overhead the two layers of abstraction (WASM3 and its middleware) bring to the proposed solution. To determine the exact overhead, the same selected algorithms (see Table 1) were run on a bare metal implementation, i.e., what vendors offer as low-level API + FreeRTOS and on the code written on C/C++ and further WASM3 compiled, where the containers execution engine is also running on FreeRTOS. The following pre-requisites were used for the performance measurements:

1. For ensuring repeatability and to be able to properly calculate the S, SD, and AVG times of the computation, 1000 measurements are planned for each experiment;
2. For each algorithm, 3 different sizes of data will be used. This will be mainly used to prove that the implementation is carried out in a proper manner and O-complexity is followed;
3. All measurements will be rounded to two digits after the comma;
4. For the CRC-16 algorithm, the sets of 100, 1000, and 10,000, 32-bit signed integers will be used;
5. For the bubble sort algorithm, the sets of 100, 500, and 1000, 32-bit signed integers will be used bubble sort sets with the worst possible condition are used—the data vector to be sorted was filled with numbers placed in a back-sorted order.
6. For the Fast Fourier Transform algorithm, the sets of 128, 256, and 512 samples will be used.

Table 1. The algorithms planned for the performance tests.

Algorithm Name	Time Complexity in Big O Notation	Notes
Checksum Calculation (CRC-16)	$O(N)$	Linear Time
bubble sort (worst case)	$O(N^2)$	Quadratic Time
Fast Fourier Transform	$O(N * logN)$	Linearithmic Time

3. Results

After the successful adoption of the algorithms to the C and C++ languages, the porting of the code to the WASM3 containers was performed too. To obtain the results, the telemetry channel was used in the look of a hyper-terminal, where the data of the calculation duration were obtained via the regular "printf()" C-function. The duration of the execution of the algorithm was obtained by using the microprocessor's system timer with a precision of 1 μS.

The first experiment was made on the entire implementation of pure C language and processor-dependent APIs w/o FreeRTOS and other 3rd party libraries.

The second experiment was carried out with the written C-language container, compiled for the WASM3 container engine, and running under FreeRTOS. The container image was uploaded via the SD-card image and transferred to it from the PC.

Comparative Analysis and Pre-Conclusoins

For the comparison of the two hardware platforms and native C (bare metal) versus WASM3 on the performance of FreeRTOS containers, the test results were chosen on a given number of test data. To provide a good graphical representation of the results (especially for the Native C/bare-metal results), the arithmetic sum of 1000 measurements was used as a basis for comparison.

The following size of the data sets was used for the final performance comparison:

- Sum of 1000 measurements for the CRC-16: 1000 × 32-bit signed integers;
- Sum of 1000 measurements for bubble sort: 1000 × 32-bit signed integers;
- Sum of 1000 measurements for the FFT: 512 samples.

The results can be represented as the numeric data in the following Table 2 and graphically on the Figures 21–23:

Table 2. Comparative results data at given samples per algorithm—sum of 1000 measurements. All numbers are in ms.

Algorithm Name	Rasperry Pi Pico: Native C	Falco CDHM 0.1: Native C	Rasperry Pi Pico: WASM3	Falco CDHM: WASM3
CRC-16 (1000 × 32-bit integers)	2426.82	572.71	66,075.18	15,540.95
Bubble Sort, (1000 × 32-bit integers)	291,932.8	60,856.55	6,039,676.61	1,818,025.11
Fast Fourier Transform (512 samples)	7237.35	303.79	47,259.05	13,252.53

Graphically, such a difference in three selected algorithms could be represented as follows:

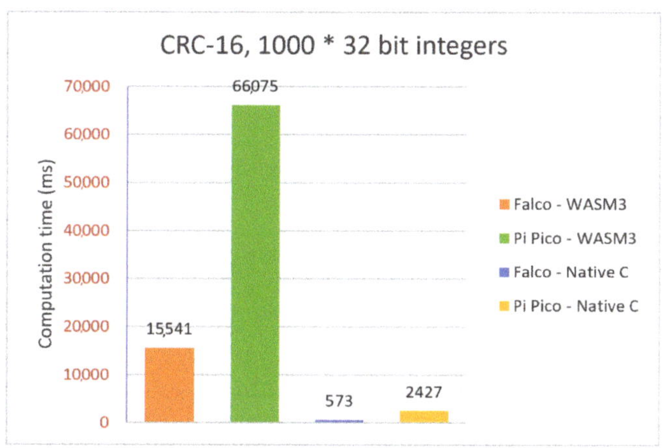

Figure 21. Comparative analysis—CRC-16 (platforms, implementations).

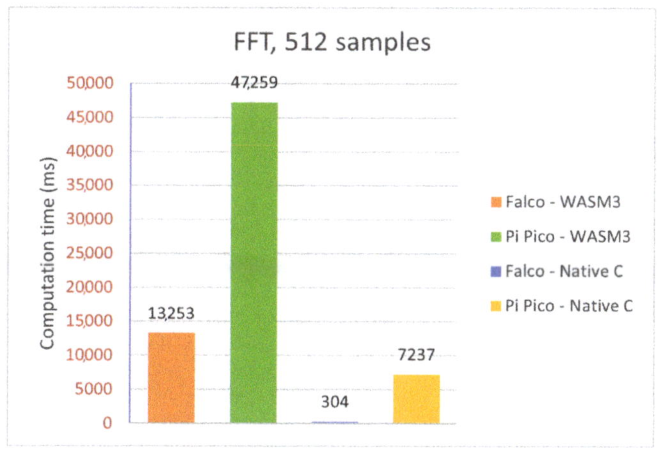

Figure 22. Comparative analysis—FFT (platforms, implementations).

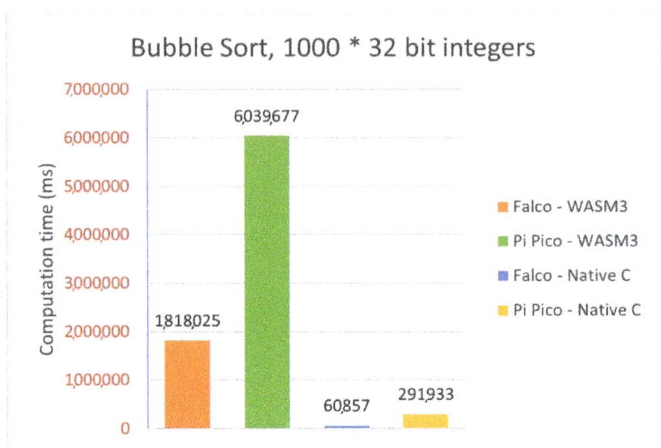

Figure 23. Comparative analysis—Bubble Sort (platforms, implementations).

For further detailed analysis please refer to Appendix A.

4. Discussion

Porting WASM3 to both hardware platforms and running and executing the tests has brought the authors to the desired results and knowledge.

Performance: It is quite clear from Table 2 that there is a pretty visible performance difference when running the same algorithms on the native C implementation and the WASM3 implementation. The simple and yet true performance difference is in the range of 30 times, which means that the native C implementation is 30 times faster than the WASM3-one. However, the reader shall remember that it is rather unfair to compare the Native C implementation (which is very close to bare metal digital machine implementation) and the RTOS-based high-level implementation. The obtained difference in performance in the size of two orders (and in some cases one order) is expected and might sound big. However, knowing the typical calculation tasks for the CubeSats, such a performance difference is not a lifesaver and can be accepted.

Implementation Complexity: During the implementation of the algorithms to Native C and WASM3 on the FreeRTOS platforms, it was found that the differences in the required skill set and the speed of implementation are quite different. For the simple algorithms bubble sort and CRC-16, the implementation was rather simple, and thus there was no major difference in the development speed, while for the FFT implementation, it was rather clear that the WASM3 implementation, for which the hardware capabilities are not really taken into account, is much simpler. It was also found that the implementation by the regular undergraduate student is very straightforward even for a student with rather basic programming and data science skills.

Time at the Hardware and Debug: During the implementation of a Native C algorithm, it can be seen that the hardware access was a MUST. It is very typical for embedded software engineers to run the Debug process on a target, even for the hardware-decoupled algorithms we used in the work. At the same time, for the WASM3 implementation, the hardware was not a need as during the development it is clear that the cross-compilation is required to upload the software to the target platform. So, as stipulated in the correct development model research of this article, the demand for the hardware availability is rather low and could help student teams to work simultaneously on the creation and debugging of a CubeSat.

Falco CDHM as the Low-cost and Powerful CubeSat Platform: One of the secondary tasks in this work was to prove that the selected low-cost automotive grade Microchip SAMV71Q21 (Cortex-M7) microprocessor can be a good basement for the student's CubeSat "KhAI-1 spacecraft" being developed by the National Aerospace University "Kharkiv Aviation Institute". As the results of this synthetic performance testing demonstrate a substantial performance, the Microchip SAMV71 microprocessor is recommended for future use. It is low-cost, automotive grade and demonstrates outstanding performance.

Future Work: During this research and experiments, quite a few conceptual and practical questions and tasks have been raised. These topics can be stated as follows:

- Further development of the ported WASM3 engine so it can be further optimized and support the concurrent container's execution. This shall allow simple yet powerful orchestration, for instance, on the basis of event-driven architecture (EDA) and/or the use of the "Saga" design pattern;
- Further performance optimization shall be carried out, and the real performance penalty overhead sources shall be found;
- Research and implement (if required) safe yet performance-optimal hardware low-level access to the microprocessor's peripherals. The existing implementation of the WASM3 port does not allow that; if there is no easy way to solve it via the WASM3 approach, propose a new one that will be a combination of HAL, FreeRTOS, and WASM3 facilities;
- Research low-power modes of such a WASM3 implementation of a CubeSat CDHM as the power consumption requirements for the spacecraft are very constrained;
- Research and implement hardware debug facilities that will help to find and fix complex hardware-related issues when the software to be verified is fully "packed" into containers.

Author Contributions: Conceptualization, O.L. and I.T.; methodology, O.L.; software, O.L.; validation, L.V. and O.L.; formal analysis, O.L.; investigation, O.L.; resources, L.V.; data curation, I.T.; writing—original draft preparation, L.V.; writing—review and editing, O.L., I.T. and V.P.; visualization, L.V.; and supervision, V.P. All authors have read and agreed to the published version of the manuscript.

Funding: This research received no external funding.

Data Availability Statement: All of the archived datasets analyzed and generated during the study, as well as the source code for the experiments, can be obtained upon request.

Acknowledgments: The authors acknowledge the help of engineering company Ektos-Ukraine LLC for their support with borrowing hardware platforms and helping with the porting toolchain setup and fine-tuning. Visit https://ektos.net/ (accessed on 14 August 2023) for more details.

Conflicts of Interest: The authors declare no conflict of interest.

Appendix A. Detailed Performance Test Results

The following tests were performed for both of the selected hardware platforms (Falco OBC/CDHM, based on Microchip SAMV71Q21, and Raspberry Pi Pico, based on RP2040) and for both Native C (Bare Metal) and for the WASM3 + FreeRTOS container implementation:

- For the CRC-16 Checksum calculation algorithm with 32-bit signed integers: 100/1000/ 10,000 elements;
- For the Fast Fourier Transform: 128/256/512 samples;
- For bubble sort (with the worst case condition, back-sorted) with 32-bit signed integers: 100/500/1000 elements.

For each algorithm, 1000 measurement cycles were made. This was done to increase the precision of the measurements, especially with the RTOS-based implementation where, due to the system task scheduling, you normally see a relatively high measurement jitter. To be able to look into the quality of the obtained statistical data and its dispersion, σ and σ^2 were calculated for each data set of 1000 measurements. In order to compare potentially small numbers for the small amount of elements data sets, there was a decision to compare the arithmetical sum of 1000 measurements time. The following sub-sections will represent the obtained results.

Appendix A.1. Falco OBC/CDHM, Native C-SAMV71 @ 300 Mhz

For the Microchip SAMV71, the following results were obtained:

Appendix A.1.1. Falco OBC/CDHM, Native C, CRC-16

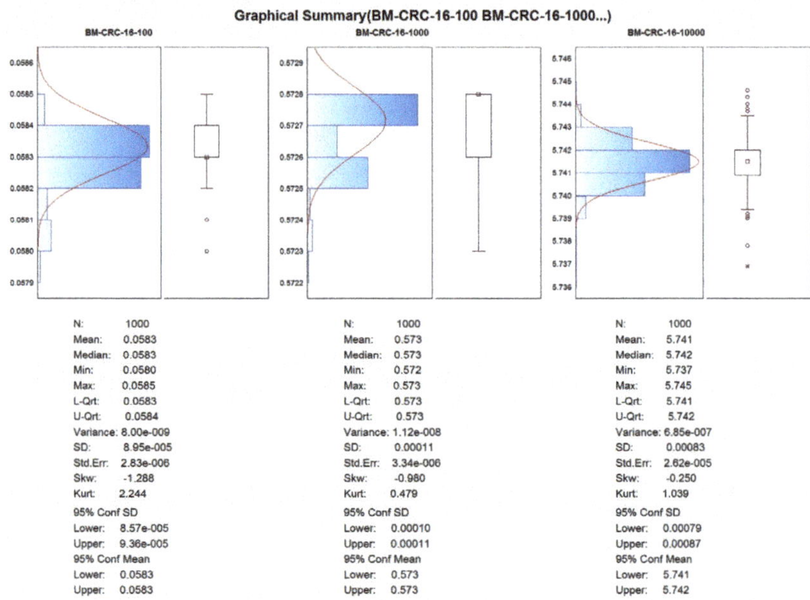

Figure A1. Native C/Falco CDHM: CRC-16 performance at different numbers of test samples, 1/1000 measurements.

Table A1. Native C/Falco CDHM test results: CRC-16 Algorithm.

Data-Set Size (Elements)	Min (ms)	Max (ms)	M (ms)	σ^2	σ	$\sum$ (ms)
100	0.058	0.058	0.058	0.0	0.0001	58.33
1000	0.572	0.573	0.573	0.0	0.0001	572.71
10,000	5.707	5.745	5.742	0.0	0.0008	5741.47

Appendix A.1.2. Falco OBC/CDHM, Native C, and FFT

Disclaimer: It is important to mention that Falco's microprocessor, Atmel SAMV71Q21, has an FPU (floating-point unit) that is used for floating-point arithmetic. As the FFT implementation contains both the float data type and trigonometry functions as *sin* and *cos*, where the floating point arithmetic is used, a performance much stronger than RP2040's performance was expected.

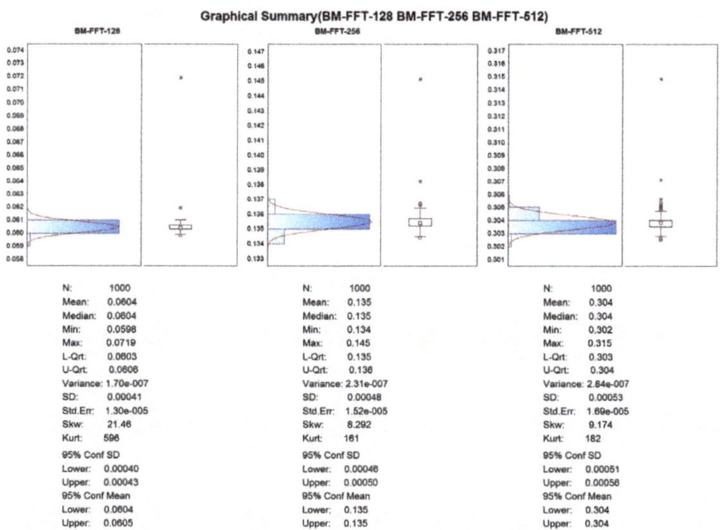

Figure A2. Native C/Falco: FFT performance at different numbers of test samples, 1/1000 measurements.

Table A2. Native C/Falco CDHM test results: FFT algorithm.

Data-Set Size (Samples)	Min (ms)	Max (ms)	M (ms)	σ^2	σ	$\sum$ (ms)
128	0.0598	0.072	0.06	0.0	0.0004	60.44
256	0.134	0.145	0.13	0.0	0.0005	135.46
512	0.302	0.315	0.30	0.0	0.0005	303.79

Appendix A.1.3. Falco OBC/CDHM, Native C, and Bubble Sort

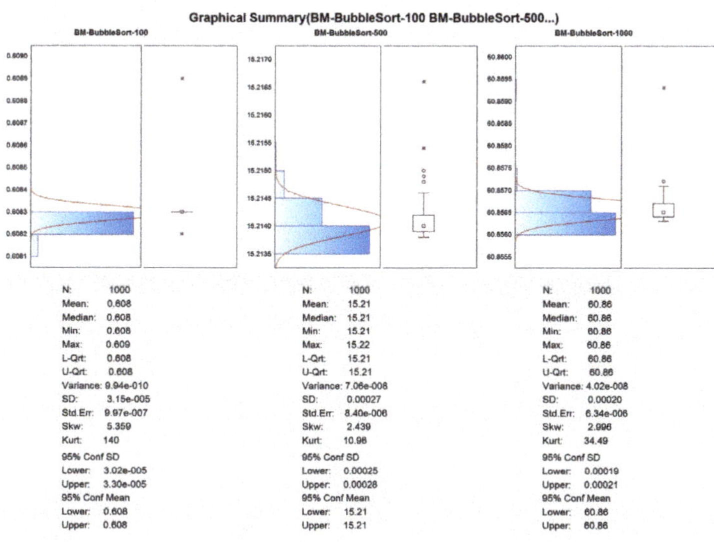

Figure A3. Native C/Falco: bubble sort performance at different numbers of test samples, 1/1000 measurements.

Table A3. Native C/Falco CDHM test results: Bubble sort.

Data-Set Size (Elements)	Min (ms)	Max (ms)	M (ms)	σ^2	σ	$\sum$ (ms)
100	0.6082	0.6089	0.6083	0.0	0.0	608.29
500	15.21	15.22	15.21	0.0	0.0003	15,214.08
1000	60.86	60.86	60.86	0.0	0.0002	60,856.55

Appendix A.2. Raspberry Pi Pico, Native C: RP2040 @ 133Mhz

For the Raspberry Pi Pico, the following results were obtained (Statistica analysis Figure A4, look Table A4): (Please bear in mind that due to the limitations of the system timer resolution, the data precision is lower by a factor of one digit).

Appendix A.2.1. Raspberry Pi Pico, Native C, and CRC-16

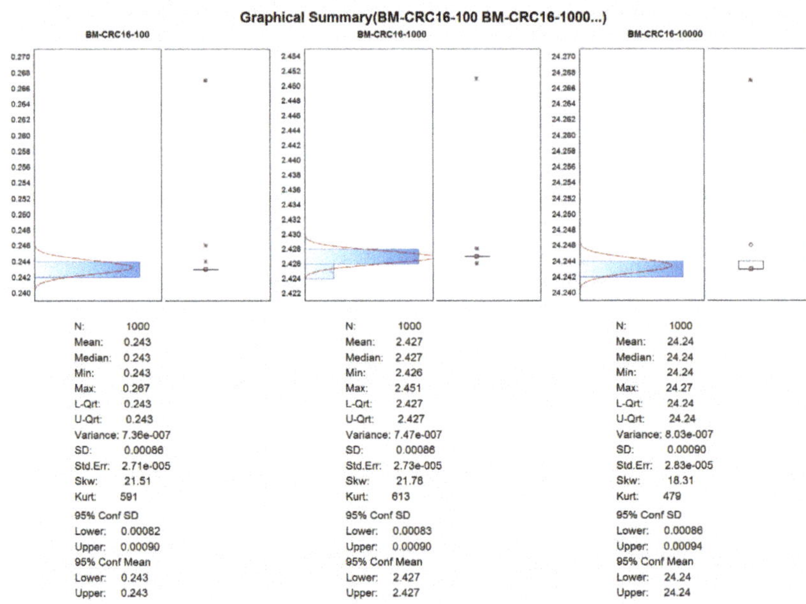

Figure A4. Native C/Pi Pico: CRC-16 performance at different numbers of test samples, 1/1000 measurements.

Table A4. Native C/Pi Pico test results: CRC16 Algorithm.

Data-Set Size (Elements)	Min (ms)	Max (ms)	M (ms)	σ^2	σ	$\sum$ (ms)
100	0.243	0.267	0.2432	0.0	0.0009	243.23
1000	2.426	2.451	2.4268	0.0	0.0009	2426.82
10,000	24.243	24.267	24.2434	0.0	0.0009	24,243.43

Appendix A.2.2. Raspberry Pi Pico, Native C, and FFT

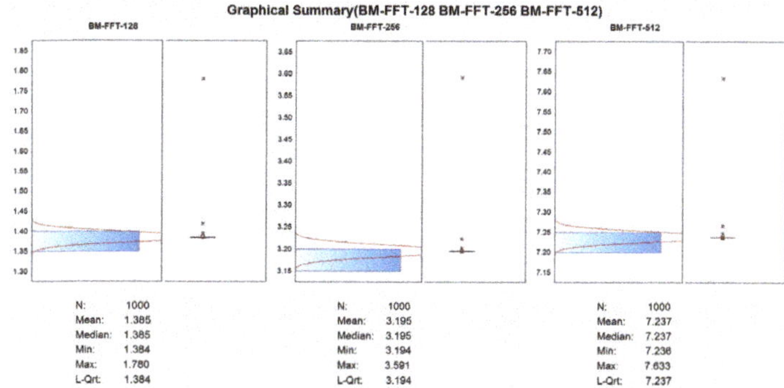

Figure A5. Native C/Pi Pico: FFT performance at different numbers of test samples, 1/1000 measurements.

Table A5. Native C/Pi Pico test results: FFT algorithm.

Data-Set Size (Elements)	Min (ms)	Max (ms)	M (ms)	σ^2	σ	$\sum$ (ms)
128	1.384	1.78	1.385	0.0	0.0126	1385.16
256	3.194	3.591	3.195	0.0	0.0126	3195.08
512	7.236	7.633	7.237	0.0	0.0126	7237.35

Appendix A.2.3. Raspberry Pi Pico, Native C, and bubble Sort

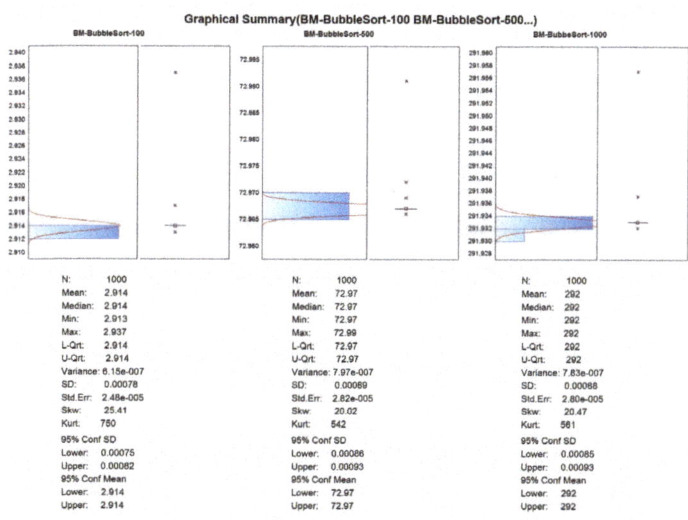

Figure A6. Native C/Pi Pico: bubble sort performance at different numbers of test samples, 1/1000 measurements.

Table A6. Native C/Pi Pico test results: bubble sort.

Data-Set Size (Elements)	Min (ms)	Max (ms)	M (ms)	σ^2	σ	$\sum$ (ms)
100	2.913	2.937	2.914	0.0	0.0008	2913.947
500	72.966	72.991	72.967	0.0	0.0009	72,966.8
1000	291.932	291.957	291.933	0.0	0.0009	291,932.8

Appendix A.3. Falco OBC/CDHM, WASM3, Single Container Configuration: SAMV71 @ 300 MHz

For the Microchip SAMV71 and WASM3, the following results were obtained:

Appendix A.3.1. Falco OBC/CDHM, WASM3, and CRC-16

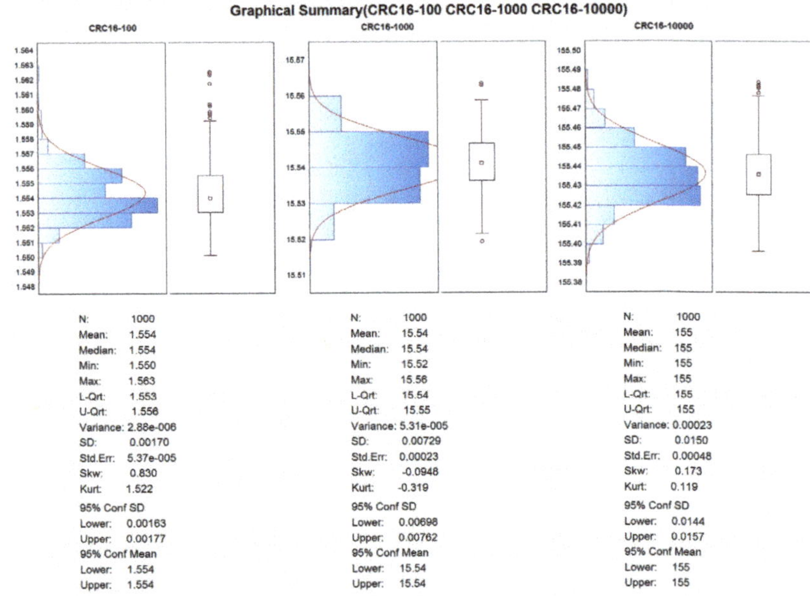

	CRC16-100	CRC16-1000	CRC16-10000
N:	1000	1000	1000
Mean:	1.554	15.54	155
Median:	1.554	15.54	155
Min:	1.550	15.52	155
Max:	1.563	15.56	155
L-Qrt:	1.553	15.54	155
U-Qrt:	1.556	15.55	155
Variance:	2.88e-006	5.31e-005	0.00023
SD:	0.00170	0.00729	0.0150
Std.Err:	5.37e-005	0.00023	0.00048
Skw:	0.830	-0.0948	0.173
Kurt:	1.522	-0.319	0.119
95% Conf SD			
Lower:	0.00163	0.00698	0.0144
Upper:	0.00177	0.00762	0.0157
95% Conf Mean			
Lower:	1.554	15.54	155
Upper:	1.554	15.54	155

Figure A7. WASM3/Falco: CRC-16 performance at different numbers of test samples 1/1000 measurements.

Table A7. WASM3/Falco CDHM: CRC16 Algorithm, 1000 measurements.

Data-Set Size (Elements)	Min (ms)	Max (ms)	M (ms)	σ^2	σ	$\sum$ (ms)
100	1.55	1.56	1.55	0.0	0.002	1554.29
1000	15.52	15.56	15.54	0.0001	0.007	15,540.95
10,000	155.40	155.48	155.44	0.0002	0.015	155,436.47

Appendix A.3.2. Falco OBC/CDHM, WASM3, and FFT

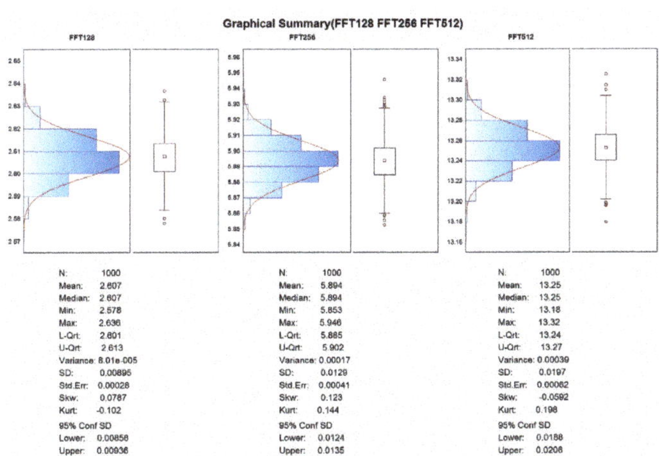

Figure A8. WASM3/Falco: FFT performance at different numbers of test samples, 1/1000 measurements.

Table A8. WASM3/Falco CDHM: FFT algorithm, 1000 measurements.

Data-Set Size (Elements)	Min (ms)	Max (ms)	M (ms)	σ^2	σ	$\sum$ (ms)
128	2.58	2.64	2.61	0.0001	0.009	2607.25
256	5.85	5.95	5.89	0.0002	0.013	5893.53
512	13.18	13.32	13.25	0.0004	0.02	13,252.53

Appendix A.3.3. Falco OBC/CDHM, WASM3, and Bubble Sort

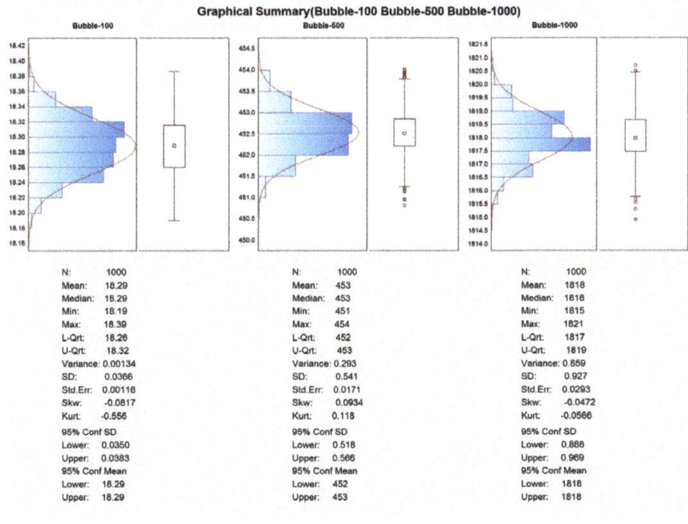

Figure A9. WASM3/Falco: bubble sort performance at different numbers of test samples, 1/1000 measurements.

Table A9. WASM3/Falco test results: bubble sort.

Data-Set Size (Elements)	Min (ms)	Max (ms)	M (ms)	σ^2	σ	$\sum$ (ms)
100	18.19	18.39	18.29	0.001	0.037	18,287.40
500	450.82	453.99	452.52	0.29	0.54	452,521.56
1000	1814.90	1820.7	1818.03	0.86	0.93	1,818,025.11

Appendix A.4. Raspberry Pi Pico, WASM3, and Single Container Configuration: RP2040 @ 133 MHz

For the Raspberry RP2040 and WASM3, the following results were obtained:

Appendix A.4.1. Raspberry Pi Pico, WASM3, and CRC-16

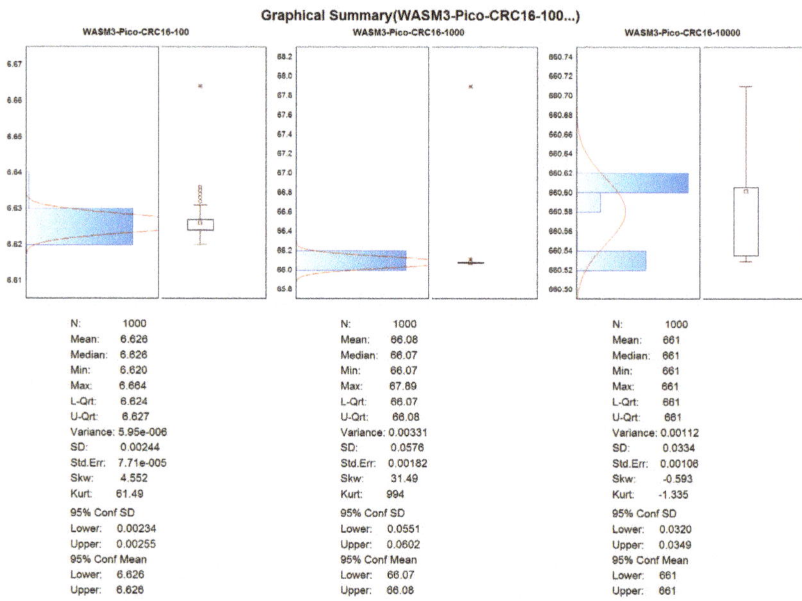

Figure A10. WASM3/Pi Pico: CRC-16 performance at different numbers of test samples, 1/1000 measurements.

Table A10. WASM3/Pi Pico test results: CRC16 Algorithm, 1000 measurements.

Data-Set Size (Elements)	Min (ms)	Max (ms)	M (ms)	σ^2	σ	$\sum$ (ms)
100	6.62	6.664	6.625	0.0	0.0024	6625.76
1000	66.067	67.891	66.075	0.0033	0.0576	66,075.18
10,000	660.53	660.71	660.58	0.0011	0.0334	660,580.33

Appendix A.4.2. Raspberry Pi Pico, WASM3, and FFT

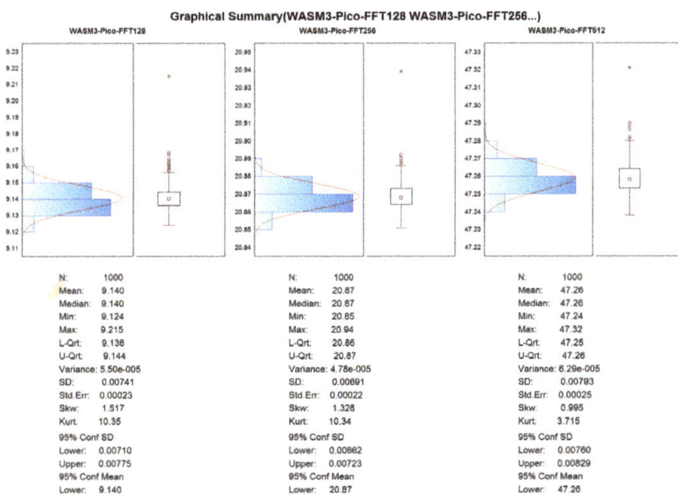

Figure A11. WASM3/Pi Pico: FFT performance at different numbers of test samples, 1/1000 measurements.

Table A11. WASM3/Pi Pico test results: FFT algorithm, 1000 measurements.

Data-Set Size (Elements)	Min (ms)	Max (ms)	M (ms)	σ^2	σ	$\sum$ (ms)
128	9.124	9.215	9.14	0.0001	0.0074	9140.41
256	20.851	20.939	20.14	0.0	0.0069	20,868.49
512	47.238	47.321	47.259	0.0001	0.0079	47,259.05

Appendix A.4.3. Raspberry Pi Pico, WASM3, and Bubble Sort

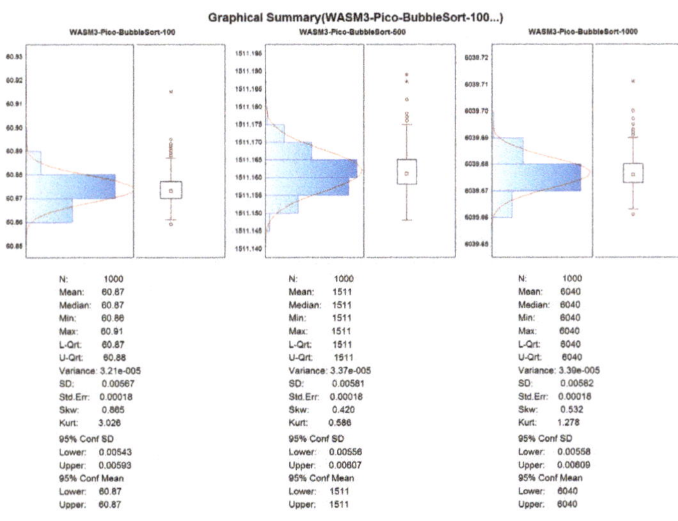

Figure A12. WASM3/Pi Pico: Bubble sort performance at different numbers of test samples, 1/1000 measurements.

Table A12. WASM3/Pi Pico test results: bubble sort.

Data-Set Size (Elements)	Min (ms)	Max (ms)	M (ms)	σ^2	σ	$\sum$ (ms)
100	60.86	60.91	60.87	0.0	0.0057	60,873.67
500	1511.15	1511.19	1511.16	0.0	0.0058	1,511,161.81
1000	6039.66	6039.71	6039.68	0.0	0.0058	6,039,676.61

References

1. CubeSat.org. Cubesat Design Specification Rev 14.1 (by the CubeSat Program). 2022. Available online: https://www.cubesat.org/cubesatinfo (accessed on 15 April 2023).
2. Cappelletti, C.; Robson, D. 2-CubeSat missions and applications. In *Cubesat Handbook*; Academic Press: Cambridge, MA, USA 2021; pp. 53–65. [CrossRef]
3. Shkolnik, E.L. On the verge of an astronomy CubeSat revolution. *Nat. Astron.* **2018**, *2*, 374–378. [CrossRef]
4. Crusan, J.; Galica, C. NASA's CubeSat Launch Initiative: Enabling broad access to space. *Acta Astronaut.* **2019**, *157*, 51–60. [CrossRef]
5. ESA. European Space Agency. Fly Your Satellite Program Intro. 2018. Available online: https://www.esa.int/Education/CubeSats_-_Fly_Your_Satellite/Fly_Your_Satellite!_programme (accessed on 22 August 2023).
6. CalPoly. P-POD User Guide. California Polytechnic State University. 2014. Available online: https://static1.squarespace.com/static/5418c831e4b0fa4ecac1bacd/t/5806854d6b8f5b8eb57b83bd/1476822350599/P-POD_MkIIIRevE_UserGuide_CP-PPODUG-1.0-1_Rev1.pdf (accessed on 24 August 2023).
7. Brycetech. Smallsats by the Numbers 2023. 2023. Available online: https://brycetech.com/reports/report-documents/Bryce_Smallsats_2023.pdf (accessed on 5 June 2023).
8. Kang, J.; Gregory, J.; Temkin, S.; Sanders, M.; King, J. Creating Future Space Technology Workforce Utilizing CubeSat Platforms Challenges, Good Practices, and Lessons Learned. In Proceedings of the AIAA Scitech 2021 Forum, Virtual Event, 11–15 & 19–21 January 2021; pp. 1–12. [CrossRef]
9. EXA. Cubesat Market. KRATOS 1U Platform. 2021. Available online: https://www.cubesat.market/kratos1uplatform (accessed on 26 August 2023).
10. Reznik, S.; Reut, D.; Shustilova, M. Comparison of geostationary and low-orbit "round dance" satellite communication systems *IOP Conf. Ser. Mater. Sci. Eng.* **2020**, *971*, 052045. [CrossRef]
11. Swartwout, M. Sant Louis University "Cubesat Database". 2022. Available online: https://sites.google.com/a/slu.edu/swartwout/cubesat-database (accessed on 5 June 2023).
12. Kulu, E. NewSpace Index "Nanosats Database". 2022. Available online: https://www.nanosats.eu/database (accessed on 5 June 2023)
13. NASA. NASA CubeSat 101: Basic Concepts and Processes for First-Time CubeSat Developers. 2018. Available online: https//www.nasa.gov/sites/default/files/atoms/files/nasa_csli_cubesat_101_508.pdf (accessed on 25 June 2023).
14. El Allam, A.K.; Jallad, A.H.M.; Awad, M.; Takruri, M.; Marpu, P.R. A Highly Modular Software Framework for Reducing Software Development Time of Nanosatellites. *IEEE Access* **2021**, *9*, 107791–107803. [CrossRef]
15. Bocchino, R.L., Jr.; Canham, T.K.; Watney, G.J.; Reder, L.J.; Levison, J.W. F Prime: An Open-Source Framework for Small-Scale Flight Software Systems. In Proceedings of the SSC-18-XII-04 32nd Annual AIAA/USU Conference on Small Satellites, Logan UT, USA, 4–9 August 2018; pp. 110–119.
16. Paiva, D.; Lima, R.; Carvalho, M.; Mattiello-Francisco, F.; Madeira, H. Enhanced software development process for CubeSats to cope with space radiation faults. In Proceedings of the 2022 IEEE 27th Pacific Rim International Symposium on Dependable Computing (PRDC), Beijing, China, 28 November–1 December 2022; pp. 78–88. [CrossRef]
17. Liubimov, O.; Turkin, I. Data Model and Methods for Ensuring the Reliability and Relevance of Data for the CubeSat Projects. In Proceedings of the 2022 12th International Conference on Dependable Systems, Services and Technologies (DESSERT), Athens Greece, 9–11 December 2022; pp. 1–7. [CrossRef]
18. Goyal, T.; Aggarwal, K. Simulator for Functional Verification and Validation of a Nanosatellite. In Proceedings of the 2019 IEEE Aerospace Conference, Big Sky, MT, USA, 2–9 March 2019; pp. 1–8. [CrossRef]
19. Batista, C.L.G.; Martins, E.; de Fátima Mattiello-Francisco, M. On the use of a failure emulator mechanism at nanosatellite subsystems integration tests. In Proceedings of the 2018 IEEE 19th Latin-American Test Symposium (LATS), Sao Paulo, Brazil, 12–14 March 2018; pp. 1–6. [CrossRef]
20. Paiva, D.; Duarte, J.M.; Lima, R.; Carvalho, M.; Mattiello-Francisco, F.; Madeira, H. Fault injection platform for affordable verification and validation of CubeSats software. In Proceedings of the 2021 10th Latin-American Symposium on Dependable Computing (LADC), Florianópolis, Brazil, 22–26 November 2021; pp. 1–11. [CrossRef]
21. *ISO/IEC 15408-1:2022*; Information Security, Cybersecurity and Privacy Protection—Evaluation Criteria for IT Security—Part 1 Introduction and General Model. International Organization for Standardization (ISO): Geneve, Switzerland, 2022. Available online: https://www.iso.org/standard/72891.html (accessed on 26 August 2023).

22. Potii, O.; Illiashenko, O.; Komin, D. Advanced Security Assurance Case Based on ISO/IEC 15408. In *Theory and Engineering of Complex Systems and Dependability, Proceedings of the Tenth International Conference on Dependability and Complex Systems DepCoS-RELCOMEX, Brunów, Poland, 29 June–3 July 2015*; Springer International Publishing: New York, NY, USA, 2015; pp. 391–401. [CrossRef]
23. Falco, G.; Viswanathan, A.; Santangelo, A. CubeSat Security Attack Tree Analysis. In Proceedings of the 2021 IEEE 8th International Conference on Space Mission Challenges for Information Technology (SMC-IT), Pasadena, CA, USA, 26–30 July 2021; pp. 68–76. [CrossRef]
24. Tanaka, K. *Embedded Systems: Theory and Design Methodology*; IntechOpen: London, UK, 2012; pp. 101–120. [CrossRef]
25. Siewert, S.; Rocha, K.; Butcher, T.; Pederson, T. Comparison of Common Instrument Stack Architectures for Small UAS and CubeSats. In Proceedings of the 2021 IEEE Aerospace Conference (50100), Big Sky, MT, USA, 6–13 March 2021; pp. 1–17. [CrossRef]
26. *MIL-STD-499*; Military Standard: System Engineering Management. Defense Logistics Agency: Fort Belvoir, VA, USA, 2017. Available online: http://everyspec.com/MIL-STD/MIL-STD-0300-0499/MIL-STD-499_10376/ (accessed on 26 August 2023).
27. *MIL-STD-1521B*; Military Standard: Technical Reviews and Audits for Systems, Equipments, and Computer Software. Defense Logistics Agency: Columbus, OH, USA, 1995. Available online: http://everyspec.com/MIL-STD/MIL-STD-1500-1599/MIL_STD_1521B_1503/ (accessed on 26 August 2023).
28. *ISO/IEC/IEEE 15288:2023*; Systems and Software Engineering—System life Cycle Processes. International Organization for Standardization (ISO): Geneve, Switzerland, 2023. Available online: https://www.iso.org/standard/81702.html (accessed on 26 August 2023).
29. *IEC 61508 Ed. 2.0 en:2010 CMV*; Functional Safety of Electrical/Electronic/Programmable Electronic Safety-Related Systems—Parts 1 to 7 Together with a Commented Version (See Functional Safety And IEC 61508). International Organization for Standardization (ISO): Geneve, Switzerland, 2021. Available online: https://webstore.ansi.org/standards/iec/iec61508eden2010cmv (accessed on 26 August 2023).
30. *IEC 62304 Ed. 1.1 b:2015*; Medical Device Software—Software Life Cycle Processes. International Organization for Standardization (ISO): Geneve, Switzerland, 2020. Available online: https://webstore.ansi.org/standards/iec/iec62304ed2015 (accessed on 26 August 2023).
31. *ISO 14971:2019*; Medical Devices—Application of Risk Management to Medical Devices. International Organization for Standardization (ISO): Geneve, Switzerland, 2019. Available online: https://www.iso.org/standard/72704.html (accessed on 26 August 2023).
32. *ISO 26262-6:2018*; Road Vehicles—Functional Safety—Part 6: Product Development at the Software Level. International Organization for Standardization (ISO): Geneve, Switzerland, 2018. Available online: https://www.iso.org/standard/68388.html (accessed on 26 August 2023).
33. Edmunds, A.; Olszewska (Plaska), M.; Waldén, M. Using the Event-B Formal Method for Disciplined agile Delivery of Safety-critical Systems. In Proceedings of the Second International Conference on Advances and Trends in Software Engineering—SOFTENG 2016, Lisbon, Portugal, 21–25 February 2016.
34. Al-Debagy, O.; Martinek, P. Extracting Microservices' Candidates from Monolithic Applications: Interface Analysis and Evaluation Metrics Approach. In Proceedings of the 2020 IEEE 15th International Conference of System of Systems Engineering (SoSE), Budapest, Hungary, 2–4 June 2020; pp. 289–294. [CrossRef]
35. IBM. What Is Containerization? 2018. Available online: https://www.ibm.com/topics/containerization (accessed on 12 April 2023).
36. IBM. Containers In the Enterprise. 2020. Available online: https://www.ibm.com/downloads/cas/VG8KRPRM (accessed on 12 April 2023).
37. Tamanaka, G.T.B.; Aroca, R.V.; de Paula Caurin, G.A. Fault-tolerant architecture and implementation of a distributed control system using containers. In Proceedings of the 2022 Latin American Robotics Symposium (LARS), 2022 Brazilian Symposium on Robotics (SBR), and 2022 Workshop on Robotics in Education (WRE), São Bernardo do Campo, Brazil, 18–21 October 2022; pp. 1–6. [CrossRef]
38. Wang, S.; Du, C.; Chen, J.; Zhang, Y.; Yang, M. Microservice Architecture for Embedded Systems. In Proceedings of the 2021 IEEE 5th Information Technology, Networking, Electronic and Automation Control Conference (ITNEC), Xi'an, China, 15–17 October 2021; Volume 5, pp. 544–549. [CrossRef]
39. Shymanskyy, V. WASM3 GitHub Page. 2021. Available online: https://github.com/wasm3/wasm3 (accessed on 28 June 2023).
40. RaspberryPi. RP2040 Microprocessor Page. 2020. Available online: https://www.raspberrypi.com/products/rp2040/ (accessed on 28 June 2023).
41. Liubimov, O. Falco Engineering. 2023. Available online: https://www.falco.engineering/ (accessed on 28 August 2023).
42. Microchip. ATSAMV71Q21 Microprocessor Page. 2020. Available online: https://www.microchip.com/en-us/product/ATSAMV71Q21 (accessed on 28 June 2023).
43. Microchip. COTS-to-Radiation-Tolerant and Radiation-Hardened Devices. 2019. Available online: https://www.microchip.com/en-us/solutions/aerospace-and-defense/products/microcontrollers-and-microprocessors/cots-to-radiation-tolerant-and-radiation-hardened-devices (accessed on 28 June 2023).

Article

Analysis of Effectiveness of Combined Surface Treatment Methods for Structural Parts with Holes to Enhance Their Fatigue Life

Olexander Grebenikov [1], Andrii Humennyi [1,*], Serhii Svitlychnyi [2,*], Vasyl Lohinov [3] and Valerii Matviienko [4]

[1] Department of Airplanes and Helicopters Design, Faculty of Aircraft Engineering, National Aerospace University "Kharkiv Aviation Institute", 61070 Kharkiv, Ukraine; o.grebenikov@khai.edu
[2] Department of Theoretical Mechanics, Mechanical Engineering and Robotic Systems, Faculty of Aircraft Engine, National Aerospace University "Kharkiv Aviation Institute", 61070 Kharkiv, Ukraine
[3] JSC FED, 61023 Kharkiv, Ukraine
[4] JSC "Ukrainian Research Institute of Aviation Technology", 04080 Kyiv, Ukraine; cmti@ukrniat.com
* Correspondence: a.gumennyy@khai.edu (A.H.); s.svetlichniy@khai.edu (S.S.)

Abstract: The typical and most widespread stress concentrators in the lower wing panels of aircraft are the drain holes located on the stringer vertical ribs. These are prime sources for the initiation and development of fatigue cracks, which lead to early failure of the wing structure. Therefore, improving fatigue life in these critical areas is one of the significant issues for research. Two combined methods of surface plastic treatment in the location around drain holes are discussed in this paper. Using the finite element method and ANSYS software, we created a finite element model and obtained nonlinear solution results in the case of tension in a plate with three holes. In addition, the development of residual stress due to the surface plastic treatment of the hole-adjacent areas was taken into account. In this paper, it is shown that after surface treatment of the corresponding areas of the holes, residual stress, which exceeds the yield stress for the plate material, is induced. When combined with alternative tensile stress, these reduce the amplitude of the local stresses, thus increasing the number of stress cycles before failure. The benefits of this technology were confirmed by fatigue test results, which include the fatigue failure types of the plates. Graphs showing the impact of applicable surface treatment combined methods on the number of cycles to failure were also plotted.

Keywords: drain holes; fatigue life; surface plastic treatment; hole cold expansion; rolling; residual stress; fatigue cracks; finite element analysis

Citation: Grebenikov, O.; Humennyi, A.; Svitlychnyi, S.; Lohinov, V.; Matviienko, V. Analysis of Effectiveness of Combined Surface Treatment Methods for Structural Parts with Holes to Enhance Their Fatigue Life. *Computation* **2024**, *12*, 8. https://doi.org/10.3390/computation12010008

Academic Editors: Victor Calo, Mykola Nechyporuk and Dmitriy Kritskiy

Received: 1 November 2023
Revised: 9 December 2023
Accepted: 19 December 2023
Published: 8 January 2024

1. Introduction

Aircraft structural components experience cyclic loading during operation that causes damage accumulation, fatigue crack initiation and development, and eventually failure. Stress and strain concentration in these components is the major factor which defines the strength and fatigue life of the structural elements. The typical and most widespread stress concentrators in wing structural components are holes. All holes made in structural elements can be conventionally divided into two classes: free unused holes and fastener holes, which are either free of loads or carrying loads. The former class are drain holes, which are going to be considered in this paper.

To ensure the reliability and the assigned service life of wing panels in the area of drain holes, the stress and strain concentration must be reduced. Consequently, a current challenge in designing new structures or modifying existing ones is in developing and applying design and technological methods for improving structural elements' fatigue life in stress concentration areas.

Surface plastic deformation (SPD) methods have proved to be effective in enhancing the fatigue life of wing panels around drain holes, as demonstrated by the results of fatigue tests [1–5]. The primary concept underlying these methods is to create residual compressive stresses near the stress concentrator. These stresses, when combined with cyclic tensile stress, reduce the amplitude of local alternating stresses, thereby increasing the number of stress cycles before failure. Additionally, these surface treatment methods have the positive effect of smoothing microroughness and waviness on the surface, thereby enhancing surface roughness and accuracy.

A brief description of the publications focusing on the enhancement of fatigue life in structural components within areas of stress concentration, along with the significant results, is provided below.

Sasan Faghih et al. [6] conducted a series of fatigue tests to find the optimal degree of expansion that resulted in the greatest enhancement in fatigue performance of cold expanded samples. The study revealed that a 6% cold expansion was the optimum expansion level for the investigated material, resulting in substantial plastic deformation around the hole without causing any macro- or micro-structural damage under the processing conditions. A digital image correlation (DIC) technique was adopted for crack detection and monitoring to examine the effect of the cold expansion process on crack growth behavior. It was found that cold expansion not only postponed the crack initiation but also considerably decreased the crack growth rate in cold-expanded specimens.

Qi Li, Qichao Xue et al. [7] performed a three-dimensional finite element analysis of cold expansion strengthening of the 7050 aluminum alloy orifice plate to study the distribution of von Mises and residual stress around the hole after cold expansion. The influence of different parameters such as extrusion amounts and plate thickness on the residual stress distribution around the hole was analyzed. The relationship between extrusion amount, plate thickness, and residual stress was obtained. An X-ray diffraction (XRD) technique was employed to measure the residual stress on the surface of the extrusion entrance and to validate the finite element modeling results. The study results indicated that the cold expansion process of slotted bushing holes can enhance the residual stress distribution around the hole and form a strong residual stress layer. As the amounts of extrusion increased, the residual stress tended to increase, reaching the maximum when the amounts of extrusion were 4%.

Wuzhu Yan et al. [8] proposed a bi-directional cold expansion procedure to study the homogenization of residual stress in the thickness direction of the cold-expanded hole, which enabled further improvement of the antifatigue performance of the cold-expanded hole. To investigate the effectiveness of the bi-directional cold expansion procedure and find the optimal process parameters, a series of finite element (FE) simulations were carried out. The results showed that the optimized bi-directional cold expansion process generated a more uniform distribution of residual circumferential compressive stress in the thickness direction compared to the simplified bi-directional cold expansion process using a single mandrel.

The authors of the work [9] applied finite element modeling and fatigue testing to compare residual hoop stresses induced by the cold expansion process using multiple balls with those arising from the cold expansion using a single ball and a tapered mandrel. The results showed that the use of three incremental balls significantly reduced the magnitude of non-conforming residual hoop stresses and the extension of this detrimental zone.

Ambriz R. et al. [10] evaluated the influence of variable load spectra on the fatigue crack growth of cold-expanded specimens. The results of the fatigue tests showed a 3.5 times improvement in the fatigue life of cold-expanded holes compared to untreated holes.

Zuccarello B. and Franco G Di. proposed a numerical experimental method for the analysis of residual stresses in cold-expanded holes [11]. The idea of the proposed method was to determine the residual strains through the thickness of the specimen near the hole in circumferential direction at different values of interference fit. In this case, rectangular grooves were successively milled on both sides of the plate. Strain gages attached to both sides of the plate were used to measure residual strains. The measured values of residual strains were used to calculate the residual stresses through the thickness of the specimen near the hole using the integral method, which involved computation of the influence matrix. In order to calculate influence coefficients, a numerical simulation was carried out using the boundary elements method and the commercial code BEASY 9.0.

Karuppanan S., Hashim M. H., and Wahab A. A. presented the results of a numerical study of the influence of the proximity of hole location and plate thickness on the distribution of residual stresses in a plate with two holes made of 7075-T6 aluminum alloy in the case of cold expansion of the hole [12]. Two-dimensional and three-dimensional models were created in ANSYS to obtain the results. It was shown that as the distance between adjacent holes increased, the residual stresses decreased. As the thickness of the plate increased, stresses decreased. Moreover, the residual stresses at the mid-thickness of the plate were higher than the results at the entrance and exit faces.

V. Archard et al. [13] assessed the influence of the cold expansion ratios on the fatigue strength of Ti-6Al-4V tensile specimens by conducting an experimental and numerical study. The aim of the study was to understand the impact of the high-expansion ratio in titanium holes and the influence on fatigue performance. The authors argued that a very good correlation between the experimental and numerical results was observed.

Gao Y. and Zhong Z. pointed out that residual stresses have a significant effect on material fatigue, stress corrosion cracking, and other fracture parameters [14]. The authors presented a set of parameters describing the residual stresses in the specimen for different types of treatment such as shot peening, hole cold expansion, laser shock peening, and ultrasonic peening. A series of experiments was used to study the effect of residual stresses on the initiation and propagation of fatigue cracks in specimens of aluminum alloys, steel, and titanium.

Yang Z., Lee Y., He S., Jia W., and Zhao J. introduced parameters such as dent size, the distance between shots, and shot numbers for a quantitative description of shot peening coverage [15]. A finite element simulation of shot peening was carried out using the commercial finite element program, ABAQUS, to estimate the surface treatment intensity by comparing results of the finite element simulation with the results of the Almen test.

Zarutskiy A. proposed a method for residual stress accounting, which was produced during barrier compression by introducing the concept of additional dummy compressive stresses [16]. Experimental verification of the proposed method to predict fatigue life was carried out using a series of pre-compressed specimens made of the D16AT (Д16АТ) and V95pchT2 (В95пчТ2) alloys. The influence of the compression depth, swage width, and sweep angle on variations in residual stress was analyzed.

Calif-Chica J., Marin M. M., Rubio E. M., Teti R., and Segreto T. presented the results of their numerical parametric simulation of the thick-walled cylinder hole cold expansion process obtained for an axisymmetric model using the finite element method and ANSYS v.19 software [17]. The influence of geometric parameters of a mandrel on the value of residual stresses and forces during hole cold expansion was investigated.

Aid A., Semari Z., and Benguediab M. proposed a method for studying the effect of the radial interference value during hole cold expansion on the growth rate of a fatigue crack in a strip of the 6082A T6 aluminum alloy [18]. The method involved carrying out numerical simulations and a full-scale experiment with subsequent verification of the results. The finite element analysis was used to simulate the fatigue crack growth in a plate with a central hole using the ANSYS software. Based on the simulation results, using Irwin's theory, the stress intensity factor was calculated. The Paris equation was used to estimate the fatigue crack growth rate. The experimental part of this work was subdivided into three steps: a quasi-static tensile test to determine the mechanical characteristics of the material, fatigue tests, and cold expansion with different degrees of expansion. The cold working expansion process was realized by forcing a hard steel ball of 6 mm inside a predrilled hole.

P. Yasniy, O. Dyvdyk et al. [19] adopted an experimental technique to examine the impact of cold expansion holes in D16chT aluminum alloy plates with a pre-existing quarter-elliptical crack on fatigue crack growth and residual life. It was found that cold expansion of the hole with an expansion ratio of 2.7% significantly retarded the fatigue crack growth rate and increased residual lifetime by three times.

Vorobiov Y., Voronko V., and Stepanenko V. presented the results of a comparative analysis of quasi-static low-speed, high-speed, and pulse hole cold expansion processes. For comparison, the authors used a multivariate full-scale test [20]. The test results showed that in the case of impulse hole cold expansion, the hourglass shape of the hole decreased by 4–5% in comparison with quasi-static low-speed hole cold expansion. It was also noted that galling on the supporting surface decreased during impulse hole cold expansion with varying degrees of interference.

Sementsov V. and Vasylevskyi E. analyzed the influence of some of the parameters of technological tools used for barrier compression and the loading level on the characteristics of the local stress–strain state of a plate with a hole during its tension [21]. It was found numerically that the use of barrier compression with a depth of 0.3 mm helped to reduce the maximum principal stresses of an equivalent repeated cycle by 1.5–2.2 times in the section along the axis of hole and by 1.4–1.7 times in the compressed zone, with respect to stresses for a plate with an untreated hole.

The aim of this work is to estimate the possible benefits of using surface treatment-combined methods for enhancing the fatigue life of lower wing panel structural elements in the area of a drain hole by carrying out numerical simulations and fatigue testing.

2. Initial Data

A specimen of a plate with three holes of 8 mm in diameter was used to simulate operation under loading of the real part of the structure which contains drain holes.

A rectangular plate (Figure 1) has the following geometric dimensions: the plate thickness is 5 mm, width in sections along the axes of holes is 48 mm, the hole diameter is 8 mm, ratio of plate width to diameter (B/d) is 6, and the distance between the centers of the holes is 12 mm. The edges of the holes are chamfered with dimensions of 0.5 × 45°.

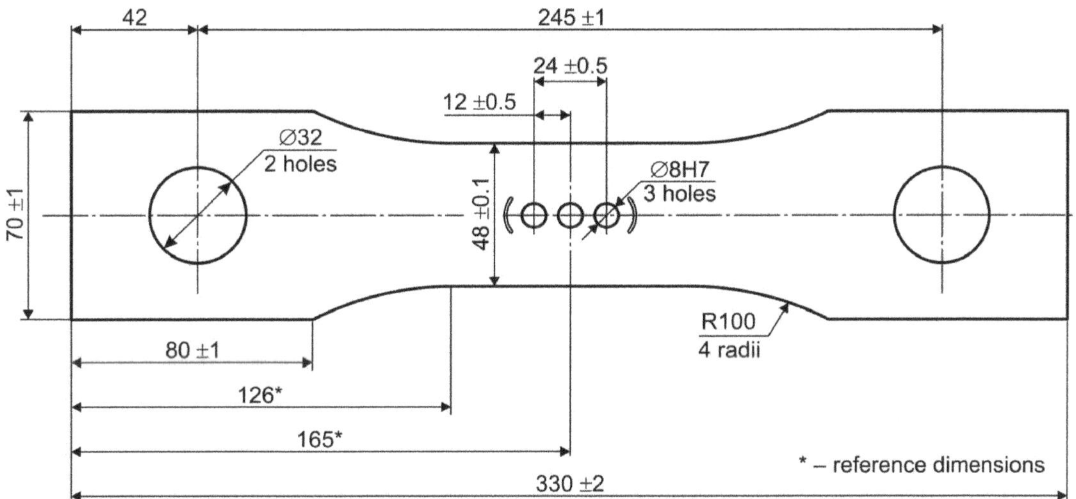

Figure 1. Specimen of plate with three holes of 8 mm in diameter.

All of the dimensions shown in Figures 1–3 are in millimeters.

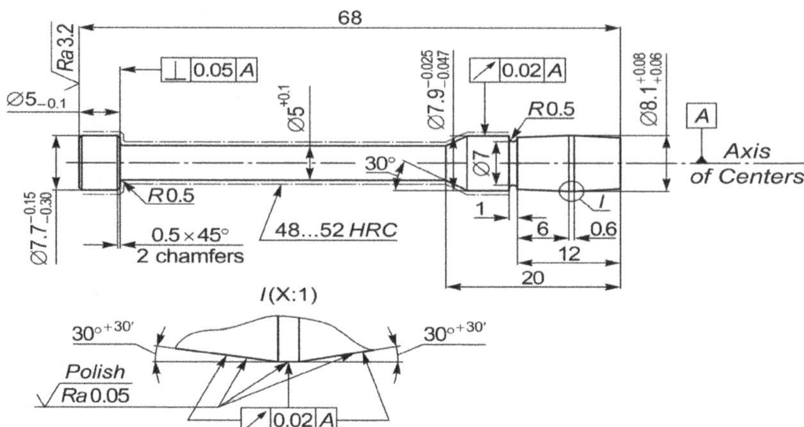

Figure 2. Geometry and main dimensions of mandrel.

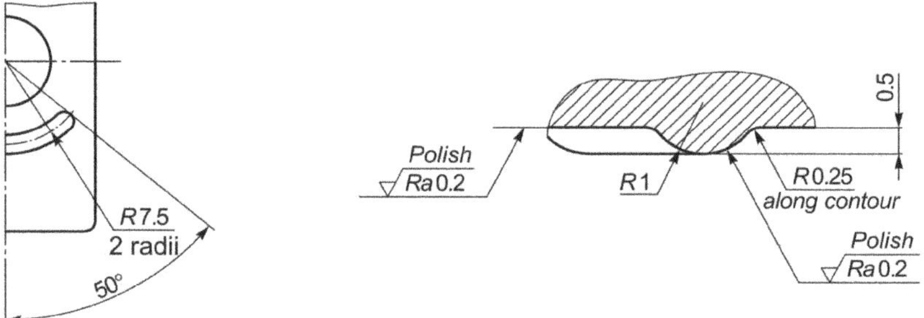

Figure 3. Geometry and main dimensions of swage for stamping of segment dimples.

The two combined methods of surface plastic treatment in the location around drain holes are:

1. holes rolling with an interference of 0.4% and barrier segment compression around the holes with a depth of 0.3 mm;
2. hole cold expansion with an interference of 2% and barrier segment compression around the holes with a depth of 0.3 mm.

Figures 2 and 3 show the geometry and main dimensions of the technological tools used for treatment.

The plate material is the D16T (Д16Т) aluminium alloy (Figure 1), which has the following physical and mechanical characteristics: Young's modulus E = 72,000 MPa, Poisson's ratio ν = 0.3, yield stress $\sigma_{0.2}$ = 285 MPa, and ultimate stress σ_B = 430 MPa [22]. As the local stresses on the plate in the vicinity of the holes exceed the yield stress resulting in developing permanent plastic deformations, nonlinear structural analysis is conducted which requires the definition of the stress–strain curve. Figure 4 shows a true stress–strain diagram of the D16T (Д16Т) aluminum alloy.

The mandrel (Figure 2) is made of steel KhVG (ХВГ) according to GOST 5950-73 [23] with the following characteristics: Young's modulus E = 200,000 MPa, yield stress $\sigma_{0.2}$ = 1400 MPa, ultimate tensile stress is σ_B = 1600 MPa, and Poisson's ratio is ν = 0.3 [24].

The swages (Figure 3) used for barrier compression are made of the U8A (У8А) steel according to GOST 1435-90 [25] with the following characteristics: Young's modulus E = 200,000 MPa, yield stress $\sigma_{0.2}$ = 750 MPa, ultimate tensile stress σ_B = 1200 MPa, and Poisson's ratio ν = 0.3 [24].

The finite element method and ANSYS software are used to calculate the local stress–strain state in the plate.

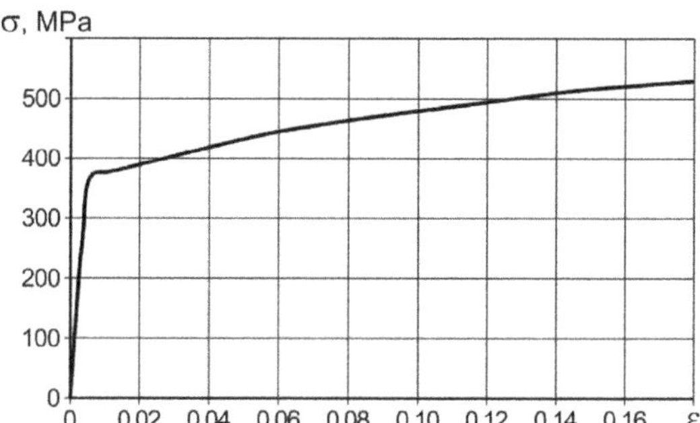

Figure 4. True stress–strain diagram of specimen from D16T (Д16Т) aluminum alloy.

To carry out nonlinear structural analysis, the boundary and contact conditions are applied as follows:

- **Boundary conditions.**

  Symmetry condition:

$$U_x = 0, \ X_i \in \partial\Gamma_{U_1}; \ U_y = 0, \ X_i \in \partial\Gamma_{U_2}; \ U_z = 0, \ X_i \in \partial\Gamma_{U_3}. \tag{1}$$

  Force boundary condition:

$$\sigma_{ij}n_i = \sigma_{gross}, \ X = L/2. \tag{2}$$

- **Contact conditions.**

$$\left(\sigma_{ij}^+ - \sigma_{ij}^-\right) \cdot n_i = 0, \ \left(x_i^+ - x_j^-\right) \leq g, \ X_i \in \partial\Gamma_{C_1} \cup \partial\Gamma_{C_2} \tag{3}$$

Here, U_x, U_y and U_z are the components of the displacement vector in the global Cartesian coordinate system; ∂_{u_1}, ∂_{u_2}, and ∂_{u_3} are boundaries on which symmetry condition is applied; σ_{ij} are stress tensor components; n_i is the unit normal vector; x_i^+ and x_j^- are the coordinates of nodes on target and contact surfaces; g is the gap which defines permissible interpenetrations of nodes; ∂_{C_1} and ∂_{C_2} are surfaces on which contact conditions are applied.

Figure 5 shows the plate with three holes with a visualization of the boundary conditions and applied loads.

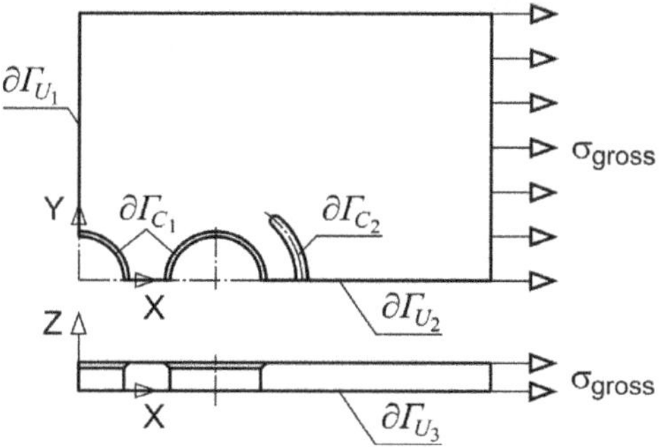

Figure 5. Plate with three holes with visualization of boundary conditions and applied loads.

3. Finite Element Modeling

Because of the symmetry of the model and applied loads, only 1/8 of the plate is used for the three-dimensional modeling in ANSYS, with appropriate boundary conditions set on the symmetry planes. Due to the fact that the rigidity of the technological tool used for treatment is three times higher than that of the plate, and that in this study stress and strain in the tool are outside the area of interest, we then consider the tool to be a rigid body. This allows us to reduce the overall dimensions of the finite element model and computational costs.

The finite element model of the plate (Figure 6) consists of a set of the three-dimensional elements of the solid deformable body, SOLID45 [26]. The element is defined by eight nodes having three degrees of freedom at each node: translations in the nodal x, y, and z directions.

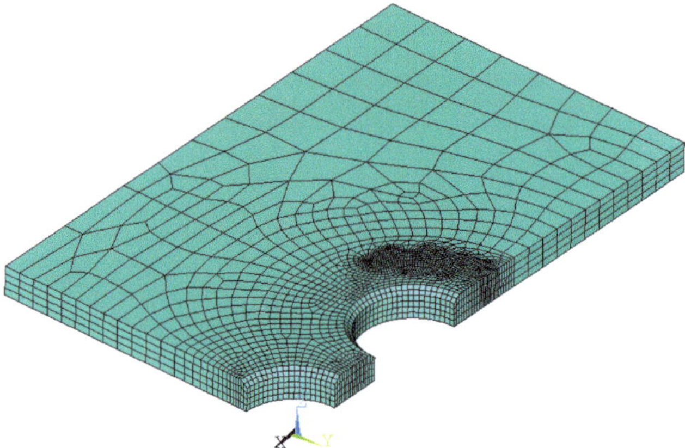

Figure 6. Finite element model of plate with three holes.

To accurately capture the local stresses in the stress concentration regions, a fine mesh was generated around the holes and in the area of the segment dimples; meanwhile, in the other regions which were far away from the stress risers, the model had a coarse mesh (Figures 6 and 7).

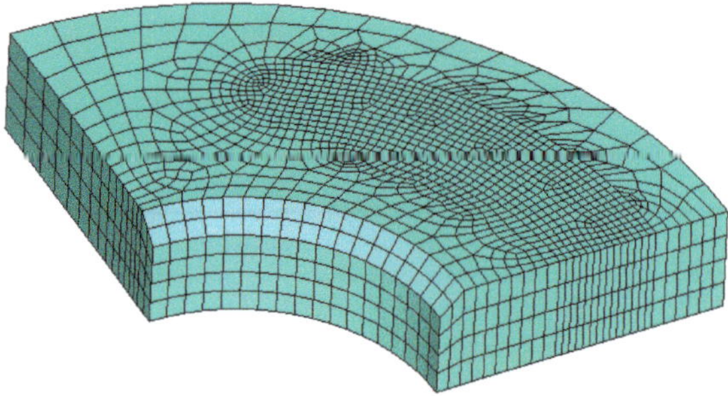

Figure 7. Portion of finite element model of plate with three holes in the area of segment dimple stamping.

Figures 8–10 show the finite element models of the roller, mandrel, and swage which were created with the conditions described in this section.

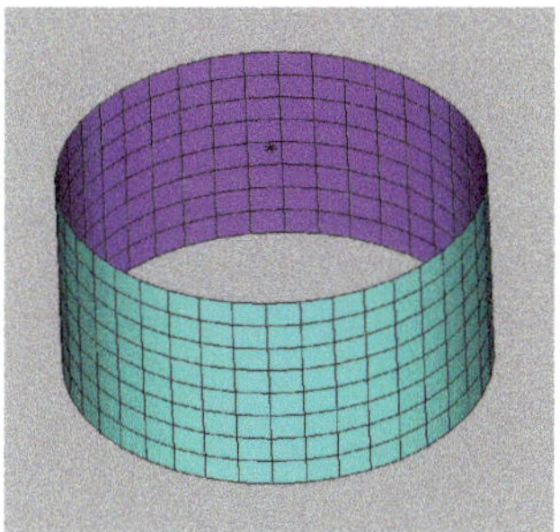

Figure 8. Finite element model of roller.

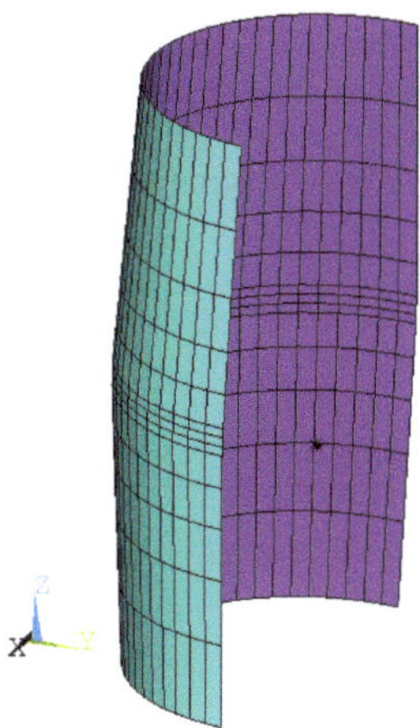

Figure 9. Finite element model of mandrel.

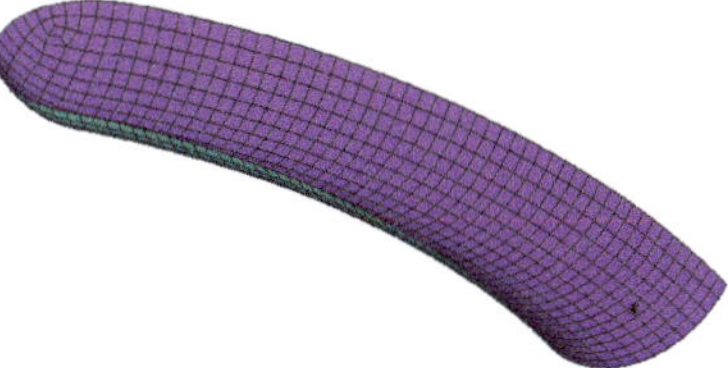

Figure 10. Finite element model of the swage for stamping of segment dimples.

Contact Impact Model

A general "Surface-to-Surface" contact model was used to simulate contact interaction between the technological tool and respective areas of the plate [27]. The "Rigid-to-Flexible" type of contact was used for this model [27], where the technological tool is considered to be a rigid body and the plate is considered to be a deformable body.

Contact surfaces are defined via a set of the TARGE170 target elements [26] and the CONTA173 contact elements [26].

The target elements are overlaid on the part that defines the technological tool geometry, while the contact elements are overlaid on the respective surfaces of the plate. To include friction in the contact algorithm, the Coulomb model of friction was used by setting the friction factor as f = 0.6. The pilot node has six degrees of freedom and is used to control six independent motions of the entire target surface of the rigid body: three translations along the X, Y, and Z direction and three rotations around X, Y, and

Z. Interference occurring during rolling and hole cold expansion processes was modeled geometrically by considering an oversized roller or mandrel that was pushed through the hole to yield it and to generate residual compressive stress. The initial penetration of the target elements through the contact surface was included in the contact algorithm. The motion of the mandrel and swage was modeled by setting respective translations in pilot nodes in a negative Z direction.

Nonlinear analysis was carried out taking into account geometric and material nonlinearities. In this case, the level of design loads was set equal to four numbers of gross stress σ_{gross} = 0, 100, 130, and 150 MPa.

4. Results

The following section presents the numerical simulation and fatigue test results. The main results of the finite element modeling involve von Mises stress distributions at the different stages of the surface treatment and provide graphs showing the impact of loading level and treatment method on the variation in maximum principal stresses, strains, and specific strain energy of the repeated stress cycle in the plate. The fatigue test results are fatigue failure types of the specimens tested and graphs are provided illustrating the number of cycles to failure for the plate specimens with and without surface treatments.

4.1. Numerical Modeling Results

Using the finite element model created in ANSYS software, we carried out a step-by-step simulation of the surface treatment process in the areas surrounding the holes. Then, we followed this with tensile loading of the plate in order to study the effect of application of the considered treatment methods on the fatigue life of the plate with holes. Local stress–strain state parameters were calculated in two areas:

1. cross-section along the extreme hole axis;
2. stamping area of segment dimples.

The amplitude of the principal stress, strain, and specific strain energy were calculated in the cross-section along the extreme hole axis and the stamping area of segment dimples, using the following equations:

$$\sigma_{1a} = \frac{\sigma_{1max} - \sigma_{1min}}{2}; \; \varepsilon_{1a} = \frac{\varepsilon_{1max} - \varepsilon_{1min}}{2}; \; w_{1a} = \frac{1}{2}\sigma_{1a} \cdot \varepsilon_{1a}, \tag{4}$$

in which:

- σ_{1max}, ε_{1max}, and w_{1max} are the maximum principal tensile stresses, strains, and specific strain energy. We took the value of the local parameter (stress, strain, or specific strain energy) in the corresponding area for the state when external loads are applied as the maximum value.
- σ_{1min}, ε_{1min}, and w_{1min} are the minimum principal tensile stresses, strains, and specific strain energy. We took the value of the specific parameter in the corresponding area for the state corresponding to the load release as the minimum value.

When calculating strain amplitude, elastic strain was considered, because it significantly changes during cyclic loading.

The initial loading cycle (Figure 11) was converted to an equivalent repeated cycle using Oding's formula:

$$\sigma_{01\,max} = \sqrt{2\sigma_{1a} \cdot \sigma_{1\,max}}; \; \varepsilon_{01\,max} = \sqrt{2\varepsilon_{1a} \cdot \varepsilon_{1\,max}}; \tag{5}$$

$$w_{01\,max} = \frac{1}{2}\sigma_{01\,max} \cdot \varepsilon_{01\,max}. \tag{6}$$

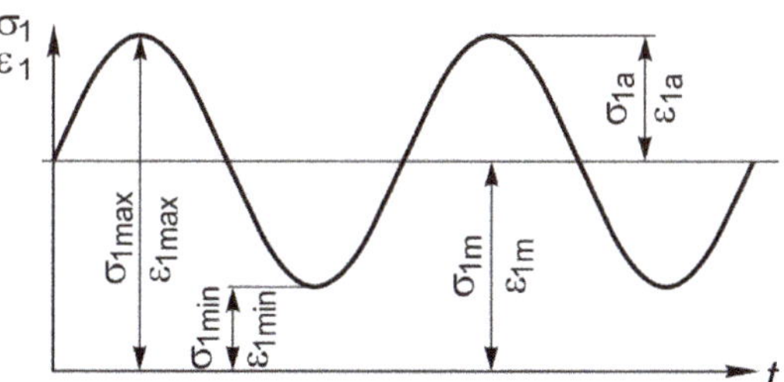

Figure 11. Loading cycle and its parameters.

The mechanism of the plastic strain area formation was studied and distribution of the von Mises stress in the plate with three holes was analyzed step by step in the case of the treatment using the hole cold expansion method with an interference of 2% and following stamping of the segment dimples (Figure 12). The unit of the von Mises stress is MPa.

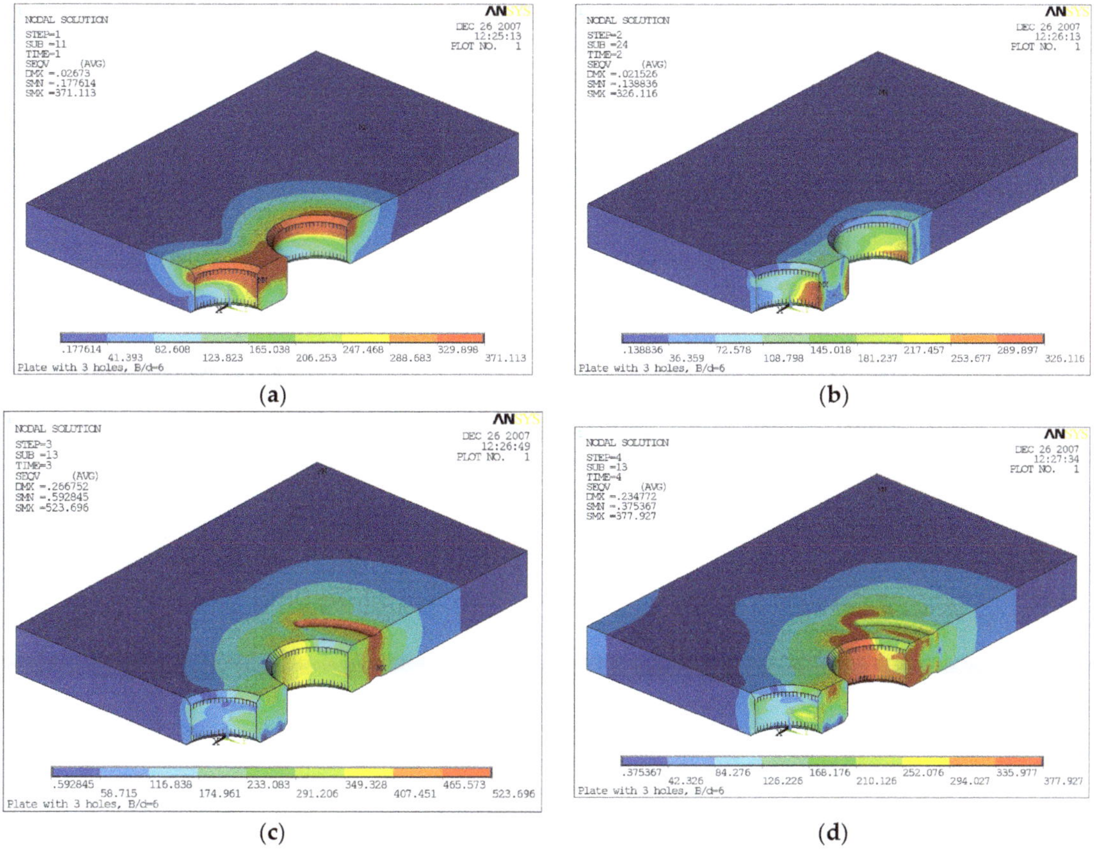

(**a**)

(**b**)

(**c**)

(**d**)

Figure 12. Distribution of the von Mises stress in a plate with three holes in the case of using the hole cold expansion method with an interference of 2% and following barrier stamping with a depth of 0.3 mm: (**a**) initial step is penetration of mandrel; (**b**) after hole cold expansion; (**c**) after hole cold expansion and following stamping with a depth of 0.3 mm; (**d**) after surface treatment when the tools are removed.

4.2. Experimental Results

To carry out experimental studies to determine the characteristics of fatigue life, standard specimens were designed as follows: a plate with a central hole (B/d = 6) (Figure 13), a plate with three holes of 8 mm in diameter (distance between hole centers is 12 mm) without treatment (Figure 14); and with surface treatment in the hole areas (Figure 15). The shape and dimensions of the test specimen were designed in accordance with [28].

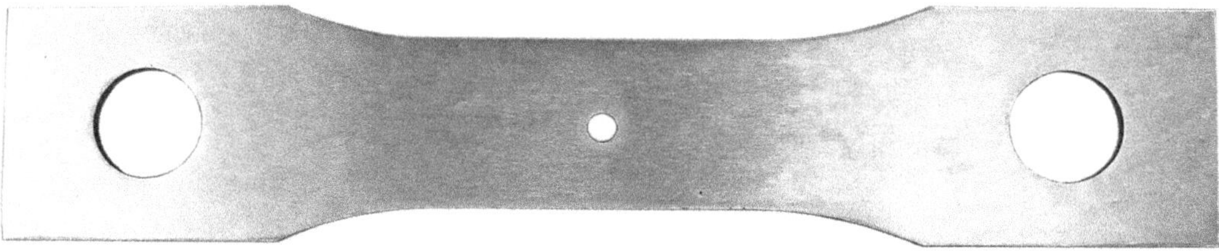

Figure 13. Test specimen of plate with central hole.

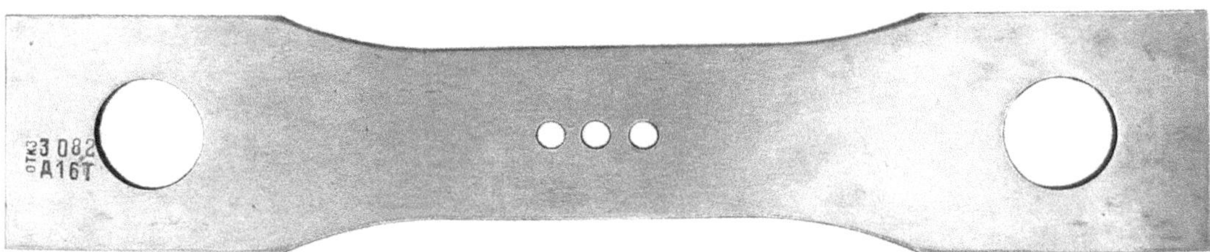

Figure 14. Test specimen of plate with three holes.

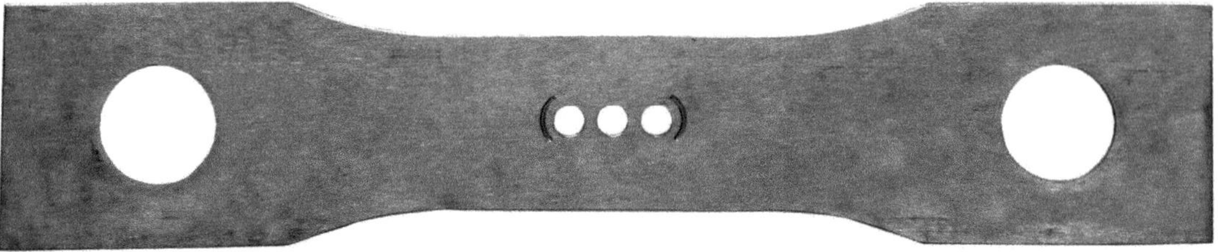

Figure 15. Test specimen of plate with three holes treated around holes.

The tests of specimens of plates with holes were carried out in the "STRENGTH" research laboratory of the National Aerospace University's "Kharkiv Aviation Institute" using the UMM-02 electromechanical test machine. Specimens of plates were loaded with uniaxial alternating cyclic loads with a frequency of 12 Hz. Specimens of plates were tested at the level of cyclic loading of $\sigma_{gross\,min}$ = 0 MPa and $\sigma_{gross\,maz}$ = 150 MPa.

Three specimens of the plates with a central hole were tested at this loading level. The fatigue failure of specimens took place in a section along the axis of the hole. Fatigue cracks were initiated on the edge between the conical surface of chamfer and the cylindrical surface of the hole. Figure 16 shows the regions and fatigue failure types.

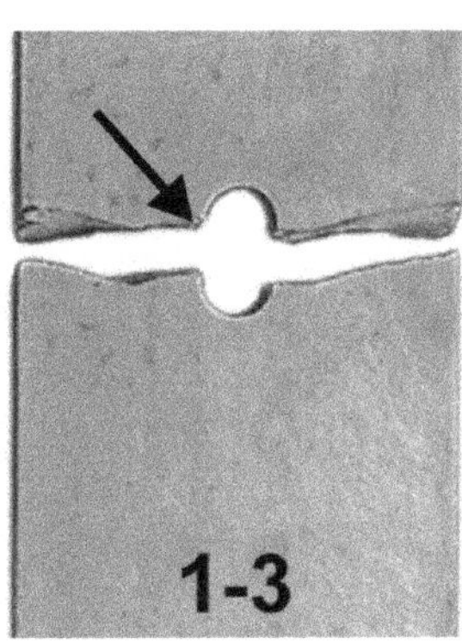

Figure 16. Fatigue failure of specimens of plates with central holes fabricated from pressed section (D16T (Д16) aluminum alloy) with a loading spectrum of $\sigma_{gross\ min} = 0$ MPa and $\sigma_{gross\ maz} = 150$ MPa; arrow in figure (⟶) is the area where fatigue crack is initiated; 1-1, 1-2, 1-3 are numbers of specimens.

Three specimens of plates with three untreated holes were tested. The fatigue failure of specimens occurred in a section along the axis of the extreme hole. Fatigue cracks were initiated on the edge between the conical surface of the chamfer and the cylindrical surface of the hole. Figure 17 shows the areas and fatigue failure types.

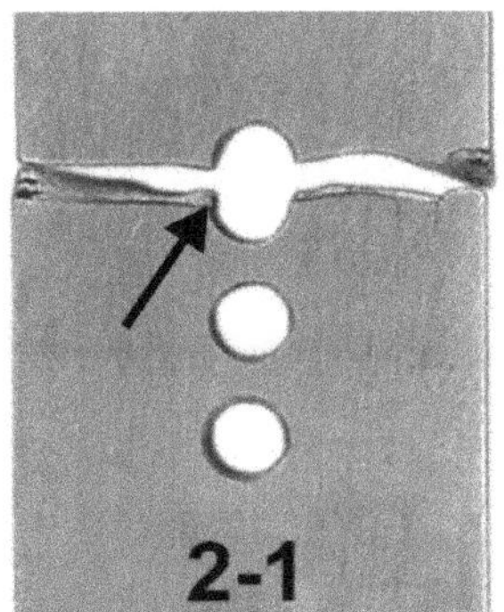
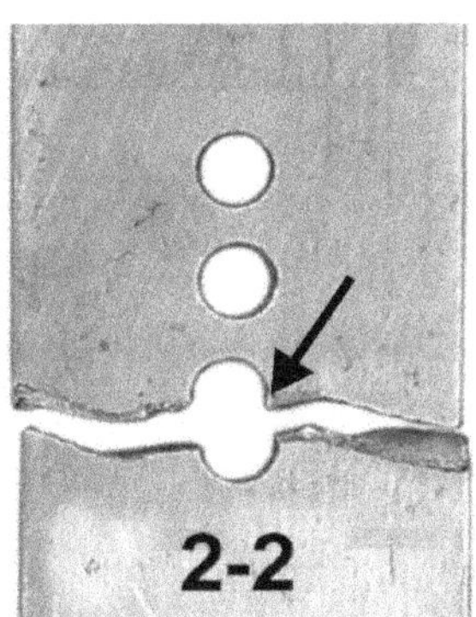

Figure 17. Fatigue failure of specimens of plates with three untreated holes fabricated from pressed section (D16T (Д16) aluminum alloy) with a loading spectrum of $\sigma_{gross\ min} = 0$ MPa and $\sigma_{gross\ maz} = 150$ MPa; arrow in figure (⟶) is the area where fatigue crack is initiated; 2-1, 2-2, 2-3 are numbers of specimens.

Three specimens of plates with three holes were tested, which were treated using the rolling method with an interference of 0.4% and barrier compression. Crack propagation in specimen No. 3-1 started from the surface in the area of the extreme hole, in specimen No. 3-2 crack propagation started from the surface in the area of the extreme hole and segment dimple, and specimen No. 3-3 failed at the grips location due to fretting corrosion. Figure 18 shows the areas and fatigue failure types of the specimens tested.

Three specimens of the plates with three holes were tested, which were treated using the hole cold expansion method with an interference of 2% and barrier compression. Specimen No. 4-1 failed at the grips location due to fretting corrosion. Crack propagation in specimen No. 4-2 started from the surface in the fillet area. Crack propagation in specimen No. 4-3 started from the surface in the section of the extreme hole on the edge between the conical surface of chamfer and the cylindrical surface of the hole. Figure 19 shows the areas and fatigue failure types of the specimens tested.

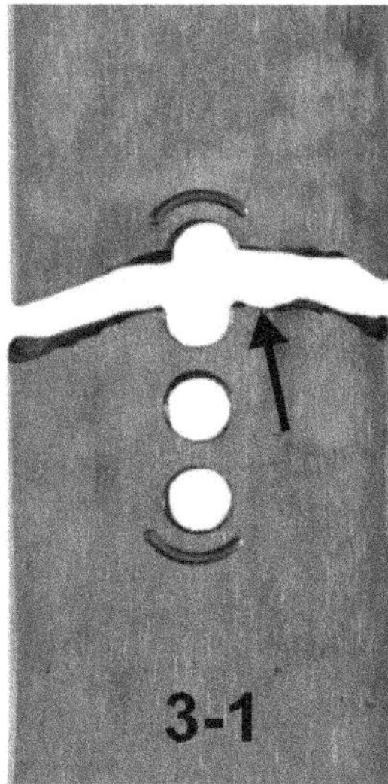
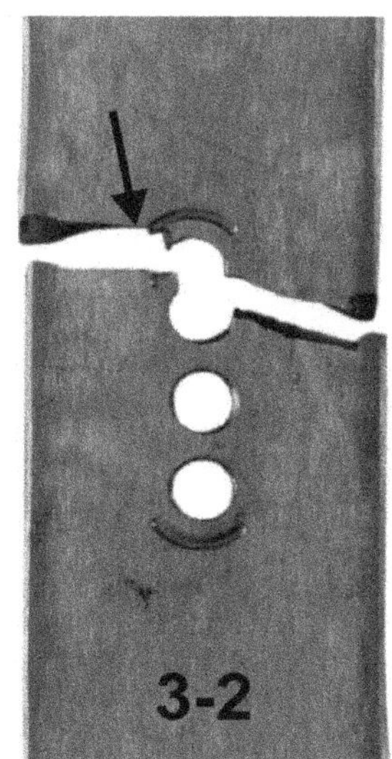
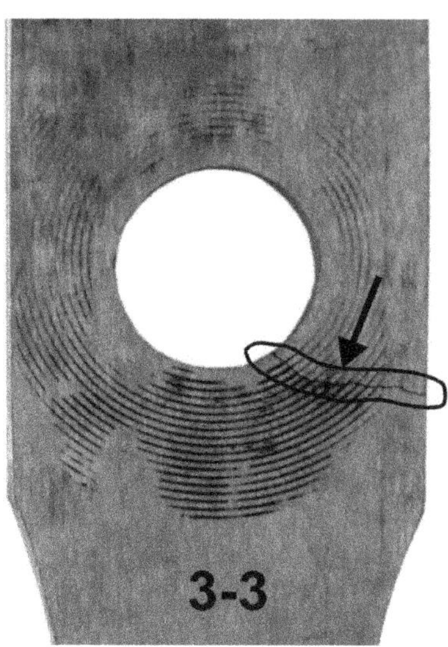

Figure 18. Fatigue failure of specimens of D16 (Д16) aluminum plates with three holes treated using rolling with an interference of 0.4% and barrier compression with a loading spectrum of $\sigma_{gross\,min}$ = 0 MPa and $\sigma_{gross\,maz}$ = 150 MPa; arrow in figure (➤) is the area where fatigue crack is initiated; 3-1, 3-2, 3-3 are numbers of specimens.

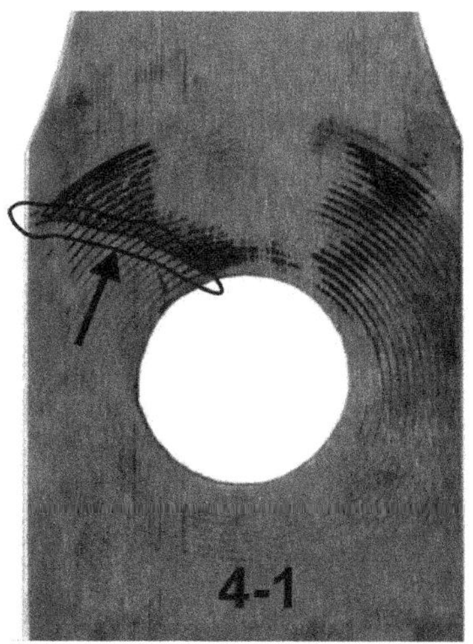
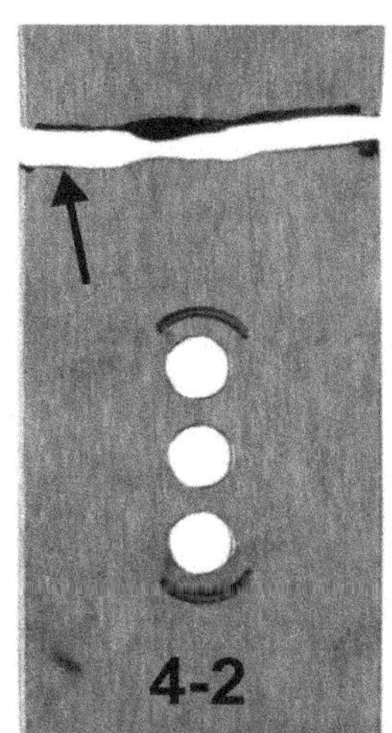
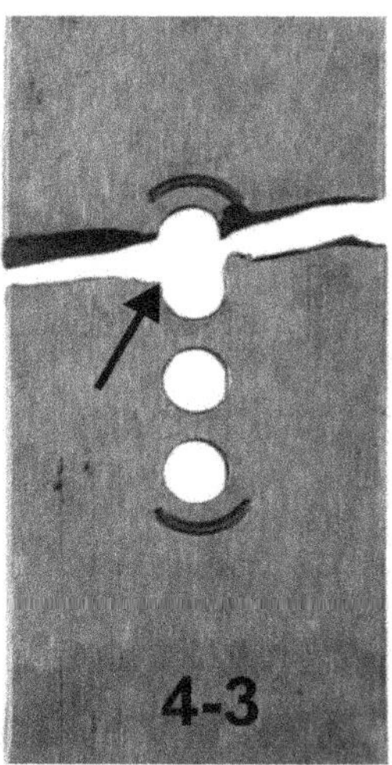

Figure 19. Fatigue failure of specimens of D16 (Д16) aluminum plates with three holes treated using hole cold expansion with an interference of 2% and barrier compression with a loading spectrum $\sigma_{gross\,min}$ = 0 MPa and $\sigma_{gross\,max}$ = 150 MPa; arrow in figure (➤) is the area where fatigue crack is initiated; 4-1, 4-2, 4-3 are numbers of specimens.

5. Discussion

In this section, based on the results from the previous sections, the following are discussed: the distribution of principal tensile stress, strain, and specific strain energy of an equivalent repeated stress cycle; the impact of loading level and treatment method on variations in maximum principal stress, strain, and specific strain energy of equivalent repeated cycle in the plate; and the impact of hole number and surface treatment methods on the fatigue life of plates with holes.

The analysis of the stress distribution revealed that in the area to be treated, the local von Mises stress exceeded the yield stress which indicated the development of plastic strain regions in the material during its treatment. The regions are localized in close proximity to the areas of the plate to be treated. In the case of the segment stamping, the plastic strain region is distributed through the entire thickness directly under the dimple formed by treatment. The obtained results indicate that during surface treatment of corresponding areas of the plate, residual compressive stresses are produced which in combination with working tensile stresses lead to a reduction in local stress level in the areas. This, in turn, enhances the fatigue life.

To analyze the impact of loading level and treatment method on variations in maximum principal stress, strain, and specific strain energy of the equivalent repeated cycle in the plate, the graphical dependencies (Figures 20–22) between calculated values (see Section 3) of $\sigma_{01\,max}$ and σ_{gross}, $\varepsilon_{01\,max}$ and σ_{gross}, and $w_{01\,max}$ and σ_{gross} were constructed.

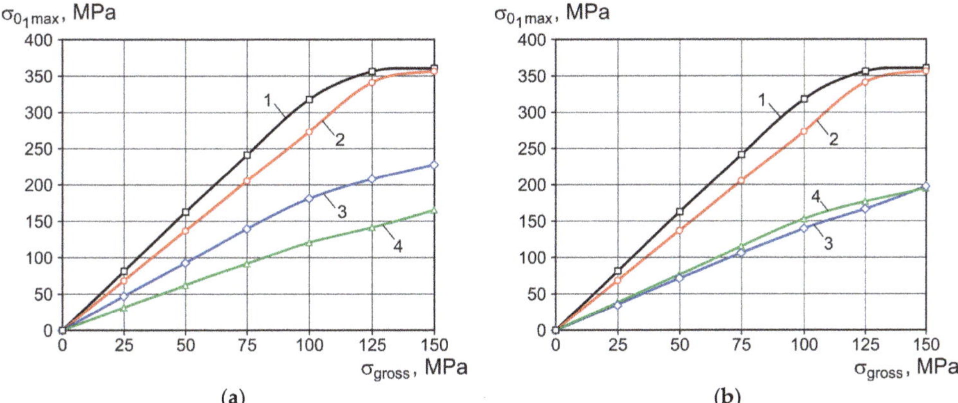

Figure 20. Impact of loading level and treatment method on variations in maximum principal stress of repeated cycles in plate: (**a**) along the axis of the extreme hole; (**b**) stamping region.

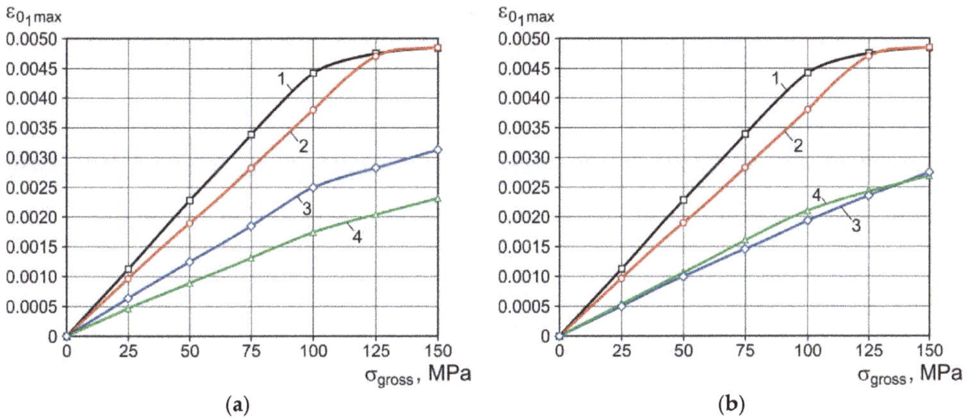

Figure 21. Impact of loading level and treatment method on variations in maximum principal strain of repeated cycles in plate: (**a**) along the axis of the extreme hole; (**b**) stamping region.

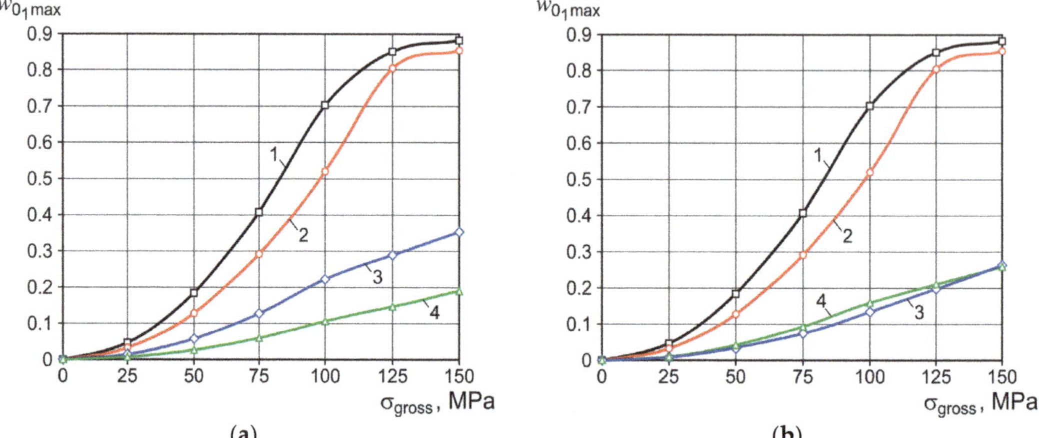

Figure 22. Impact of loading level and treatment method on variations in maximum principal specific strain energy of repeated cycles in plate: (**a**) along the axis of the extreme hole; (**b**) stamping region.

There are four curves on each graphic, which correspond to the following:

- Number 1 represents the plate with a hole of 8 mm in diameter and a chamfer of $0.5 \times 45°$ without surface treatment;
- Number 2 represents the plate with three holes of 8 mm in diameter and a chamfer of $0.5 \times 45°$; the distance between holes is 12 mm without surface treatment;
- Number 3 represents the plate with three holes of 8 mm in diameter and a chamfer of $0.5 \times 45°$; the distance between holes is 12 mm in the case of hole cold expansion with an interference of 2% and stamping of segment dimples with a depth of 0.3 mm;
- Number 4 represents the plate with three holes of 8 mm in diameter and a chamfer of $0.5 \times 45°$; the distance between holes is 12 mm, in the case of rolling with an interference of 0.4% and stamping of segment dimples with a depth of 0.3 mm.

Analysis of the obtained results showed that the stamping region is more dangerous since the characteristics of the local stress–strain state in this region exceed the corresponding characteristics in the area along the axis of the extreme hole.

In the case of combined application of the rolling method with an interference of 0.4% and stamping of segment dimples with a depth of 0.3 mm, the level of the maximum principal tensile stresses of the equivalent repeated cycles in the section along the axis of the extreme hole decreased by 2.16–2.4 times, the maximum principal strains by 2.13–2.33 times and the specific strain energy by 2.42–3.07 times compared to these characteristics for a plate with three holes without any surface treatment of the holes.

The combined application of hole cold expansion with an interference of 2% and stamping of segment dimples with a depth of 0.3 mm results in decreasing the maximum principal tensile stresses of the equivalent repeated cycles by 1.84–2.06 times, the maximum principal strain by 1.79–2.01 times, and the specific strain energy by 2.51–3.01 times compared to these characteristics for a plate with three holes without surface treatment of holes.

The chart in Figure 23 demonstrates the influence of surface treatment methods on the fatigue life of plates with holes, where the designations (numbers in the rectangles) correspond to the following:

- Number 1 represents the plate with a central hole of 8 mm in diameter (B/d = 6);
- Number 2 represents the plate with three holes of 8 mm in diameter (B/d = 6) (distance between holes' center is 12 mm) without surface treatment;

- Number 3 represents the plate with three holes of 8 mm in diameter (B/d = 6) (distance between holes' center is 12 mm) treated using rolling with an interference of 0.4% and barrier compression;
- Number 4 represents the plate with three holes of 8 mm in diameter (B/d = 6) (distance between holes' center is 12 mm) treated using hole cold expansion with an interference of 2% and barrier compression.

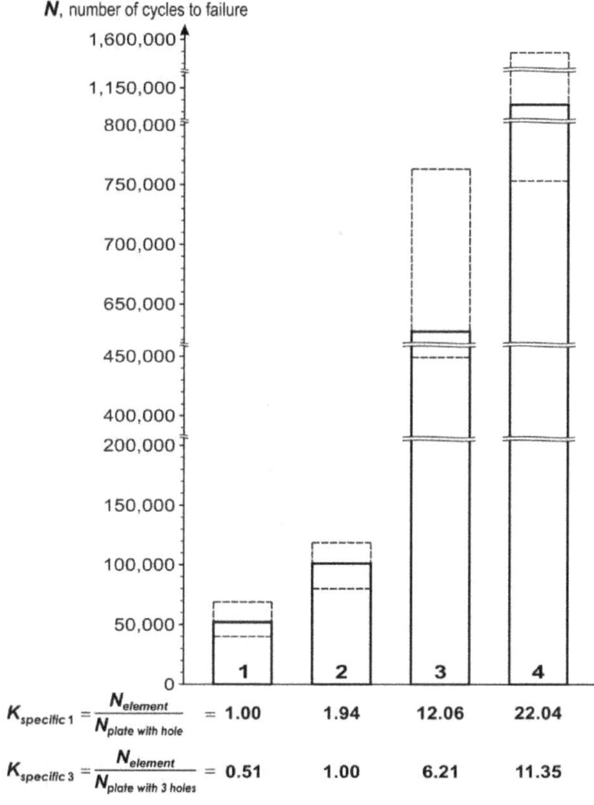

$$K_{\text{specific 1}} = \frac{N_{element}}{N_{plate\ with\ hole}} = 1.00 \qquad 1.94 \qquad 12.06 \qquad 22.04$$

$$K_{\text{specific 3}} = \frac{N_{element}}{N_{plate\ with\ 3\ holes}} = 0.51 \qquad 1.00 \qquad 6.21 \qquad 11.35$$

Figure 23. Impact of hole numbers and surface treatment methods on fatigue life of plates with holes at the loading level of $\sigma_{0\ gross\ max}$ = 150 MPa.

Figure 23 provides the following notations: $K_{\text{specific1}}$ is the ratio of number of cycles of specimens 1, 2, 3, or 4 to the number of cycles of specimen 1 (plate with the hole of 8 mm in diameter); $K_{\text{specific2}}$ is the ratio of the number of cycles of specimens 1, 2, 3, or 4 to the number of cycles of specimen 2 (plate with three holes of 8 mm in diameter without surface treatment).

The following results were obtained during fatigue tests: the fatigue life of the plate with three holes treated by rolling with an interference of 0.4% and barrier compression is higher by 12.06 times than the fatigue life of the plate with a central hole. The same enhancement in fatigue life compared to fatigue life of the plate with three untreated holes is by 6.21 times. The fatigue life of the plate with three holes, which are treated by hole cold expansion with an interference of 2% and barrier compression, is higher by 22.04 times than the fatigue life of the plate with a central hole. The same enhancement of fatigue life compared to the fatigue life of the plate with three untreated holes is by 11.35 times.

6. Conclusions

In this paper, the effectiveness of two combined methods of surface plastic treatment to enhance the fatigue life of structural parts with holes is analyzed using numerical simulation and fatigue testing. These combined methods are hole rolling with an interference of 0.4% and barrier segment compression around the holes with a depth of 0.3 mm, and hole cold expansion with an interference of 2% and barrier segment compression with a depth of 0.3 mm.

Initially, we created a finite element model of the plate with three holes of 8 mm in diameter and the corresponding technological tools used for the surface treatment of the hole-adjacent areas to perform a step-by-step simulation of the surface treatment and subsequent tension of the plate. In order to accurately capture the local stresses in the regions of stress concentration, a fine hexahedral mesh was generated there; meanwhile, other regions, which are far from stress rises, have a coarse mesh. Because of the symmetry of the plate and the applied load, only 1/8 of the plate was used for three-dimensional modeling in ANSYS with the application of appropriate boundary conditions on the symmetry planes. Taking into account that the rigidity of the technological tool is three times higher than that of the plate and that in this study, stress and strain in the tool are outside the area of interest, we, therefore, considered the tool to be a rigid body. This enabled us to reduce the overall dimensions of the finite element model and computational costs.

A nonlinear solution of the problem was obtained in ANSYS software with respect to the following nonlinearities: finite deformations, plasticity of the plate material, and contact interaction with friction. Then, distributions of major principal stress, strain, and specific strain energy along the extreme hole axis and stamping area of segment dimples were plotted for two load steps: when the tensile load was applied to obtain the maximum value and after surface treatment of hole-adjacent areas to calculate the minimum value of the local parameters. Based on distributions of the maximum and minimum values of major principal stress, strain, and specific strain energy in these two areas of interest, the amplitude of the local parameters was calculated. After that, the initial loading cycle was converted to an equivalent repeated cycle using Oding's formula to obtain the maximum values of major principal stress, strain, and specific strains of the equivalent repeated cycle.

Finally, in order to estimate the effectiveness of applicable combined methods of surface plastic treatment, graphs showing variations in the maximum values of major principal stress, strain, and specific strains of equivalent repeated cycles versus operating loading were plotted. The results showed that application of combined surface treatment of the areas in the vicinity of the holes significantly reduced the corresponding values of the maximum major principal stress, strain, and specific strains of the equivalent repeated cycles as compared to these characteristics for a plate with three holes without surface treatment of holes. Consequently, reduction in the corresponding values of stress, strain, and specific energy leads to fatigue life enhancement of the surface-treated specimens.

Concurrently, in order to confirm the numerical simulation results, a series of fatigue tests were carried out. The results of the test comprise fatigue failure of the specimens and a chart showing the impact of the surface treatment methods on fatigue life of plates with holes.

Concerning the fatigue failure of the specimens, the results showed that in the case of using hole cold expansion with an interference of 2% and barrier compression, only one of the three specimens failed in the stress concentration area, whereas all three specimens failed in the stress riser region for the plate with a single and three untreated holes. This may indirectly prove the effectiveness of the application of combined surface treatments.

To estimate the benefits of using combined methods of surface treatment, the number of cycles to failure of tested specimens was compared and the ratio of the respected number of cycles for the plates with treated holes to the number of cycles for plates with untreated holes was calculated.

The corresponding values of $K_{specific1}$ = 12.06 and $K_{specific3}$ = 6.21 were obtained for the plate with three holes treated using rolling with an interference of 0.4% and barrier compression, and $K_{specific1}$ = 22.04 and $K_{specific3}$ = 11.35 were obtained for the plate with three holes treated using hole cold expansion with an interference of 2% and barrier compression, respectively.

Overall, based on the numerical simulation and fatigue test results, hole cold expansion with an interference of 2% and barrier compression with a depth of 0.3 mm was chosen as the most effective method for improvement in the fatigue life of structural elements in the vicinity of drain holes.

Author Contributions: Conceptualization, O.G. and A.H.; methodology, O.G., A.H. and V.M.; software, S.S.; validation, V.L. and A.H.; formal analysis, O.G. and S.S.; investigation, S.S.; resources, O.G., A.H., V.L. and V.M.; data curation, A.H.; writing—original draft preparation, S.S.; writing—review and editing, O.G.; visualization, A.H.; supervision, V.M. All authors have read and agreed to the published version of the manuscript.

Funding: This research received no external funding.

Data Availability Statement: Data are contained within this article.

Acknowledgments: We are grateful to the administration of the National Aerospace University's "Kharkiv Aviation Institute" for the use of the "STRENGTH" research laboratory and the UMM-02 electromechanical test machine. Thanks are given to the administration and engineers of the Kharkiv State Aviation Manufacturing Company for their help in producing specimens for tests.

Conflicts of Interest: Author Vasyl Lohinov was employed by the company FED. Author Valerii Matviienko was employed by the company Ukrainian Research Institute of Aviation Technology. The remaining authors declare that the research was conducted in the absence of any commercial or financial relationships that could be construed as a potential conflict of interest.

References

1. Ol'kin, B.I.; Givanova, N.I.; Rodchenko, T.S.; Masyuk, A.M.; Shcheglova, N.I.; Barten'eva, G.F. Eksperimental'nye issledovaniya vliyaniya konstruktivnykh i tekhnologicheskikh faktorov na vynoslivost' tipovykh elementov aviatsionnykh konstruktsii. *Tr. TsAGI* **1980**, 1–86.
2. Rychik, V.P.; Litvinenko, E.A.; Ostapenko, N.T.; Vasilevskii, E.T. Issledovanie vynoslivosti nizhnikh panelei kryla s otverstiyami dlya peretekaniya topliva. *Tr. TsAGI* **1976**, 100–103.
3. Yucan, F.; Ende, G.; Honghua, S.; Jiuhua, X.; Renzheng, L. Cold expansion technology of connection holes in aircraft structures: A review and prospect. *Chin. J. Aeronaut.* **2015**, *28*, 961–973. [CrossRef]
4. Grebenikov, A.G.; Movchan, Y.A.; Grebenikov, V.A. Analiz kharakteristik lokal'nogo NDS s pomoshch'yu sistemy ANSYS v plastine s otverstiyami, podverzhennoi uprochneniyu dornovaniem ili glubokogo plasticheskogo deformirovaniyu i rastyazheniyu. *Vopr. Proekt. I Proizv. Konstr. Letatel'nykh Appar.* **2003**, *32*, 124–138.
5. Maharjan, N.; Chan, S.Y.; Ramesh, T.; Nai, P.; Tanako Ardi, D. Fatigue Performance of Laser Shock Peened Ti6Al4V and Al6060-T6 Alloys. *Fatigue Fract. Eng. Mater. Struct.* **2021**, *44*, 733–747. Available online: https://www.researchgate.net/publication/344856398_Fatigue_performance_of_laser_shock_peened_Ti6Al4V_and_Al6061-T6_alloys (accessed on 27 October 2023). [CrossRef]
6. Faghih, S.; Behzad, S.; Sugrib, B.; Shaha, K.; Jaheb, H. Effect of Split Sleeve Cold Expansion on Fatigue and Fracture of Rolled AZ31B Magnesium Alloy. *Theor. Appl. Fract. Mech.* **2023**, *123*, 103715. [CrossRef]
7. Li, Q.; Xue, Q.; Song, T.; Wang, Y.; Li, S. Cold Expansion Strengthening of 7050 Aluminum Alloy Hole: Structure, Residual Stress and Fatigue life. *Int. J. Aerosp. Eng.* **2022**, *2022*, 4057898. [CrossRef]
8. Yan, W.; Wang, Y.; Liang, S.; Huo, S. Study on the Residual Stress Distribution of Bi-directional Cold Expansion Process Performed on Open Holes. *J. Press. Vessel. Technol.* **2023**, *145*, 041501. [CrossRef]
9. Curto-Cardenas, D.; Calaf-Chica, J.; Bravo Diez, P.M.; Preciado Calzada, M.; Garcia-Tarrago, M.-J. Cold expansion Process with Multiple Balls—Numerical Simulation and Comparison with Single Ball and Tapered Mandrels. *Materials* **2020**, *13*, 5536. [CrossRef] [PubMed]
10. Ambriz, R.; Garcia, C.; Rodriguez-Reyna, S.L.; Ramos-Azpeitia, M.; Jaramillo, D. Synergy Effects in the Fatigue Cracks Growth of Hole Cold Expanded Specimens under Variable Cyclic Loading. *Int. J. Fatigue* **2020**, *140*, 105807. [CrossRef]
11. Zuccarello, B.; Franco, G.D. Numerical-experimental Method for the Analysis of Residual Stresses in Cold-expanded Holes. *Exp. Mech.* **2013**, *53*, 679–686. [CrossRef]
12. Karuppanan, S.; Hashim, M.H.; Wahab, A.A. Finite Element Simulation of Residual Stresses in Cold-expanded Plate. *Asian J. Sci. Res.* **2013**, *6*, 518–527. [CrossRef]

13. Archard, V.; Daidie, A.; Paredes, M.; Chirol, C. Cold Expansion Process on Hard Alloy Holes—Experimental and Numerical Evaluation. *Mech. Ind.* **2016**, *17*, 303. [CrossRef]
14. Gao, Y.; Zhong, Z. Residual Stresses Induced by Surface Enhancement Processes. In Proceedings of the 13th International Conference on Fracture, Beijing, China, 16–21 June 2013.
15. Yang, Z.; Lee, Y.; He, S.; Jia, W.; Zhao, J. Analysis of Influence of High Peening Coverage on Almen Intensity and Residual Compressive Stress. *Appl. Sci.* **2020**, *10*, 105. [CrossRef]
16. Zarutskii, A.V. Vliyanie parametrov predvaritel'nogo obzhatiya materiala v zone otverstiya na ustalostnuyu dolgovechnost' elementov aviatsionnykh konstruktsii. In Proceedings of the IV Mizhnarodna Naukovo-Tekhnichna Konferentsiya "Poshkodzhen nya Materialiv pid Chas Ekspluatatsiï, Metodi Iogo Diagnostuvannya i Prognozuvannya", Vid-vo TNTU Imeni Ivana Pulyuya Ternopil, Ukraine, 21–24 September 2015.
17. Calaf-Chica, J.; Marín, M.M.; Rubio, E.M.; Teti, R.; Segreto, T. Parametric Analysis of the Mandrel Geometrical Data in a Cold Expansion Process of Small Holes Drilled in Thick Plates. *Materials* **2019**, *12*, 4105. [CrossRef] [PubMed]
18. Aid, A.; Semari, Z.; Benguediab, M. Finite Element Method Investigation of the Effect of Cold Expansion Process on Fatigue Crack Growth in 6082 Aluminum Alloy. *Model. Numer. Simul. Mater. Sci.* **2014**, *10*, 25–31. [CrossRef]
19. Yasniy, P.; Dyvdyk, O.; Semenets, O.; Iasnii, V.; Antonov, A. Fatigue Crack Growth in Aluminum Alloy from Cold expanded Hole with Preexisting Crack. *Sci. J. Ternopil Natl. Tech. Univ.* **2020**, *3*, 5–16. [CrossRef]
20. Vorob'ev, Y.A.; Voron'ko, V.V.; Stepanenko, V.N. Sravnitel'nyi analiz sposobov dornovaniya otverstii. *Sist. Obrobki Informatsii* **2007**, *5*, 35–38.
21. Sementsov, V.F.; Vasilevskii, E.T. Vliyanie bar'ernogo obzhatiya na kharakteristiki lokal'nogo NDS polosy s otverstiem pri ee rastyazhenii. *Otkrytye Informatsionnye Komp'yuternye Integrirovannye Tekhnologii* **2015**, *68*, 86–92.
22. Shalina, R.E. Alyuminievye i berillievye splavy. In *Aviatsionnye Materialy*; ONTI: Moscow, Russia, 1983; Volume 4.
23. *GOST 5950-73*; Bars and Strips of Tool Alloyed Steel. Specifications. Mezhgosudarstvenyi Sovet po Standartizatsii, Metrologii I Sertifikatsii: Minsk, Belarus, 1973.
24. Shalina, R.E. Konstruktsionnye stali. In *Aviatsionnye Materialy*; ONTI: Moscow, Russia, 1983; Volume 1.
25. *GOST 1435-90*; Bars, Strips and Reels of Tool Unalloyed Steel. General Specifications. Mezhgosudarstvenyi Sovet po Standartizat sii, Metrologii I Sertifikatsii: Minsk, Belarus, 1990.
26. Element References. Section 3.1. Element Library. 001087. In *ANSYS Element Guide*, 4th ed.; ANSYS, Inc.: Canonsburg, PA, USA, 1983; Volume 3.
27. Surface-to-Surface Contact. 001087. In *ANSYS Contact Technology Guide*, 4th ed.; ANSYS, Inc.: Canonsburg, PA, USA, 1983; Volume 3.
28. *GOST 25.502-79*; Calculations and Tests for Strength in Mechanical Engineering. Mechanical Test Methods for Metals: Fatigue Test Methods. Standartinform Publ.: Moscow, Russia, 1981.

 *computation*

Article

The Use of IoT for Determination of Time and Frequency Vibration Characteristics of Industrial Equipment for Condition-Based Maintenance

Ihor Turkin, Viacheslav Leznovskyi, Andrii Zelenkov *, Agil Nabizade, Lina Volobuieva and Viktoriia Turkina

Software Engineering Department, National Aerospace University "Kharkiv Aviation Institute", 61070 Kharkiv, Ukraine; i.turkin@khai.edu (I.T.); lieznovskiy@gmail.com (V.L.); agilnabizade@gmail.com (A.N.); l.volobuieva@khai.edu (L.V.); v.turkina@khai.edu (V.T.)
* Correspondence: a.zelenkov@khai.edu; Tel.: +38-066-447-5807

Abstract: The subject of study in this article is a method for industrial equipment vibration diagnostics that uses discrete Fourier transform and Allan variance to increase precision and accuracy of industrial equipment vibration diagnostics processes. We propose IoT-oriented solutions based on smart sensors. The primary objectives include validating the practicality of employing platform-oriented technologies for vibro-diagnostics of industrial equipment, creating software and hardware solutions for the IoT platform, and assessing measurement accuracy and precision through the analysis of measurement results in both time and frequency domains. The IoT system architecture for industrial equipment vibration diagnostics consists of three levels. At the autonomous sensor level, vibration acceleration indicators are obtained and transmitted via a BLE digital wireless data transmission channel to the second level, the hub, which is based on a BeagleBone single-board microcomputer. The computing power of BeagleBone is sufficient to work with artificial intelligence algorithms. At the third level of the server platform, the tasks of diagnosing and predicting the state of the equipment are solved, for which the Dictionary Learning algorithm implemented in the Python programming language is used. The verification of the accuracy and precision of the vibration diagnostics system was carried out on the developed stand. A comparison of the expected and measured results in the frequency and time domains confirms the correct operation of the entire system.

Keywords: internet of things; digital platform; vibration diagnostics; calibration; accelerometer; industrial equipment; Allan variance

Citation: Turkin, I.; Leznovskyi, V.; Zelenkov, A.; Nabizade, A.; Volobuieva, L.; Turkina, V. The Use of IoT for Determination of Time and Frequency Vibration Characteristics of Industrial Equipment for Condition-Based Maintenance. *Computation* **2023**, *11*, 177. https://doi.org/10.3390/computation11090177

Academic Editor: Francesco Cauteruccio

Received: 27 June 2023
Revised: 31 July 2023
Accepted: 1 September 2023
Published: 5 September 2023

1. Introduction

Technological advancements in information technologies are progressively transforming our lives [1]. These changes demand an accelerated pace of management decision making [2,3]. As stated by the authors in [4], in the current scenario, producing an innovative product (or providing a service) that meets user requirements typically involves the integration of resources and competencies from multiple companies. The main finding from article [5] highlights the necessity for research and development in several crucial areas, including digital equipment maintenance and end-to-end automation, in order to enhance industries' preparedness for future problems.

Digital technologies and the Internet of Things (IoT) offer data homogeneity, distribution, editability, and the ability to self-reflect and reprogram, as stated in [6]. These features enable the implementation of multiple inheritance in distributed software applications, where no single owner possesses the entire design hierarchy or dictates the platform's core. As a result, products become open to new uses after manufacturing, as they can be arbitrarily combined through standardized interfaces [7].

The concept of condition-based maintenance (CBM) of industrial equipment [8] allows the determination of maintenance requirements and offers numerous benefits. CBM improves equipment credibility and dependability, and reduces maintenance resource costs compared to a late-scheduled maintenance approach. Under the CBM approach, maintenance is carried out only when specific metrics indicate declining performance or faults. The main problem with CBM is the need to spend significant resources on implementing equipment condition-monitoring tools, which usually include such non-invasive methods as visual inspection, and measurement of power consumption, noise, temperature, and vibration. Paper [9] proposes an integrated framework, which takes a broad perspective on CBM implementation, and integrates technological, organizational, and user-related elements. This study contributes to the field of CBM with a comprehensive view of implementation challenges and solutions in real-world implementations, from the original equipment manufacturer's (OEM's) point of view. Of the solutions proposed in article [9] for current research, we chose to prioritize the following in our work:

- Use a state-of-the-art IoT platform for development;
- Define modular project-level software decisions;
- Outline methods for collaboration between hardware and software specialists.

The authors of paper [10] emphasize the difference between Condition-based Maintenance (CBM) and Predictive-based Maintenance (PM) as two effective and complementary maintenance methods: CBM monitors the current condition; PM uses the CBM results to generate a future prediction for a machine.

Digital IoT platforms and end-to-end automation allow us to overcome the substantial resource consumption of the CBM concept. Well-known publications [11,12] extensively explore the development of cost-effective hardware and software solutions for vibration diagnostics using microelectromechanical systems (MEMS). A platform-oriented approach creates new possibilities for equipment fault diagnosis and state forecasting.

Smart sensors play a crucial role in CBM systems. According to the IEEE 1451.0-2007 standard [13], sensors with functions beyond the minimum required for measurements are classified as intelligent. Along with the digital interface and self-testing capabilities, these sensors have redundant functionality that simplifies their integration into networked applications.

In various mechanical systems, vibration diagnostics are an essential method for assessing the condition of mechanical systems, holding significant importance across multiple fields of application. Vibration is a highly versatile parameter that considers almost all aspects of a unit's state, allowing operating modes to determine the technical condition of the equipment.

Vibration acceleration is the vibration value directly related to the force that caused the vibration. Vibration acceleration characterizes the power interaction dynamic of elements inside the unit, which causes this vibration. The use of vibration acceleration is theoretically ideal since the accelerometer specifically measures the acceleration, which does not need to be specially converted. The disadvantage is that there are no practical developments regarding norms and threshold levels, and no generally accepted physical or spectral interpretations of the features of the manifestation of vibration acceleration.

Usually, for the vibration diagnostic, it is necessary to measure the vibration velocity. Vibration velocity is the speed of movement of the controlled point of the equipment during its precession along the measurement axis. Standard ISO 20816-1:2016 establishes general conditions and procedures for measuring and evaluating vibration, using measurements of vibration velocity root mean square value (RMS). The physical essence of the vibration velocity RMS parameter is the equality of the energy impact on the machine supports of an actual vibration signal and a fictitious constant, numerically equal in value to the RMS.

The accelerometer manufacturer establishes the output characteristics following extensive testing, typically encompassing the influence of various operating conditions, such as temperature changes and magnetic fields. Paper [14] proposes a method for estimating the thermal behavior of capacitive MEMS accelerometers and compensating for their drifts in order to reduce orientation and temperature effects. It is a necessary solution, but insufficient for solving the general problem of compensating for accelerometer errors during regular operation.

The accelerometer metric of displacement at 0 g holds significant importance as it sets the baseline for measuring actual acceleration. Mounting the system with an accelerometer introduces additional measurement errors, which can arise due to stresses in the printed circuit board and the application of various compounds during mounting. As recommended in [15], we will calibrate after system assembly to exclude these errors.

The ISO 16063 series standards [16] set the modern requirements for vibration sensors and their calibration methods. Usually, a MEMS accelerometer calibration involves averaging the measurement values using a calibration scheme, where the accelerometer system is positioned to have one axis, typically the Z axis, experiencing a 1 g gravitational field, while the other axes, X and Y, remain in a 0 g field. After installation at a specific location, additional calibration is conducted by comparing the measurement results with those of a reference accelerometer [17,18].

Although recommendations to use simultaneous analysis of vibration parameters in the time and frequency domains have been known since the standard [19], modern publications [20,21] do not attempt to combine analysis methods and the capabilities of modern IoT technologies into one whole.

Known hardware and software solutions for vibration diagnostics have the following disadvantages:

- Use of outdated sensors for measuring acceleration [22,23];
- Use of energy-inefficient wireless communication protocols between the sensor and the host [23];
- Use of suboptimal data processing algorithms that meet the standard [19] but are insufficient to ensure energy efficiency, affecting the duration of autonomous operation [12].

The study's objective is to offer a new resource-efficient IoT-oriented wireless solution and technology for vibration diagnostics, which utilizes the contact method of vibration measurement using MEMS accelerometers according to the standard [19].

To achieve the goal, we carry out the following tasks:

- Justify the choice of a cloud solution;
- Develop the architecture of a digital platform for vibration diagnostics, including a relational database model for storing measurement results;
- Determine the procedure for calibrating linear acceleration sensors;
- Propose a method for processing measurement results based on analysis in the frequency and time domains, which allows the application of a standard for assessing the current state of industrial equipment;
- As a result, prove that combining state-of-the-art information technologies and data analysis methods in the time and frequency domains allows us to achieve solutions qualitatively better than those known before.

2. Materials and Methods

2.1. IoT Platform Software and Hardware Solutions

Diagnostic results are processed and formed in a system with a three-level architecture (Figure 1). The main criterion in the development was to reduce installation and operation costs while ensuring high energy efficiency for extended autonomous operations. The system architecture consists of three levels:

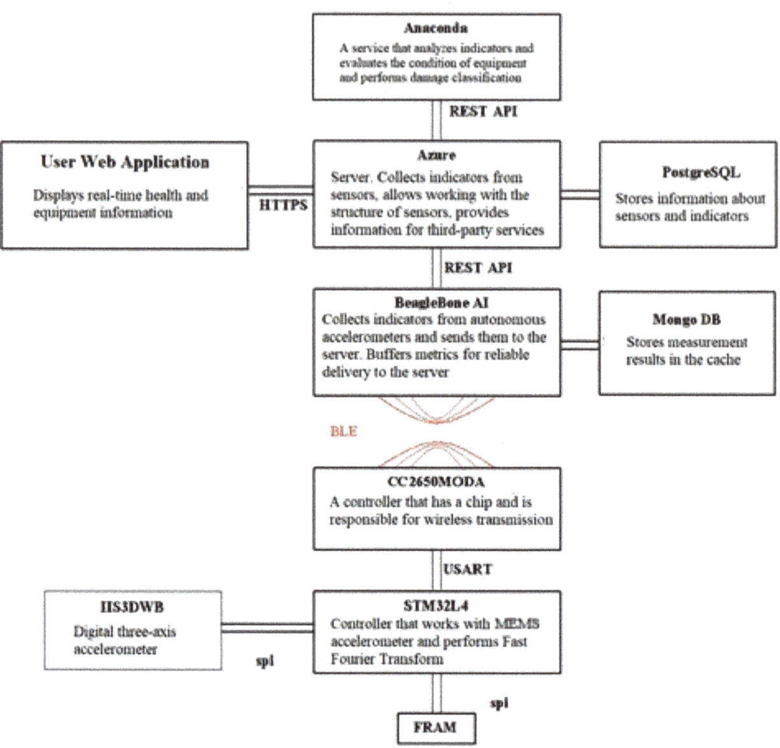

Figure 1. The architecture of the IoT system for vibration diagnostics.

1. The autonomous sensor level is responsible for reading vibration acceleration indicators. It is based on an STM32L476 microcontroller and a three-axis digital accelerometer from STMicroelectronics called IIS3DWB [24]. The IIS3DWB capacitive accelerometer is mounted on the monitoring object and connected to the microcontroller via the Serial Peripheral Interface (SPI). It has low power consumption, high resolution (16 bits), and a reprogrammable measurement range of ±2 g, ±4 g, ±8 g, and ±16 g. The measurement result is read byte by byte in the form of 16-bit data. IIS3DWB has a bandwidth ranging from 0.05 to 6000 Hz, enabling it to capture vibrations with frequencies up to 1000 Hz. The IIS3DWB accelerometer saves the resulting calibration values in its OFFSET registers for automatic error compensation. Each OFFSET register's content is added to the measured acceleration value along the respective axis, and the resulting values are then stored in the DATA registers. Depending on the measurement frequency, these sensors are designed for autonomous operation for 6–12 months. We use BLE (Bluetooth Low Energy) digital wireless data transmission technology to transmit data from the sensor to the hub level, ensuring low energy consumption.

2. The hub layer is deployed on a device using a single-board microcomputer called the BeagleBone® AI, which is designed to run artificial intelligence algorithms. This layer receives data from the autonomous sensor layer and transmits it to the server layer. Depending on the selected analysis algorithm, it may pre-process data before transmission, significantly reducing the server's load.

3. The server layer offers an API that allows clients and third-party services to interact with the sensor structure and data. The Microsoft cloud platform Azure IoT Suite [25] provides infrastructure for creating and managing applications in the cloud. The Azure Internet of Things Suite is a comprehensive service that leverages the full

capabilities of Azure to establish connections with devices. It effectively captures a diverse range of data generated by these devices. Through seamless integration and organization, the suite manages and analyzes the data, presenting it in a format that facilitates informed decision-making. We selected Microsoft Azure IoT for further implementation because this platform provides well-established solutions, and its budget requirements are feasible for startups in the initial phase.

2.2. Digital Platform Database

The PostgreSQL DBMS was selected to ensure long-term data storage in the digital platform for vibration diagnostics of industrial equipment at the second level of the hub. PostgreSQL has been in development since 1996. PostgreSQL boasts advanced features, including Multi-Version Concurrency Control, asynchronous replication, and nested transactions (savepoints). At the third level of the server platform, we use the NoSQL database MongoDB for caching data on the BeagleBone AI Mini PC.

The ER model of the digital platform database (Figure 2) characterizes the relationships between the following entities: sensors, IoT devices in which they are included, and data on the results of measurements by these sensors.

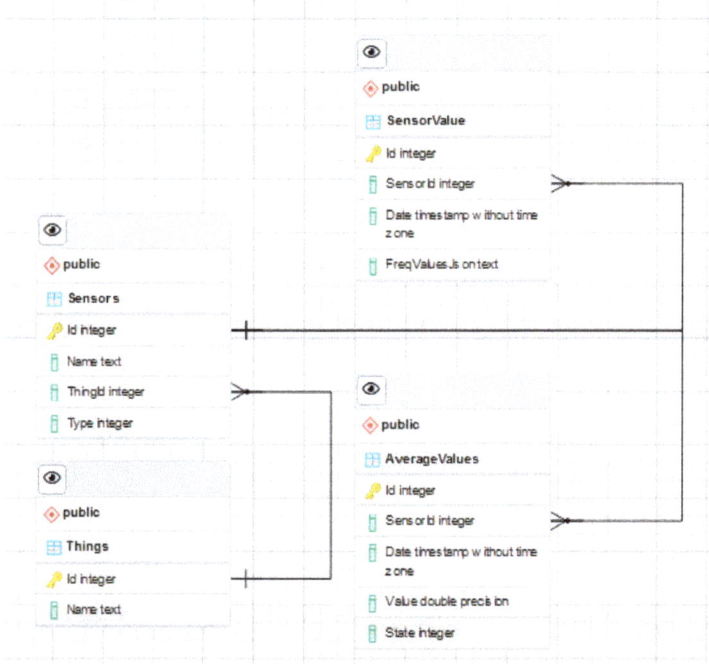

Figure 2. ER model of the database of the digital platform.

Entities in the database, which are the basis for business logic, contain the following data according to the IEEE 1451.0-2007 standard:

- AverageValue—average values of measurement results by sensors, such as average rotation speed;
- Sensor—data about sensors: sensor identifier, name;
- SensorValue—data on current and historical measurement results: timestamp (Timestamp) and result (Value);
- Things—data about the equipment on which the sensors are installed: name and location.

- The definitions of relationships between entities are presented as follows:
- The «Sensor» entity is related to the «AverageValue» entity by the ratio «1: N»—each sensor has many numerical indicators;
- The «Sensor» entity is related to the «SensorValue» entity by a «1: N» relationship—each sensor has many measurement results;
- The «Thing» entity is related to the «Sensor» entity by a «1: N» relationship—each piece of industrial equipment can have many sensors.

2.3. The Bench Equipment and Measurement Algorithm

The adequacy of the system, the accuracy, and the correctness of the work of the components were evaluated using the bench equipment shown in Figures 3 and 4. This setup replicated an electric motor and generator interconnected by a coupling.

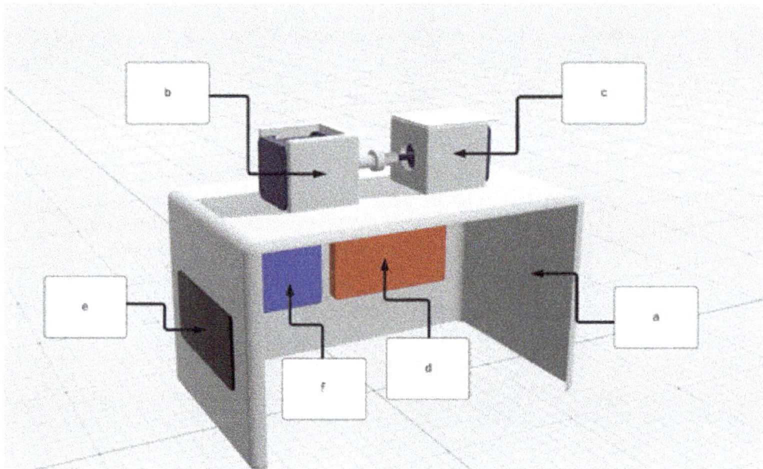

Figure 3. A 3D-model of the testbed (a—Testbed base, b—Load node, c—Node with motor and vibration sensor mount, d—Power supply unit, e—User control panel, f—Testbed controller unit).

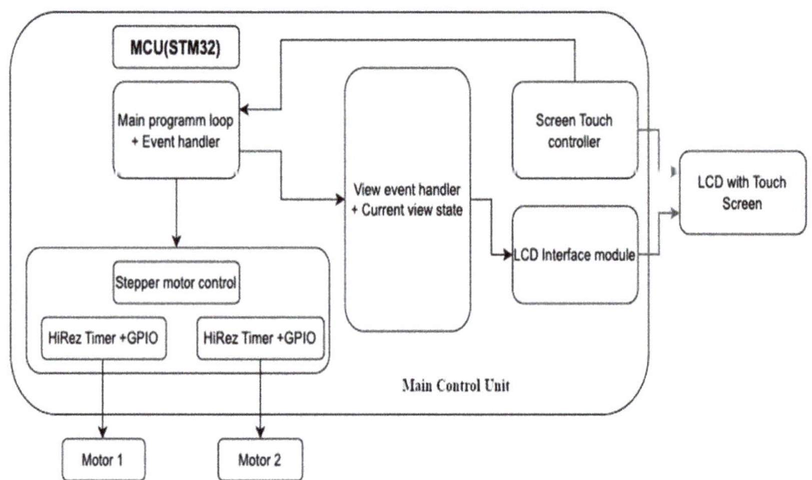

Figure 4. Schema of the testbed electronics.

The testbed replicates the operational conditions of the pumping unit at the enterprise. The testbed consists of:

(a)   A testbed base;
(b)   A load node that replicates the behavior of the generator;
(c)   A test node that replicates the behavior of the electric motor, with a vibration sensor mount;
(d)   A power supply unit which provides power for the whole testbed;
(e)   A user control panel with a sensor display for visualization and control of motor speed;
(f)   A testbed controller unit, which controls motors and the user control panel.

The central controller is the STM32F429 chip; the primary actuators are stepper motors in conjunction with stepper motor controllers. High-precision timers are used as part of the microcontroller to generate pulse sequences. The use of hardware timers allows for minimization of the load on the central core of the controller since, after the initial initialization, it will be sufficient for the controller to change the value of the divider register when a speed change event occurs.

The operation of the microcontroller is based on a mechanism for responding to events coming from the stand operator when interacting with the touch screen. When registering a click on an interface element, the main loop will update the current state of the GUI as well as the value of the speed variables of the motors.

Figure 5 shows a measurement system algorithm.

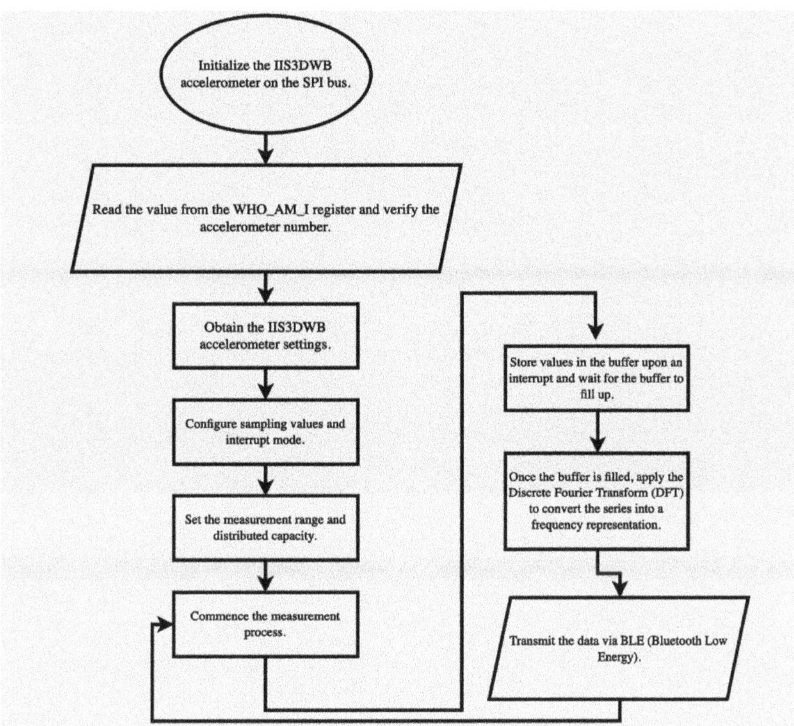

Figure 5. Measurement algorithm.

According to the algorithm, the most energy-consuming steps will be steps of the measurement process and the transmission of data via BLE.

2.4. Sensor Calibration Procedure

Sensor calibration is divided into two steps:

- Measuring energy consumption in all potential work conditions;
- Lengthy testing of sensors in field conditions.

However, we must first check the calibration with debugging using the JTAG interface. The sensor has three axes, so the calibration is carried out under conditions in which one axis is installed perpendicular to the plane of the work table.

If the Z axis is perpendicular to the plane of the desktop (Figure 6), then the measurement results at rest along the Z axis will equal 1 g, and 0 g along the X and Y axes. Two identical designs along the measurement axes can be made perpendicularly to obtain a biaxial accelerometer, but the third axis, usually vertical-Z, requires a different construction. This results in poorer performance for the third axis, reducing sensitivity and increasing error and noise [24].

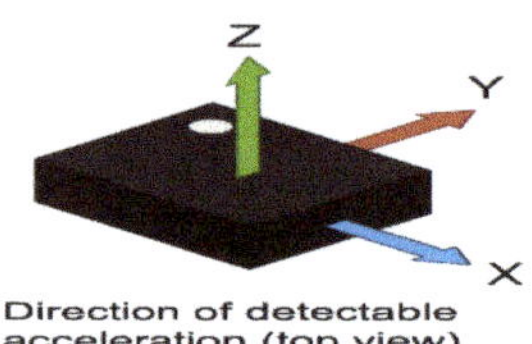

Figure 6. Scheme of the location of measurement axes.

Figure 7 presents the measurement results of the sensor calibration.

Build Analyzer Static Stack Analyzer Debug SFRs Memory Include Browser Expressions	
Name	Value
"id"	
"x"	0.021
"y"	0.014
"z"	1.015
Add new expression	

Figure 7. Measurement results during calibration.

2.5. The Method of Processing Measurement Results

The digital platform offers customers an equipment condition assessment in the form of classification, categorized as "good", "satisfactory", or "unsatisfactory". The primary goal of developing the evaluation algorithm is to fulfill the minimum training requirement by using the smallest amount of labeled data. As a result, the algorithm should be capable of recognizing patterns and learning from unlabeled input data, a technique known as System Identification. The algorithm was implemented using the Python programming language. Python allows for the use of various tools and libraries for machine learning, data processing, and signal analysis. The equipment condition evaluation algorithm was split to separate services to provide greater scalability and flexibility for the whole system.

Accelerometer noise analyses using the Allan variance method include the following steps:

1. Primary statistical data processing.

For statistical processing of the initial information, the correspondence of the data to the normal distribution or Gaussian distribution of the probability density distribution is checked at the first stage.

For all points, the arithmetic mean measurement at each position is calculated. We start by using the average. The scatter is indicated by the standard deviation (SD). The mean deviation is the spread around the mean and shows the initial instability.

$$x = \frac{1}{n}\sum\nolimits_{j=1}^{n} x_j, SD = \sqrt{\frac{1}{n}\sum\nolimits_{j=1}^{n}(x_j - x)^2}, \tag{1}$$

where n—the total number of measurements;

x_j—measurement results of j points.

For groups of positions in the runs, the overall average across all runs and the average variability between runs are calculated, considering the exclusion of anomalous measurements. To assess compliance with the Gaussian distribution, we utilize the Shapiro–Wilk test [26]. This test employs the sum of squared deviations between the characteristic functions derived from sample data and a normal distribution.

2.   Allan deviation analysis provides excellent differentiation between the particular noise types. We use the Allan transform to quickly check whether the measurement results are consistent with «white» noise, as proposed in the paper [26]. The formula for calculating the Allan variance $\sigma_A^2(\tau)$ under the condition of a uniform polling step Δt is as follows:

$$\sigma_A^2(\tau) = \frac{\sum_{n=0}^{L-2l}\left(\sum_{i=1}^{l}((x(t_{n+i}) - x(t_{n+l+i})))\right)^2}{2\tau^2(L - 2l + 1)} \tag{2}$$

where L—the total number of measurements;

l—total number of measurements in the averaging interval $1 < l \leq L/2$;

$x(t_k)$—measurement result at time $t_k = k \bullet \Delta t$.

3.   For noise spectral density analysis:

$$f(x) = a_0 + \sum_{n=1}^{\infty}\left(a_n \cos \cos \frac{n\pi x}{L} + b_n \sin \sin \frac{n\pi x}{L}\right), \tag{3}$$

where a_n, b_n—Fourier series coefficients.

We use Discrete Fourier Transform (DFT), which realizes transformations:

$$X_k = \sum_{j=0}^{n-1} x_j\left(\cos\left(\frac{2\pi}{n}kj\right) - i \bullet \sin\left(\frac{2\pi}{n}kj\right)\right) \tag{4}$$

We realized the Discrete Fourier Transform (DFT) at the second (hub) level to enhance determination of the frequency composition of input signals. The external software library (FFTW) was used to compute the frequency vibration characteristics. The DFT recorded spectral information represents the vibration acceleration signals of each axis. To improve the accuracy of the Fourier transform, we used the adaptive filtering method for the MEMS gyroscope with a dynamic noise model [15].

4.   One of the main reasons for the accelerometer's systematic errors is the temperature-dependence of its characteristics [24]. When integrating to calculate the vibration speed, these errors quickly accumulate and give an utterly unreliable result. To determine the root-mean-square of the vibration velocity (RMS_{VV}), as recommended by standard ISO 20816-1:2016, we use the integration of the measured vibration acceleration (VA) with a correction for the moving average of the vibration acceleration (MA_{AV}) and vibration velocity (MA_{VV}):

$$MA_{VA_k} = g \bullet \sum_{j=k-n}^{k+n} x_j/(2 \bullet n + 1), \; VA_k = g \bullet x_k - MA_{VA_k},$$

$$V_k = V_{k-1} + (VA_{k-1} + 4VA_k + VA_{k+1})\Delta\tau/6 \tag{5}$$

$$MA_{VV_k} = \sum_{j=k-n}^{k+n} V_j/(2 \bullet n + 1), \; VV_k = V_k - MA_{VA_k}$$

where g—gravitational acceleration, g = 9.8 m/s^2;

$x_j, x_{k-1}, x_k, x_{k+1}$—raw acceleration data, g;

VA_{k-1}, VA_k, VA_{k+1}—calculated values of the vibration acceleration with acceleration offset compensation, m/s^2;

V_k, V_j—calculated values of the vibration velocity without velocity offset compensation, m/s^2;

VV_k, VV_j—calculated values of the vibration velocity with velocity offset compensation, m/s^2;

n—variable that specifies the half-width of the time window for which averaging and offset compensation is performed.

With the known dependence of the vibration velocity on time $VV(\tau)$, root-mean-square value over a time interval RMS_{VV} can be calculated as:

$$RMS_{VV}(\tau) = \sqrt{\frac{1}{T}\int_{\tau-T/2}^{\tau+T/2} VV^2(t)dt} \approx \sqrt{\sum_{j=k-n}^{k+n} VV_j^2/(2n+1)}. \qquad (6)$$

2.6. Comparison with Analogs

We compared our decision with those previously offered by viewing the following three similar systems, each allowing several accelerometers to connect with a high-level system.

1. The adaptive Kalman filter and Allan variance method used for Inertial Measurement with Unit MPU6050. This method is based on the dynamic noise model [22].
2. The compact and low-powered MEMS accelerometer and microcontroller with wireless connectivity and a run time of approximately eight hours [23].
3. A budget-friendly vibration measurement system dedicated to assessing the condition of construction structures [12].

3. Results

The sensors are calibrated to assess the potential of the hardware part for maintenance-free use. During calibration, we compared the results from the visualization tools and the debugger with the information displayed by the software.

1. Using Allan variation, it is easy to eliminate the systematic error in estimating the statistical characteristics of the original series, while for uncorrelated data, the variance estimate will be unbiased. The presence of any periodic function will be displayed on the plot of the Allan variance versus time (Figure 8).

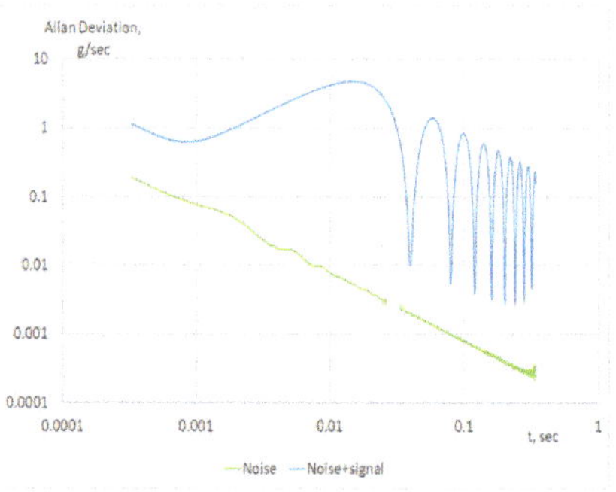

Figure 8. Simulation results: Allan deviation for white Gaussian noise and the additive mixture of 25 Hz harmonic signal and white Gaussian noise (signal-to-noise ratio is 40 dB).

Practically, the offset of the fitting straight line and the standard error of the deviation from the straight-line equation are reliable indicators of the magnitude of the vibration (Figure 9). As a result, we get visual evidence that the measurement noise is white noise as the slope of the characteristic on a log–log scale is -1. The problem with this solution is the lack of generally accepted standards linking the Allan deviation measurement results and the technical condition of industrial equipment.

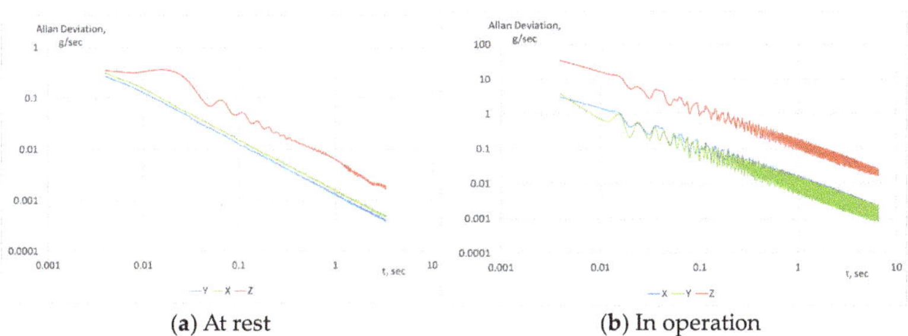

(**a**) At rest (**b**) In operation

Figure 9. Allan deviation acceleration data at rest and during the operation of industrial equipment.

2. The noise spectral density analysis allows us to study the problem in the frequency domain. In the numerical simulation, the dependence of the results of the Fourier transform in the presence of an additive mixture of a single valid periodic signal (50 Hz) and noise of various intensities looks like Figure 10.

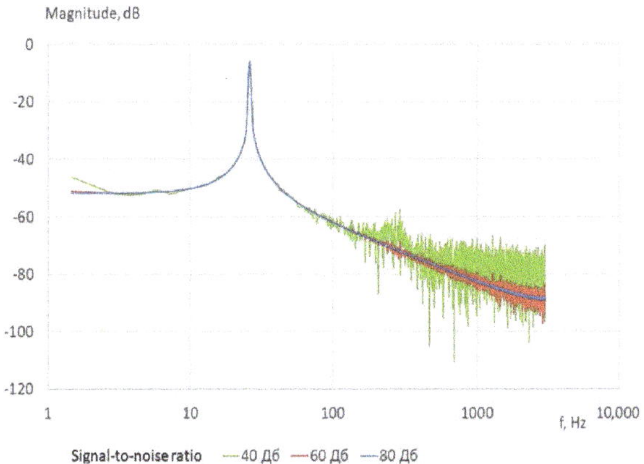

Figure 10. Outcomes of the simulation: The vertical acceleration frequency response under varying noise-to-signal ratios.

In contrast to theoretical model calculations, the results of practical measurements demonstrate the presence of many harmonics in the frequency range of 20–100 Hz (Figure 11). As a result, we concluded that the width of the 5–20 ms time window, determined by n in Equation (5), is the most suitable for calculating the moving average.

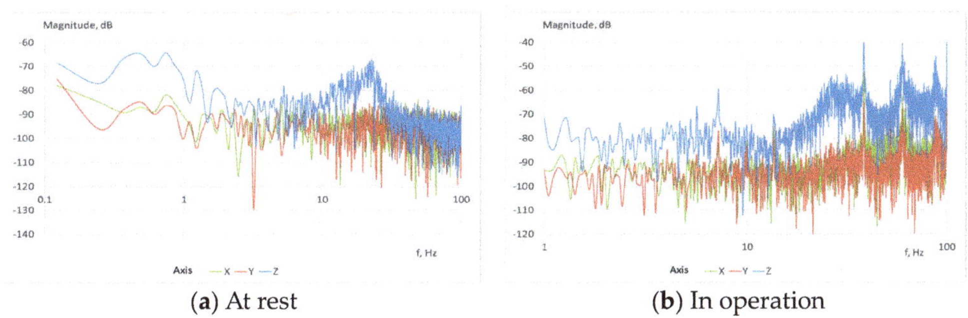

(**a**) At rest (**b**) In operation

Figure 11. Frequency response of-3-axis accelerations with the different signal-to-noise ratios.

3. Despite pre-calibration, the raw data contains a significant bias error. With the standard placement of the MEMS accelerometer, the sensor along the z-axis, which is the least accurate due to the technological constraints of production, measures the value of the proper acceleration (Figure 12a). This error is integrated without compensation (5), leading to unreliable results. Moving average compensation removes this bias (Figure 12b).

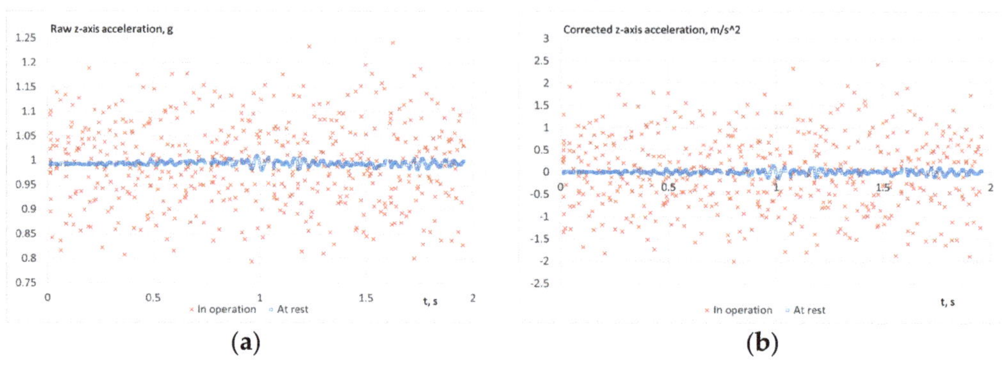

(a) **(b)**

Figure 12. Z-axis acceleration data at rest and during the operation of industrial equipment: (**a**) row measurements: x_k (5), (**b**) calculated values with acceleration bias compensation: VA_k (5)).

Similar to compensating for the acceleration measurement error by using a moving average, compensating for the velocity calculation bias makes it possible to obtain more accurate vibration velocity estimates (Figure 13).

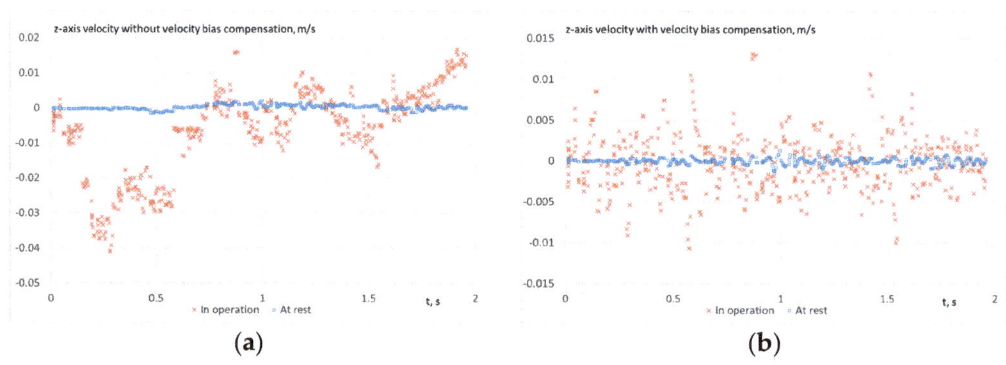

(a) **(b)**

Figure 13. Velocity data at rest and during the operation of industrial equipment: (**a**) without velocity bias compensation, (**b**) with velocity bias compensation.

4. The vibration velocity RMS calculation according to Equation (6) gives the following results. The algorithm given in Section 2.5 confirms that the controlled industrial equipment is in good condition. The calculated RMS value of the vibration velocity corresponds to the recommendations for determining the boundaries of the vibration state zones according to the standard ISO 20816-1:2016. As a result, we have the following:

- The standard sets the limit of normal operation for low-power systems as "vibration velocity no more than 0.6 mm/s";
- With a time window width of $T = 0.1$ s, the measuring system fixes velocity at no more than 0.05 mm/s at rest and no more than 0.35 mm/s during operation (Figure 14);

- Extending the time window width T from 0.1 s to 1.6 s does not significantly improve the vibration velocity estimations (Table 1). It can be assumed that the time window length should be within 0.1–1 s. The experimental estimates of Allan's variation (Figure 9) and the results of the Fourier transform (Figure 11) fully correspond to this decision.

Table 1. Dependence of the calculated maximum vibration velocity (Max(RMS), m/s) on the time window width (T), during operation.

T, s	Max(RMS), m/s
0.1	0.35
0.2	0.28
0.4	0.26
0.8	0.25
1.6	0.25

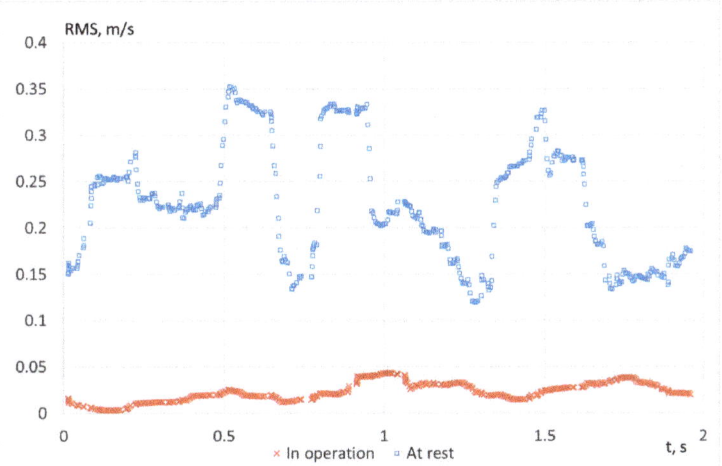

Figure 14. RMS of the vibration velocity with time window width $T = 0.1$ s.

4. Discussion

The industrial equipment condition-based and predictive-based maintenance concepts demand the complex use of multilevel digital technologies. The proposed software and hardware IoT solutions for vibration diagnostics, with the method of processing measurement results, appear to be a practical approach to accuracy and precision control of the technical state of industrial equipment.

Comparing our system with previously developed ones, which are referenced in the Introduction section, reveals the following advantages:

We achieve low-cost three-level hardware and software solutions for vibration diagnostics, with vertical integration of measuring devices with the Azure Internet of Things Suite digital platform;

- We utilize the specialized IIS3DWB smart accelerometer from STM, designed explicitly for vibration diagnostics, enabling an increased frequency range of measurements up to 6000 Hz.
- Using low energy consumption BLE and the STM32 microcontroller, we obtain a long period of autonomous work (up to one year). For comparison, prototype systems need to recharge every 8–12 h.

- We offer a method for integrating the measured vibration acceleration with a correction for the moving average of the vibration acceleration and vibration velocity, to move from the directly measured vibration acceleration to the vibration velocity RMS value recommended by the ISO 20816-1:2016 standard.

Consequently, we substantially reduce operating expenses by maintaining almost the same equipment and installation costs as in wireless analogs.

5. Conclusions

The proposed technique can effectively realize the contact method of vibration measurement using MEMS accelerometers according to the standard IEEE Recommended Practice for Inertial Sensor Test Equipment, Instrumentation, Data Acquisition, and Analysis.

Our choice of cloud solution, Microsoft Azure IoT, can be justified by the sufficient capabilities of the platform and Microsoft's pricing policy, which favors the possibility of startup projects. The developed architecture of the digital vibration diagnostics platform (Figure 1) and the relational database model (Figure 2) for storing measurement results provides an additional option for solving Predictive-based Maintenance tasks. The developed technological equipment for calibrating vibration acceleration sensors (Figures 3 and 4) allows us to reduce systematic and random accelerometer setting errors. The implemented method for the interpretation of measurement results in the frequency and time domains (Section 2.6) allows the application of the IEEE Recommended Practice for Inertial Sensor Test Equipment, Instrumentation, Data Acquisition, and Analysis requirements in assessing the current state of industrial equipment.

Thus, we proved that combining well-known and classical analysis methods in the time and frequency domains with the possibilities of their multilevel processing in the system allows us to achieve qualitatively improved results.

In the maturing of the project, we plan to maintain the limitations of the low cost of development and operation to solve the following tasks:

- Optimal distribution of solved tasks across the three levels of the IoT system;
- Reduction of the computational complexity of the algorithms in order to execute them at a lower level, increase autonomy at a low level and reduce the load on communication channels.

The upcoming stage of our plan encompasses the creation of supplementary microservices. These will facilitate the application of time series analysis techniques and state-of-the-art AI technologies, enhancing the quality of predictive maintenance. We plan to investigate, using mathematical models, the dependence of the leading indicators of Allan deviation of the vibration acceleration data on the technical condition of the equipment, measured through the root-mean-square of the vibration velocity in the theoretical part.

Author Contributions: Conceptualization, I.T. and V.L.; methodology, I.T., A.Z. and V.L.; software, V.L. and A.N.; validation, I.T. and V.T.; formal analysis, I.T. and L.V.; investigation, V.L., data curation, A.Z. and L.V.; writing—original draft preparation, V.L.; writing—review and editing, A.Z.; visualization, V.T.; supervision, I.T.; project administration, I.T. All authors have read and agreed to the published version of the manuscript.

Funding: This research received no external funding.

Data Availability Statement: Data not available due to confidentiality.

Conflicts of Interest: The authors declare no conflict of interest.

Nomenclature

x	Arithmetic means of the measurements
$\sigma_A^2(\tau)$	Allan variance
Δt	Time between measurements
L	Total number of measurements
l	Total number of measurements in the averaging interval $1 < l \leq L/2$
$x(t_k)$	Measurement result at the time $t_k = k \bullet \Delta t$
a_n, b_n	Fourier series coefficients
g	Gravitational acceleration
$x_j, x_{k-1}, x_k, x_{k+1}$	Raw acceleration data, g
VA_{k-1}, VA_k, VA_{k+1}	Calculated values of the vibration acceleration with acceleration offset compensation, m/s^2
V_k, V_j	Calculated values of the vibration velocity without velocity offset compensation, m/s^2
VV_k, VV_j	Calculated values of the vibration velocity with velocity offset compensation, m/s^2
Abbreviations	
BLE	Bluetooth Low Energy
IoT	Internet of things
CBM	Condition-based maintenance
PM	Predictive Maintenance
MEMS	Micro-electromechanical systems
RMS	Root mean square
SPI	Serial Peripheral Interface
MA	Moving average
SD	Standard deviation
I^{2}C	Inter-Integrated Circuit
VA	Vibration acceleration
ADC	Analog-to-digital converter
DBMS	Database management system
DFT	Discrete Fourier transform

References

1. Chen, X.; Han, T. Disruptive technology forecasting based on Gartner hype cycle. In Proceedings of the 2019 IEEE Technology & Engineering Management Conference (TEMSCON), Atlanta, GA, USA, 12–14 June 2019; IEEE: Piscataway Township, NJ, USA, 2019. [CrossRef]
2. Fessenmayr, F.; Benfer, M.; Gartner, P.; Lanza, G. Selection of traceability-based, automated decision-making methods in global production networks. *Procedia CIRP* **2022**, *107*, 1349–1354. [CrossRef]
3. Rajagopal, N.K.; Qureshi, N.I.; Durga, S.; Ramirez-Asis, E.H.; Huerta-Soto, R.M.; Gupta, S.K.; Deepak, S.; Ahmad, M. Future of Business Culture: An Artificial Intelligence-Driven Digital Framework for Organization Decision-Making Process. *Complexity* **2022**, *2022*, 7796507. [CrossRef]
4. Mei, J.; Zheng, G.; Zhu, L. Governance mechanisms implementation in the evolution of digital platforms: A case study of the Internet of Things platform. *RD Manag.* **2022**, *52*, 498–516. [CrossRef]
5. Umair, M.; Cheema, M.A.; Cheema, O.; Li, H.; Lu, H. Impact of COVID-19 on IoT Adoption in Healthcare, Smart Homes, Smart Buildings, Smart Cities, Transportation and Industrial IoT. *Sensors* **2021**, *21*, 3838. [CrossRef] [PubMed]
6. Neves da Rocha, F.; Pollock, N. Innovating in digital platforms: An integrative approach. In Proceedings of the 21st International Conference on Enterprise Information Systems, Heraklion, (ICEIS 2019), Crete, Greece, 3–5 May 2019; ICEIS: Crete, Greece, 2019; Volume 2, pp. 505–515. [CrossRef]
7. Tao, F.; Xiao, B.; Qi, Q.; Cheng, J.; Ji, P. Digital twin modeling. *J. Manuf. Syst.* **2022**, *64*, 372–389. [CrossRef]
8. Sánchez, R.V.; Siguencia, J.F.; Villacís, M.; Cabrera, D.; Cerrada, M.; Heredia, F. Combining Design Thinking and Agile to Implement Condition Monitoring System: A Case Study on Paper Press Bearings. *IFAC Papers OnLine* **2022**, *55*, 187–192. [CrossRef]
9. Ingemarsdotter, E.; Kambanou, M.L.; Jamsin, E.; Sakao, T.; Balkenende, R. Challenges and Solutions in condition-based maintenance implementation—A multiple case study. *J. Clean. Prod.* **2021**, *296*, 126420. [CrossRef]
10. Nata, C.; Laurence; Hartono, N.; Cahyadi, L. Implementation of Condition-based and Predictive-based Maintenance using Vibration Analysis. In Proceedings of the 2021 4th International Conference of Computer and Informatics Engineering (IC2IE), Depok, Indonesia, 14–15 September 2021; IEEE: Piscataway Township, NJ, USA, 2021; pp. 90–95. [CrossRef]

11. Yuan, X.; He, Y.; Wan, S.; Qiu, M.; Jiang, H. Remote vibration monitoring and fault diagnosis system of synchronous motor based on internet of things technology. *Artif. Intell. Edge Comput. Mob. Inf. Syst.* **2021**, *2021*, 3456624. [CrossRef]

12. Villacorta, J.J.; del-Val, L.; Martínez, R.D.; Balmori, J.-A.; Magdaleno, Á.; López, G.; Izquierdo, A.; Lorenzana, A.; Basterra, L.-A. Design and Validation of a Scalable, Reconfigurable and Low-Cost Structural Health Monitoring System. *Sensors* **2021**, *21*, 648. [CrossRef] [PubMed]

13. IEEE 1451.0-2007—IEEE Standard for a Smart Transducer Interface for Sensors and Actuators—Common Functions, Communication Protocols, and Transducer Electronic Data Sheet (TEDS) Formats//CFAT—Common Functionality and TEDS Working Group. Available online: https://standards.ieee.org/ieee/1451.0/3441/ (accessed on 14 June 2023).

14. Martínez, J.; Asiain, D.; Beltrán, J.R. Self-Calibration Technique with Lightweight Algorithm for Thermal Drift Compensation in MEMS Accelerometers. *Micromachines* **2022**, *13*, 584. [CrossRef]

15. Bai, Y.; Wang, X.; Jin, X.; Su, T.; Kong, J.; Zhang, B. Adaptive filtering for MEMS gyroscope with dynamic noise model. *ISA Trans.* **2020**, *101*, 430–441. [CrossRef] [PubMed]

16. *ISO 16063-11:1999*; Methods for the Calibration of Vibration and Shock Transducers—Part 11: Primary Vibration Calibration by Laser Interferometry. International Organization for Standardization (ISO): Geneva, Switzerland, 1999; 27p. Available online: https://www.iso.org/ru/standard/24951.html (accessed on 14 June 2023).

17. Larsonnier, F.; Rouillé, G.; Bartoli, C.; Klaus, L.; Begoff, P. Comparison on seismometer sensitivity following ISO 16063-11 standard. In Proceedings of the 19th International Congress of Metrology, Paris, France, 24–26 September 2019; p. 27003. [CrossRef]

18. Bilgic, E. Determination of Pulse Width and Pulse Amplitude Characteristics of Materials Used in Pendulum Type Shock Calibration Device. *Acta Phys. Pol.* **2017**, *132*, 857–860. [CrossRef]

19. *IEEE Std 1554-2005*; 1554-2005—IEEE Recommended Practice for Inertial Sensor Test Equipment, Instrumentation, Data Acquisition, and Analysis. IEEE: Piscataway Township, NJ, USA, 2013. [CrossRef]

20. Hayouni, M.; Vuong, T.-H.; Choubani, F. Wireless IoT universal approach based on Allan variance method for detection of artificial vibration signatures of a DC motor's shaft and reconstruction of the reference signal. *IET Wirel. Sens. Syst.* **2022**, *12*, 81–92. [CrossRef]

21. Kumari, S.; Raj, R.; Komati, R. A Thing Speak IoT Based Vibration Measurement and Monitoring System Using an Accelerometer sensor. *Int. J. Res. Appl. Sci. Eng. Technol.* **2021**, *9*, 1249–1258. [CrossRef]

22. Koene, I.; Klar, V.; Viitala, R. IoT connected device for vibration analysis and measurement. *HardwareX* **2020**, *7*, e00109. [CrossRef]

23. Villarroel, A.; Zurita, G.; Velarde, R. Development of a Low-Cost Vibration Measurement System for Industrial Applications. *Machines* **2019**, *7*, 12. [CrossRef]

24. IIS3DWB—Ultra-Wide Bandwidth, Low-Noise, 3-Axis Digital Vibration Sensor. Datasheet—Production Data. Available online: https://www.st.com/resource/en/datasheet/iis3dwb.pdf (accessed on 14 June 2023).

25. Tragos, E.Z.; Pöhls, H.C.; Staudemeyer, R.C.; Slamanig, D.; Kapovits, A.; Suppan, S.; Fragkiadakis, A.; Baldini, G.; Neisse, R.; Langendörfer, P.; et al. Building the Hyperconnected Society. In *Securing the Internet of Things*; River Publishers: Aalborg, Denmark, 2015; Available online: https://www.researchgate.net/publication/289253024_Building_the_Hyperconnected_Society (accessed on 14 June 2023).

26. Wilk, M.B. The Shapiro Wilk and Related Tests for Normality. 2015. Available online: https://math.mit.edu/~rmd/465/shapiro.pdf (accessed on 14 June 2023).

Article

A Software Verification Method for the Internet of Things and Cyber-Physical Systems

Yuriy Manzhos * and Yevheniia Sokolova *

Department Software Engineering and Busines, National Aerospace University "Kharkiv Aviation Institute", 61070 Kharkiv, Ukraine
* Correspondence: y.manzhos@khai.edu (Y.M,); y.sokolova@khai.edu (Y.S.)

Abstract: With the proliferation of the Internet of Things devices and cyber-physical systems, there is a growing demand for highly functional and high-quality software. To address this demand, it is crucial to employ effective software verification methods. The proposed method is based on the use of physical quantities defined by the International System of Units, which have specific physical dimensions. Additionally, a transformation of the physical value orientation introduced by Siano is utilized. To evaluate the effectiveness of this method, specialized software defect models have been developed. These models are based on the statistical characteristics of the open-source C/C++ code used in drone applications. The advantages of the proposed method include early detection of software defects during compile-time, reduced testing duration, cost savings by identifying a significant portion of latent defects, improved software quality by enhancing reliability, robustness, and performance, as well as complementing existing verification techniques by focusing on latent defects based on software characteristics. By implementing this method, significant reductions in testing time and improvements in both reliability and software quality can be achieved. The method aims to detect 90% of incorrect uses of software variables and over 50% of incorrect uses of operations at both compile-time and run-time.

Keywords: cyber-physical systems; internet of things; software defect model; software quality; physical dimension; physical orientation; formal verification

Citation: Manzhos, Y.; Sokolova, Y. A Software Verification Method for the Internet of Things and Cyber-Physical Systems. *Computation* **2023**, *11*, 135. https://doi.org/ 10.3390/computation11070135

Academic Editor: Benoît Caillaud

Received: 25 April 2023
Revised: 4 July 2023
Accepted: 4 July 2023
Published: 7 July 2023

1. Introduction

The Internet of Things (IoT) is a contemporary paradigm that comprises a wide range of heterogeneous inter-connected devices capable of transmitting and receiving messages in various formats through different protocols to achieve diverse goals, as noted by Bai Lan et al. [1]. Presently, the IoT ecosystem encompasses over 20 billion devices, each with a unique identifier that can seamlessly interact via existing Internet infrastructure, as noted in [2]. These devices have diverse areas of application, ranging from inside the human body to deep within the oceans and underground. The IoT refers to a network of physical devices, vehicles, buildings, and other items that are embedded with sensors, software, and other technologies to enable them to collect and exchange data. The main focus of the IoT is on enabling communication between these devices to enable automation and control.

CPS (cyber-physical systems) are similar to the IoT; however, CPS specifically refer to a system of physical, computational, and communication components that are tightly integrated to monitor and control physical processes. CPS typically involve a closed-loop feedback control system that involves sensors, actuators, and computational elements to continuously monitor and adjust physical processes in real-time.

CPS integrate physical components with software components, as noted by Buffoni et al. [3]. According to references [4,5], CPS are able to operate on different spatial and temporal scales.

Control systems coupled to physical systems are a common example of CPS, with applications in various domains such as smart grids, autonomous automobile systems, medical monitoring, industrial control systems, robotics systems, and automatic pilot avionics. CPS are becoming data-rich, enabling new and higher degrees of automation and autonomy.

New, smart CPS drive innovation and competition in a range of application domains, including agriculture, aeronautics, building design, civil infrastructure, energy, environmental quality, healthcare, personalized medicine, manufacturing, and transportation.

Despite some similarities, the primary distinction between the IoT and CPS is their focus. CPS are mainly focused on controlling physical processes, while the IoT is primarily focused on communication and data exchange between physical devices. CPS are typically used in industrial and manufacturing settings, where they facilitate real-time control of physical processes. On the other hand, the IoT has a broader range of applications, including home automation, healthcare, transportation, and other domains, where it enables seamless communication and integration of smart devices.

With the ever-increasing number of IoT and CPS devices, the need for more functional and high-quality software has become even more pressing. According to industry estimates, the global IoT market reached \$100 billion in 2017, and this figure is projected to soar to \$1.6 trillion by 2025, as noted in [6]. In 2022, enterprise spending on the IoT experienced a significant increase of 21.5%, reaching a total of \$201 billion. Back in 2019, IoT analytics had initially projected a spending growth of 24% for the year 2023. However, their growth outlook for 2023 has been revised to 18.5% according to [7], as shown in Figure 1.

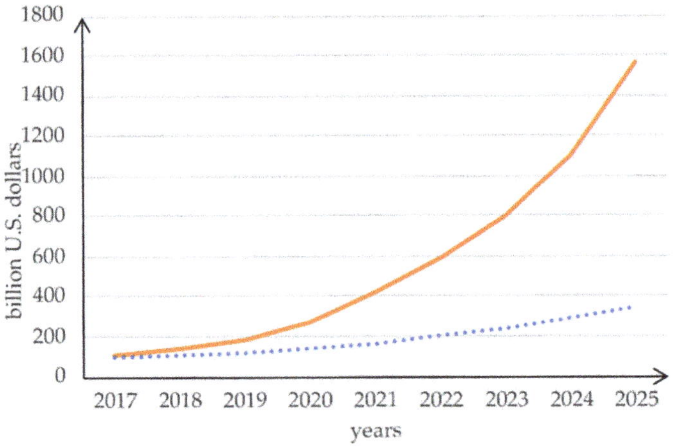

Figure 1. The global market of the Internet of Things (red solid line) and global spending on enterprise IoT technologies (blue dot line).

A model-based approach to CPS development is based on describing both the physical and software parts through models, allowing the whole system to be simulated before it is deployed.

There are several programming languages used in IoT and CPS development, including C/C++, Python, Java, JavaScript, and others. Among these, C/C++ is considered to be the most popular language for IoT development, with a popularity rate of 56.9%, according to recent research [8]. As of June 2023, GitHub has reported a total of over 53,285 IoT public repositories, with approximately 28,536 of them being C/C++ repositories, accounting for approximately 53.6% of the total number of IoT public repositories. Additionally, GitHub has reported over 22,776 CPS public repositories, out of which around 8892 are C/C++ repositories, making up approximately 39% of the total number of CPS public repositories. This is due to the fact that IoT devices typically have limited computing resources, and C/C++ is capable of working directly with the RAM while requiring minimal processing power.

CPS languages provide a unified approach to describing both the physical components and control software, making it possible to integrate modeling and simulation. Open standards such as FMI (Functional Mock-up Interface) and SSP (System Structure and Parameterization) facilitate this integration by defining a model format that utilizes the C language for behavior and XML for the interface. These standards, as specified in [9,10], enable the representation of pre-compiled models that can be exchanged between tools and combined for co-simulation.

According to the specifications outlined in [11], Modelica has been employed for the automatic generation of deployable embedded control software in C code from models. This utilization enhances the utility of Modelica as a comprehensive solution for the modeling, simulation, and deployment of CPS components.

The selection of a programming language for IoT and CPS development is highly dependent on the specific requirements of the project as well as the developer's proficiency. According to [12], utilizing C++ in embedded systems can be an effective solution, even considering the limited computing resources of microcontrollers used in small embedded applications compared to standard PCs. The clock frequency of microcontrollers may be much lower, and the amount of available RAM memory may be significantly less than that of a PC. Additionally, the smallest devices may not even have an operating system. To achieve the best performance, it is essential to choose a programming language that can be cross-compiled on a PC and then transferred as machine code to the device, avoiding any language that requires compilation or interpretation on the device itself, as this can lead to significant resource wastage.

For these reasons, C or C++ is often the preferred language for embedded systems, with critical device drivers requiring assembly language. If you follow the proper guidelines, using C++ can consume only slightly more resources than C, so it can be chosen based on the desired program structure. Overall, choosing the appropriate language for embedded systems can make a significant impact on performance and resource utilization.

The increasing number of IoT and CPS devices has resulted in a growing need for software that is both highly functional and of the utmost quality. As these devices become more ubiquitous and seamlessly integrated into our daily lives, the demand for dependable and efficient software becomes more critical than ever before. As a result, developers are constantly striving to enhance their software development methodologies and technologies to meet the ever-evolving demands of the IoT and CPS landscape.

However, given the increasing importance of the IoT and CPS as emerging technologies, it is expected that there will be more literature available on the topic of IoT and CPS software verification and quality assurance.

The typical software development life cycle (SDLC) involves several steps, including requirement analysis, design, implementation, testing and verification, and deployment and maintenance. While testing can increase our confidence in the program's correctness, it cannot prove it definitively. To establish correctness, we require a precise mathematical specification of the program's intended behavior and mathematical proof that the implementation meets the specification.

IoT verification encompasses a range of testing methodologies. These include conformance testing, as highlighted by Xie et al. [13], randomness testing, as discussed by Parisot et al. [14], statistical verification, as explored by Bae et al. [15], formal verification, as studied by Silva et al. [16], and the method known as model-based testing, as outlined by Ahmad [17]. In the specific context of the IoT, the model-checking technique, as emphasized by Clarke et al. [18], has notable representatives closely associated with it.

However, such software verification is difficult and time-consuming and is not usually considered cost-effective. In addition, modern verification methods would not replace testing in SDLC because most programs are not correct initially and need debugging before verification. The primary principle of verification involves adding specifications and invariants to the program and checking the verification conditions by proving generated lemmas based on the requirement specifications, as noted by Back [19].

However, most existing verification tools cannot detect software errors arising from incorrect usage of dimensions or units, which are commonly referred to as dimensionality errors or unit errors. These errors occur when software code or algorithms manipulate data with incompatible dimensions or units, resulting in incorrect calculations, unexpected behavior, or system failures. Such errors can have significant consequences across various domains, including engineering, finance, and scientific research:

1. Inconsistent unit conversions
2. Mixing incompatible units
3. Incorrect scaling or normalization
4. Mathematical operations on incompatible dimensions
5. Inaccurate assumptions about input units

By being aware of these potential pitfalls and implementing proper checks, validation and unit-aware programming techniques, developers can mitigate dimensionality or unit errors, ensuring accurate and reliable software functioning.

The failure of the Mars Climate Orbiter during its mission to study Mars' climate serves as a stark reminder of the consequences of navigational errors [20,21]. The spacecraft was intended to enter orbit around Mars in September 1999 but tragically entered the planet's atmosphere too low and disintegrated. This catastrophic error occurred due to a mismatch in the use of metric and imperial units, leading to incorrect calculations. Lockheed Martin, the contractor responsible for the spacecraft's navigation, used imperial units while NASA's software expected metric units. The failure resulted in a loss of $193.1 million and valuable scientific data. Lessons learned from this incident have since led to improved communication and unit conversion protocols in future space missions.

NASA's conversion concerns are particularly relevant to the constellation project which places significant emphasis on manned spaceflight [22]. Launched in 2005, the project ambitiously aims to facilitate future moon landings. However, an obstacle arises as the project's specifications and blueprints are exclusively in British imperial units. The conversion of this extensive body of work into metric units poses a considerable estimated expenditure of approximately $370 million.

In 2003, Tokyo Disneyland's Space Mountain roller coaster experienced a disruptive event when it came to a halt due to a broken axle that failed to meet design requirements [23]. The axle's excessive gap, which exceeded 1 mm instead of the required 0.2 mm, led to fractures caused by vibrations and stress. Fortunately, no injuries occurred despite the derailment. The accident resulted from discrepancies in unit systems. In 1995, the coaster's axle specifications switched to metric units, but in August 2002, an order mistakenly reverted to British imperial units, leading to 44.14 mm axles instead of the required 45 mm ones.

In 1983, an Air Canada Boeing 767 experienced fuel depletion during a Montreal to Edmonton flight [24]. Low fuel pressure warnings at 41,000 feet led to engine failures. However, the skilled captain and first officer managed to land the plane safely at an unused air force base nearby, with only a few minor passenger injuries. The incident was caused by a malfunctioning fuel indication system and an incorrect density ratio of 1.77 pounds per liter instead of the correct 0.80 kg per liter. These factors led maintenance workers to manually calculate and pump less than half the required amount of fuel, contributing to the incident.

Adding to the list of errors, in the early 1990s during the creation of the "Mir" space orbital station, another incident occurred due to incorrect usage of units of measurement. When experts from the Moscow Design Bureau sent data in kilogram-force to Khartron in Kharkiv, Ukraine (where one of the authors of this article worked), it was mistakenly interpreted as newtons. Consequently, the control system of the module, weighing approximately 20 tons, had to be reprogrammed during the flight, leading to a two-week delay in its journey to the station.

To mitigate dimensionality or unit errors, it is crucial to follow best practices, which include the following:

1. Clearly specifying and documenting the expected units and dimensions of input and output data.
2. Implementing reliable and consistent unit conversion routines.
3. Leveraging libraries or frameworks with built-in support for units and dimensions.
4. Conducting comprehensive testing, including dedicated unit tests, to validate the accuracy of calculations and conversions.
5. Validating assumptions about input units and implementing suitable checks.
6. Providing informative error messages or warnings when dimensionality or unit errors are detected.

By being mindful of dimensions and units during software development, developers can reduce the occurrence of errors and ensure the accuracy and reliability of their software.

There are several libraries and frameworks available that offer built-in support for handling units and dimensions in software development. In the following are some popular options:

As noted by Matthias Christian Schabel et al. [25], Boost.Units provides a comprehensive framework for handling physical quantities in C++ programming. It allows you to work with quantities with different units, perform arithmetic operations, and ensure dimensional consistency at compile-time. Boost.Units offers compile-time dimensional analysis, type-safe unit conversions, and supports custom unit systems. It is particularly useful in scientific and engineering applications, where precise handling of units and dimensions is crucial for accurate calculations.

As noted by Edzer Pebesma [26], UDUNITS is a flexible and extensive library primarily used in scientific and meteorological applications. It offers a comprehensive database of physical units, allowing for unit conversions, arithmetic operations, and parsing of unit expressions. UDUNITS supports a wide range of units and provides bindings for various programming languages, including Python and Java. It is a reliable solution for managing units and dimensions, especially in domains that require extensive support for physical quantities.

According to [27], Units.NET is a powerful and user-friendly library for managing physical quantities in C# applications. It provides a comprehensive set of units, supporting unit conversions, arithmetic operations, and dimensional analysis. With Units.NET, developers can work with units and dimensions in a strongly typed manner, ensuring type safety and accurate calculations. It simplifies the handling of units and dimensions in C# projects, making it convenient to work with physical quantities.

Common disadvantages of the described libraries are that they cannot utilize orientational information for checking software code.

In addition, the utilization of a specialized software language called F# enables efficient manipulation of physical units and dimensions [28]. While F# is widely recognized for its applications in general-purpose programming and data analysis, it also proves to be highly effective in the context of the IoT (Internet of Things) and CPS (cyber-physical systems) domains.

However, in the case of reusing IoT and CPS software programs that employ different physical units and orientations of physical values, which are typically implemented in languages like C++, it becomes essential to undergo additional formal verification. This verification process should incorporate orientational and dimensional information to ensure successful integration and reduce the development time of new software projects.

This article focuses on exploring a formal verification method that utilizes dimensional and orientational homogeneities and natural software invariants. Specifically, it considers the dimensions and orientations of physical quantities as defined by the International System of Units (SI), as described in references [29,30]. Additionally, it incorporates transformations of physical quantity orientations proposed by Siano [31,32] and extended by Santos et al. [33]. By leveraging these invariants, this method can effectively verify the correctness of software and detect errors that may arise due to inconsistent or incorrect use of units and dimensions, as well as incorrect usage of software operations, variables, and procedures, among other things.

The SI defines a set of base units, such as meters for length, kilograms for mass, and seconds for time, along with derived units, which are combinations of base units, such as meters per second for velocity or kilograms per cubic meter for density.

When it comes to software code, developers often need to work with and manipulate physical quantities in their programs. To ensure the correctness of such code, formal software verification methods can be applied. These methods use mathematical techniques to formally prove the properties of the software, such as its correctness, safety, or absence of certain errors.

The concept of homogeneity, derived from the SI system, plays a significant role in formal software verification. Siano proposed extending dimension homogeneity via orientation [31,32]. Now, the main concept of homogeneity states that any physical equation involving physical quantities must be both dimensionally and orientationally consistent. In other words, the units on both sides of an equation must match.

It is important to note that formal software verification involves more than just ensuring the homogeneity of physical quantities. It encompasses a broader range of techniques and methods to rigorously analyze and prove properties about software systems. However, leveraging the concept of homogeneity from the SI system can be a valuable tool in the pursuit of formal software verification, especially when dealing with physical quantities. The approach of leveraging the homogeneity of physical quantities and applying formal software verification techniques can make a significant contribution to ensuring functional and high-quality software for the IoT and CPS.

The advantages of the proposed method for software correctness and safety are that by utilizing formal software verification techniques, such as enforcing both dimensional and orientational homogeneity, developers can detect errors early in the development process and ensure that the software behaves as intended. This, in turn, reduces the potential for system failures or safety incidents.

IoT and CPS systems are typically subject to updates, maintenance, and evolution throughout their lifecycle. Enforcing both dimensional and orientational homogeneity and applying formal verification methods can enhance software maintainability and evolvability. By establishing clear and consistent units and enforcing them in the code, developers can more easily understand and modify the software, reducing the risk of introducing errors during updates or modifications. This promotes efficient maintenance and facilitates the evolution of the software as new requirements or functionalities are introduced.

Formal software verification methods, including the use of homogeneity, contribute to a rigorous quality assurance process. By systematically applying verification techniques, developers can identify and eliminate potential software defects, thereby improving the overall quality and reliability of IoT and CPS systems. This, in turn, enhances user satisfaction, reduces the risk of failures, and increases confidence in the deployed software. The thorough verification process helps ensure that the software meets the specified requirements and operates correctly in various scenarios, thereby ultimately enhancing the overall quality assurance efforts.

By combining the principles of homogeneity from the SI system with formal software verification methods, developers can create more robust, reliable, and functional software for the IoT and CPS. This approach helps mitigate risks, ensures safety, enhances interoperability, facilitates maintenance, and improves the overall quality of the software deployed in these systems.

2. The Formal Software Verification Method

Proposed is the utilization of natural software invariants, which are the physical dimensions and spatial orientation of software variables that correspond to real physical quantities. By incorporating these invariants into the program specification, it becomes possible to convert all program expressions into a series of lemmas that must be proven. This enables the verification of the homogeneity and concision of the program.

2.1. Using Dimensional Homogeneity in Formal Software Verification

According to Martínez-Rojas et al. [34], dimensional analysis is a widely used methodology in physics and engineering. It is employed to discover or verify relationships among physical quantities by considering their physical dimensions. In the SI a physical quantity's dimension is the combination of the seven basic physical dimensions: length (meter, m), time (second, s), amount of substance (mole, mol), electric current (ampere, A), temperature (kelvin, K), luminous intensity (candela, cd), and mass (kilogram, kg). Derived units are products of powers of the base units, and when the numerical factor of this product is one, they are called coherent derived units. The base and coherent derived units of the SI form a coherent set designated as the set of coherent SI units. The word "coherent" in this context means that equations between the numerical values of quantities take the same form as the equations between the quantities themselves, ensuring consistency and accuracy in calculations involving physical quantities.

Some of the coherent derived units in the SI are given specific names and, together with the seven base units, form the core of the set of SI units. All other SI units are combinations of these units. For instance, plane angle is measured in radians (rad), which is equivalent to the ratio of two lengths; solid angle is measured in steradians (sr), which is equivalent to the ratio of two areas. The frequency is measured in hertz (Hz), which is equivalent to one cycle per second. The force is measured in newtons (N), which is equivalent to $\mathrm{kg\,m/s^2}$. The pressure and stress are measured in pascals (Pa), which is equivalent to $\mathrm{kg/m\,s^2}$ or $\mathrm{N/m^2}$. The energy and work are measured in joules (J), which is equivalent to $\mathrm{kg\,m^2/s^2}$ or $\mathrm{N\,m}$.

The fundamental principle of dimensional analysis is based on the fact that a physical law must be independent of the units used to measure the physical variables. According to the principle of dimensional homogeneity, any meaningful equation must have the same dimensions on both sides. This is the fundamental approach to performing dimensional analysis in physics and engineering.

Existing software analysis tools only check the syntactic and semantic correctness of the code, but not its physical correctness. However, we can consider the program code of systems as a set of expressions consisting of operations and variables (constants). By using DA, we can verify the physical consistency of the program code and detect errors that may arise due to inconsistent or incorrect use of units and dimensions.

To check the correctness of expressions, we can use the dimensionality of program values. Preservation of the homogeneity of the expressions may indicate the physical usefulness of the expressions. Violation of homogeneity indicates the incorrect use of a program variable or program operation. Dimensional analysis provides an opportunity to check not only simple expressions but also calls to procedures and functions. The use of the physical dimension allows the verification of the software.

Dimensional analysis is a powerful tool that can be used to ensure the physical correctness of software code. By checking the dimensionality of program values, we can ensure the preservation of the homogeneity of expressions, which may indicate the physical usefulness of the code. In cases where homogeneity is violated, it may indicate an incorrect use of program variables or operations. Dimensional analysis can be applied not only to simple expressions but also to calls to procedures and functions, providing a comprehensive approach to verifying the physical consistency of the software.

We can view software as a model, and dimensional analysis can serve as a validation tool to ensure that this model adheres to the physical laws and principles governing the system it represents. By incorporating physical dimensions into the validation process, we can effectively identify and rectify errors that may arise from inconsistent or incorrect usage of units and dimensions. This approach ultimately contributes to the development of more reliable and accurate software.

In the software system, we can define it as a collection of interacting sub-systems. Each sub-system consists of interacting software units, and each unit comprises a set of operators. Operators, in turn, are ordered sets of statements or expressions. By structuring the software system in this hierarchical manner, we can effectively check the interactions and operations within the system.

To prove the homogeneity of software systems, we need to demonstrate the homogeneity of each subsystem. Similarly, to prove the homogeneity of subsystems, we need to establish the homogeneity of each software unit. Finally, to ensure the homogeneity of software units, we need to demonstrate the homogeneity of each software statement or expression. This stepwise approach allows us to systematically verify the homogeneity of the software and ensure its adherence to the specified physical dimensions.

Let us introduce a set of "multiplicative" operations {*, /, etc.} that generate new physical dimensions, while "additive" operations {+, −, =, <, ≤, >, ≥, !=, etc.} act as checkpoints to ensure dimensional homogeneity. If the source code contains variables that preserve specific physical dimensions, we can utilize this property, known as dimensional homogeneity, as a software invariant. Each additive operation serves as a source for generating lemmas. Following the principle of dimensional homogeneity, we can develop a set of lemmas to support the verification process.

By employing dimensional analysis, we can verify the physical dimensions of variables to identify errors resulting from inconsistent or incorrect usage of units and dimensions, as well as improper utilization of software operations, variables, and procedures. However, it is worth noting that certain variables may possess the same dimensions, such as moments of inertia and angular velocities. In order to detect software defects arising from the erroneous handling of such variables, careful examination of expressions involving angles, angular speed, and related quantities is required. It is important to remember that, according to the SI, angles are considered dimensionless values.

2.2. Using Orientational Homogeneity in Formal Software Verification

To address this issue, we can utilize features for transformations of angles and oriented values. In [31,32], Siano proposed an orientational analysis as an extension of dimensional analysis. This approach involves considering not only the physical dimensions but also the orientations of the quantities to enhance the analysis.

The use of orientational analysis can aid in expanding the base unit set while also ensuring dimensional and orientational consistency. Additionally, the orientational analysis technique can be applied for the formal verification of software code, allowing for a thorough evaluation of its accuracy and reliability.

Siano's proposed notation system for representing vector directions involved the use of orientational symbols l_x, l_y, l_z [31,32]. Furthermore, a symbol without orientation represented by l_0 was introduced to represent vectors that do not possess a specific orientation.

For example, a velocity in the x-direction can be represented by $V_x \doteq l_x$, while a length in the x-direction can be represented by $L_x = l_x$. Here, the symbol $=$ denotes that the quantity on the left-hand side has the same orientation as the quantity on the right-hand side. In non-relativistic scenarios, mass is considered to be a quantity without orientation.

In order for equations involving physical variables to be valid, they must exhibit orientational homogeneity, meaning that the same orientation must be utilized on both sides of the equation. Furthermore, it is crucial that the orientations of physical variables are assigned in a consistent manner. For instance, the representation of acceleration in the x-direction as $a_x = \frac{\Delta V_x}{\Delta t}$, $V_x \doteq l_x$, $\Delta t \doteq l_0$ and $a_x \doteq \frac{l_x}{l_0}$ is only valid if both sides have the same orientation.

But what about the orientation of time? From the expression $H = \frac{gt^2}{2}$ we can define the time as follows: $t = \sqrt{\frac{2H}{g}} \doteq \sqrt{\frac{l_z}{l_z}} \doteq \sqrt{l_z l_z} \doteq \sqrt{l_0} \doteq l_0$.

The physical quantity of time is considered to be without orientation, meaning it does not possess a specific orientation in space.

In order to maintain orientational homogeneity, it is necessary to introduce a characteristic length scale l_0, since time is a quantity without orientation. Therefore, l_x can be expressed as $l_x \doteq \frac{l_x}{l_0} \doteq l_x$ and $l_0 l_x \doteq l_x l_0 \doteq l_x$, $l_0^{-1} \doteq l_0$ ensuring that the orientation of l_x remains consistent. That is why $a_x \doteq l_x$.

It is essential to assign orientations to physical variables in a consistent manner. For instance, pressure is defined as force per unit area. If the force is acting in the z-direction, then the area must be normal to it etc.:

$$P = \frac{F_z}{S_{xy}} \doteq \frac{l_z}{l_x l_y} \doteq \frac{l_z}{l_z} \doteq l_0, \; P = \frac{F_x}{S_{yz}} \doteq \frac{l_x}{l_y l_z} \doteq \frac{l_x}{l_x} \doteq l_0, \; P = \frac{F_y}{S_{xz}} \doteq \frac{l_y}{l_x l_z} \doteq \frac{l_y}{l_y} \doteq l_0.$$

In order for pressure to be considered a quantity without orientation, the area S must have the same orientation as the force F. If the force F is in a particular direction, then the area S must be normal to that direction, meaning that both variables have the same orientation. Therefore, pressure can only be a quantity without orientation if this consistent orientation is maintained.

We define this as follows:

$$l_x l_y \doteq l_z, \; l_y l_z \doteq l_x, \; l_x l_z \doteq l_y, \; \frac{l_x}{l_x} \doteq l_0, \; \frac{l_y}{l_y} \doteq l_0, \; \frac{l_z}{l_z} \doteq l_0.$$

If $a \doteq l_x$, $b \doteq l_y$, $c \doteq l_z$ a volume of space, V, is a quantity without orientation:

$$V_{abc} = S_{ab} c \doteq l_z l_z \doteq l_0, \; V_{abc} = S_{bc} a \doteq l_x l_x \doteq l_0, \; V_{abc} = S_{ac} b \doteq l_y l_y \doteq l_0$$

Let us take a look at uniformly accelerated motion in the x-direction: $S_x = S_{0X} + v_x t + \frac{a_x t^2}{2}$, where S_x is the total distance, S_{0X} is the initial distance, v_x is the velocity, and a_x is the acceleration.

According to orientational homogeneity

$$S_x \doteq l_x, \; S_{0X} \doteq l_x, \; v_x t \doteq l_x l_0 \doteq l_x, \; \frac{a_x t^2}{2} \doteq \frac{l_x}{l_0}(l_0)^2 \doteq l_x$$

The orientation of derived physical variables, such as kinetic energy, can be determined by properly assigning orientation to primitive variables and applying the corresponding multiplication rules:

$$KE = \frac{mv_x^2}{2} + \frac{mv_y^2}{2} + \frac{mv_z^2}{2}, \; KE \doteq l_0 l_x l_x + l_0 l_y l_y + l_0 l_z l_z, \; KE \doteq l_0 l_0 + l_0 l_0 + l_0 l_0 \doteq l_0$$

We considered the orientation of an angle α in the x-y plane. Because $\tan(\alpha) \doteq \frac{l_y}{l_x}$ and $\lim_{\alpha \to 0}(\tan(\alpha)) = \alpha$ we deduced that $\alpha \doteq \frac{l_y}{l_x} \doteq l_z$ and an angular velocity of $\omega_{xy} = \frac{\Delta \alpha}{\Delta t} \doteq l_z$.

Let us take the following series:

$$\text{sine}(\alpha) = \alpha - \frac{\alpha^3}{3!} + \frac{\alpha^5}{5!} \ldots, \; \text{cosine}(\alpha) = 1 - \frac{\alpha^2}{2!} + \frac{\alpha^4}{4!} \ldots$$

If α has any orientation, then sine (α) would also have that orientation, while the cosine (α) would be a quantity without orientation. This is because the sine function involves the odd powers of α, while the cosine function involves the even powers of α.

Siano demonstrated that orientational symbols have an algebra defined by the multiplication table for the orientation symbols [31,32], which is as follows:

$$\begin{array}{c|cccc} & l_0 & l_x & l_y & l_z \\ \hline l_0 & l_0 & l_x & l_y & l_z \\ l_x & l_x & l_0 & l_z & l_y \\ l_y & l_y & l_z & l_0 & l_x \\ l_z & l_z & l_y & l_x & l_0 \end{array} \quad \text{and rules}: \quad \begin{array}{ll} l_o = \frac{1}{l_o} & l_x = \frac{1}{l_x} \\ l_y = \frac{1}{l_y} & l_z = \frac{1}{l_z} \end{array}$$

Based on the above, the product of two orientated physical quantities has an orientation as follows:

$$l_o l_x = l_x l_o = l_x,\; l_o l_y = l_y l_o = l_y,\; l_o l_z = l_z l_o = l_z,\; l_x l_x = l_y l_y = l_z l_z = l_0$$

If a source code contains variables that represent physical quantities with orientations, we can use the property of orientational homogeneity as a software invariant. By applying orientational homogeneity, we can transform the source code into a set of lemmas. Multiplicative operations, such as multiplication and division, introduce new physical orientations. On the other hand, additive operations such as addition, subtraction, and relational operators (e.g., =, <, $\leq$, >, $\geq$, !=) serve as checkpoints for verifying orientational homogeneity.

2.3. Examples of Software Formal Verifications

Example 1. *Consider the expression $F = ma$, where F is a force with physical units of $(kg\ m\ s^{-2})$, m is the mass in (kg), and a is the acceleration in $(m\ s^{-2})$. This expression allows us to derive the orientation and dimension of the product ma. If m is without orientation (denoted by $m \doteq l_0$) and a is in the x-direction (denoted by $a \doteq l_x$), then F has the x-orientation. If m has units of (kg) and a has units of $(m\ s^{-2})$, then the dimensions of the result are $(kg\ m\ s^{-2})$. The assignment operation "=", which is also known as the equality operator, acts as a checkpoint for our software invariants. It ensures that the physical orientation of F is equal to the physical orientation of ma and that the physical dimensions of F are equal to the physical dimensions of the result.*

Example 2. *Consider the expression $S = S_0 + vt + 0.5at^2$, where S represents the total distance, S_0 is the initial distance, v is the velocity, a is the acceleration, and t is the time. This expression generates two new physical dimensions and orientations. The second "+" operation checks the dimensions and orientations of vt and $0.5at^2$. The first "+" operation checks the homogeneity of S_0 and the result of the previous operation. Finally, the assignment operator "=" checks the homogeneity of S and the result of the previous operation. By checking the homogeneity of these variables and operations, we can ensure that the physical dimensions and orientations are consistent throughout the expression.*

Example 3. *When calling procedures and functions, it is not always possible to check the physical dimensions and orientation of the arguments. However, for function signatures such as Type1 and some Function2 (Type2 x . . .), where Type1 and Type2 have information about the physical dimensions and orientations of their arguments, it is possible to check the physical dimensions and orientations of the arguments. Each argument of a function generates a special lemma that can be used to prove dimensional and orientational homogeneities. Only after proving all the lemmas can we prove the correctness of the function call. It is important to note that real arguments of exponential and logarithmic functions must be dimensionless and without orientation to preserve dimensional and orientational homogeneities. On the other hand, arguments of trigonometric functions such as sine(x), cosine(x), and tan(x) must be orientated to preserve orientational homogeneity, while also being dimensionless to preserve dimensional homogeneity. The proposed method allows for the checking of physical dimensions and orientations in software statements and units. Repeating this check helps to ensure the correctness of the software system.*

Example 4. *Let us examine Euler's rotation equations, as described in [35], which find numerous applications in fields such as cyber-physical systems (CPS) and the Internet of Things (IoT). These equations are utilized in various contexts, including unmanned cars, helicopters, and other aerial vehicles.*

The general vector form of the equations is $I\dot{\omega} + \omega \times (I\omega) = M$, where M represents the applied torques and I is the inertia matrix. The vector $\dot{\omega}$ represents the angular acceleration.

In orthogonal principal axes of inertia coordinates the equations become

$$\begin{cases} I_x\dot{\omega}_x + (I_z - I_y)\omega_y\omega_z = M_x \\ I_y\dot{\omega}_y + (I_x - I_z)\omega_z\omega_x = M_y \\ I_z\dot{\omega}_z + (I_y - I_x)\omega_x\omega_y = M_z \end{cases} \tag{1}$$

where M_x, M_y, M_z are the components of the applied torques (kg m^2 s^{-2}); I_x, I_y, and I_z are the principal moments of inertia (kg m^2); ω_x, ω_y, and ω_z are the components of the angular velocities (s^{-1}); and $\dot{\omega}_x = \frac{d\omega_x}{dt}$, $\dot{\omega}_y = \frac{d\omega_y}{dt}$, and $\dot{\omega}_z = \frac{d\omega_z}{dt}$ have dimension (s^{-2}). However, M_k, I_k, ω_k, and $\dot{\omega}_k$ have different orientations l_k, where $k = x, y, z$.

We can verify the dimensional homogeneity of Equation (1), but it is not possible to identify all defects. This is because certain quantities have the same dimensions and cannot be distinguished solely based on their units. For example, the dimensions of the moments of inertia (I_x, I_y, I_z) are (kg m^2) and angular velocities (ω_x, ω_y, ω_z) are (s^{-1}), and the dimensions of the angular acceleration components ($\dot{\omega}_x$, $\dot{\omega}_y$, $\dot{\omega}_z$) are (s^{-2}) since angles are dimensionless. Therefore, while dimensional analysis can help identify some potential issues with the equation, it may not be able to catch all possible defects. For example, we cannot detect a defect if the expression $S = S_0 + vt + 0.5\,at^2$ does not include the initial position S_0.

In the context of Equation (1), the parameters (M_x, M_y, M_z, etc.) may have different orientations or values, which can help in detecting defects.

Furthermore, checking both the dimensional and orientational homogeneities of an equation can improve our ability to detect defects and ensure their correctness. This approach can be useful in the formal verification of CPS and IoT software, as it can help identify potential issues before they lead to real-world problems.

Let us assess the probability of detecting a software defect using both dimensional analysis and orientational analysis.

2.4. Software Defect Detection Models

2.4.1. General Software Defect Detection Model

To simplify the analysis, we assume that the software statement can only have one defect with a probability of P_{def}. The model starts with the initial event state of 'Software' and branches out into two possible outcomes at the next level: 'Software has a defect' and 'Software does not have a defect', with probabilities of P_{def} and $1 - P_{def}$, respectively.

Decision trees, as described in [36], are visual representations utilized in decision analysis and machine learning. They illustrate decisions or events along with assigned probabilities or outcomes. Decision trees offer a structured approach to analyzing intricate decision-making processes. They can be applied to predict software defect detection, facilitating the identification and prevention of software issues, as demonstrated in Figure 2.

In the state 'Software has a defect', our focus shifts to detecting the defect. At the third level, the model branches out into two possible outcomes: 'Defect detected' and 'Defect not detected', with probabilities of P_{DD} and $1 - P_{DD}$, respectively.

The revised sentence maintains clarity and correctness in grammar.

To define the conditional probability of software defect detection, we used the following formula:

$$\eta = \frac{P_{def}P_{DD}}{P_{def}P_{DD} + P_{def}(1 - P_{DD})} = P_{DD},$$

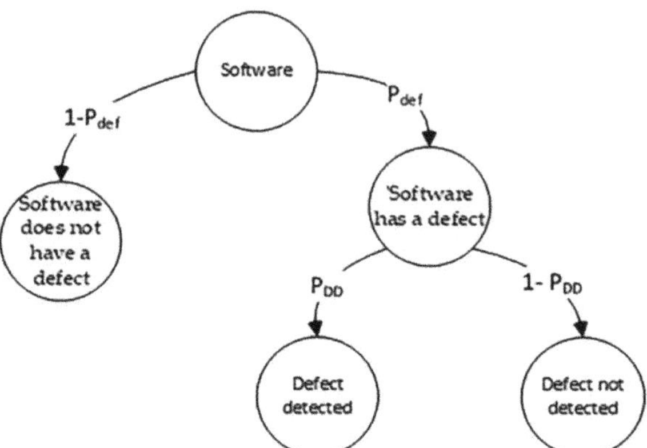

Figure 2. General software defect detection model.

Here P_{def} represents a probability of a software defect in the code; P_{DD} represents a probability of a software defect detection.

We can also introduce a more complex general software defect detection model (see Figure 3), which accounts for two types of defects: variable defects (uncorrected usage of a variable with incorrect dimension or orientation) and operation defects (incorrect usage of an operation). Despite the presence of multiple types of defects, the model still assumes that there is only one defect present in the software statement at any given time.

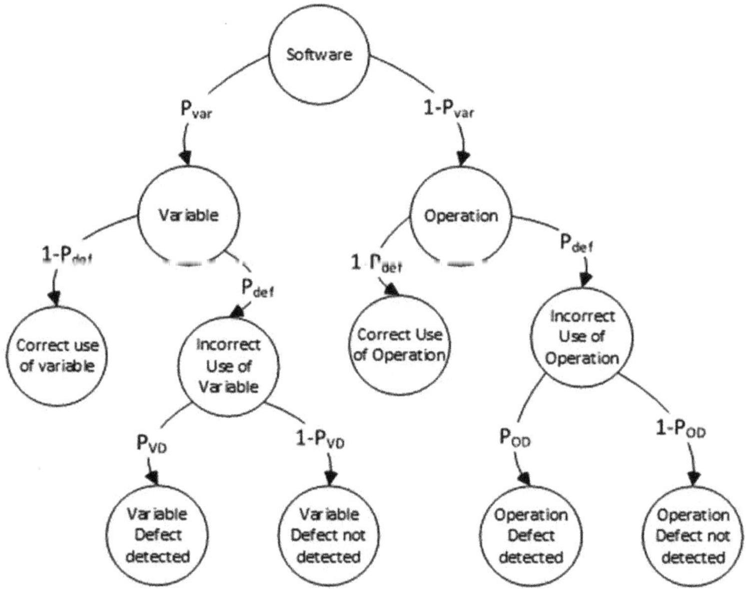

Figure 3. Complex general software defect detection model.

In this more complex model, the software statement can have two types of defects: variable defects and operation defects. A 'Variable defect' occurs when there is an incorrect usage of a variable in the code, such as using the wrong variable name. An 'Operation defect' occurs when there is an incorrect usage of an operation in the code, such as using the wrong operator symbol. Despite the presence of these two types of defects, the model still assumes that only one defect is present in the statement at any given time.

In this more complex model, the initial event state is 'Software'. At the branching point, the model expands into two possible outcomes: 'Variable' and 'Operation', with probabilities of P_{var} and $1 - P_{var}$, respectively.

The 'Variable' state has two potential outcomes at the next level: 'Correct use of variable' and 'Incorrect use of variable', with probabilities of $1 - P_{def}$ and P_{def}, respectively.

The 'Incorrect use of variable' state then branches out into two possible outcomes at the next level: 'Variable defect detected' and 'Variable defect not detected', with probabilities of P_{VD} and $1 - P_{VD}$, respectively. Here, P_{VD} represents the probability of detecting a variable defect in the source code.

In addition to the 'Variable' state, the model also has an 'Operation' state, which has two possible outcomes at the next level: 'Correct use of operation' and 'Incorrect use of operation', with probabilities of $1 - P_{def}$ and P_{def}, respectively.

The 'Incorrect use of operation' state then branches out into two possible outcomes at the next level: 'Operation defect detected' and 'Operation defect not detected', with probabilities of P_{OD} and $1 - P_{OD}$, respectively. Here, P_{OD} represents the probability of detecting an operation defect in the source code.

The conditional probability of a software defect can be defined as follows:

$$\eta = \frac{P_{variableDefectDetected} + P_{operationDefectDetected}}{P_{variableDefectDetected} + P_{variableDefectNotDetected} + P_{operationDefectDetected} + P_{operationDefectNotDetected}}$$

Here $P_{variableDefectDetected} = P_{var}P_{def}P_{VD}$, $P_{operationDefectDetected} = (1 - P_{var})P_{def}P_{OD}$
$P_{variableDefectNotDetected} = P_{var}P_{def}(1 - P_{VD})$, $P_{operationDefectNotDetected} = (1 - P_{var})P_{def}(1 - P_{OD})$

After substitution $P_{variableDefectDetected} \cdots \cdots P_{operationDefectDetected}$ in the source expression:

$$\eta = \frac{P_{var}P_{def}P_{VD} + (1 - P_{var})P_{def}P_{OD}}{P_{var}P_{def}P_{VD} + P_{var}P_{def}(1 - P_{VD}) + (1 - P_{var})P_{def}P_{OD} + (1 - P_{var})P_{def}(1 - P_{OD})}$$

$$\eta = P_{var}P_{VD} + (1 - P_{var})P_{OD} \tag{2}$$

As per Expression (2), the conditional probability of software defect detection depends on the probability of the software variables used in the source code and the conditional probabilities of detecting defects (defects of operations and defects of variables). We can determine the value of P_{var} by analyzing the software code statically, i.e., without executing the code. However, to determine the values of P_{VD} and P_{OD}, we would need to build additional software defect detection models.

2.4.2. Simple Model for Detection of Incorrect Use of Variables Based on Dimensional Analysis

Next, we introduce the simple model for the detection of incorrect use of variables based on dimensional analysis, as depicted in Figure 4.

This model has an initial state of 'Variable'. The initial state has two transitions to states 'OK' and 'Check Dimension', with probabilities $1 - P_{def}$ and P_{def}, respectively. In the state 'Check Dimension', we can evaluate the required physical dimension of the variable using dimensional analysis, such as length, mass, time, thermodynamic temperature, etc.

If the actual physical dimension is equal to the required physical dimension, we cannot detect the software defect. However, if they differ, we can identify the software defect. In this case, the probabilities are P_{dim} and $1 - P_{\text{dim}}$, where P_{dim} represents the probability of two random variables having the same physical dimension.

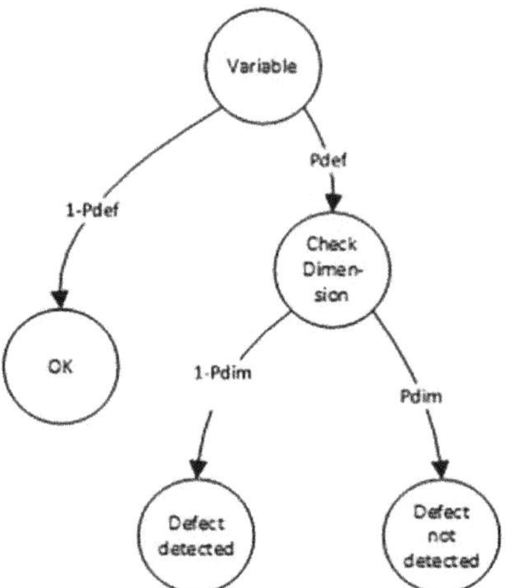

Figure 4. Simple model for the detection of incorrect use of variables based on dimensional analysis.

Let us define the conditional probability of defect detection of incorrect use of a program variable as follows:

$$P_{VD} = \frac{P_{defectDetected}}{P_{defectDetected} + P_{defectNotDetected}}.$$

Here $P_{defectDetected} = P_{def}(1 - P_{\text{dim}})$ and $P_{defectNotDetected} = P_{def}P_{\text{dim}}$

$$P_{VD} = 1 - P_{\text{dim}} \tag{3}$$

Let us consider a set of distinct software variables $\{\text{var}_1 \cdots \text{var}_{Nv}\}$ and a set of diverse physical dimensions $\{\text{dim}_1 \cdots \text{dim}_{Nd}\}$, where N_V represents the cardinality of set $\{\text{var}_i\}$ and N_D represents the cardinality of set $\{\text{dim}_j\}$.

To depict the relationship between these variables and dimensions, we can make use of an N-matrix (4):

	$\dim_1$	$\dim_2$	$\dim_3$	$\dim_4$	$\dim_5$	$\dim_6$	$\cdots$	$\dim_{N_D-1}$	$\dim_{N_D}$
var_1	n_{11}	0	0	0	0	0	$\cdots$	0	0
var_2	n_{21}	0	0	0	0	0	$\cdots$	0	0
var_3	0	n_{31}	0	0	0	0	$\cdots$	0	0
var_4	0	0	n_{43}	0	0	0	$\cdots$	0	0
var_5	0	0	0	n_{54}	0	0	$\cdots$	0	0
var_6	0	0	0	n_{64}	0	0	$\cdots$	0	0
$\vdots$	$\vdots$	$\vdots$	$\vdots$	$\vdots$	$\vdots$	$\vdots$	$\ddots$	$\vdots$	$\vdots$
var_{N_V-1}	0	0	0	0	0	0	0	0	n_{N_V-1,N_D}
var_{N_V}	0	0	0	0	0	0	0	0	n_{N_V,N_D}

$$\tag{4}$$

The equation for the total quantity of usages of all software variables that have the same j dimension can be written as follows:

$$N_{VARj} = \sum_{i=1}^{N_V} n_{ij} \tag{5}$$

where n_{ij} represents the total quantity usage of i-variable which has a j-physical dimension, and N_V is the cardinality set of software variables.

Equation (5) shows the total number of variable usages in the code:

$$N_{VAR} = \sum_{i=1}^{N_V} \sum_{j=1}^{N_D} n_{ij} \tag{6}$$

To define the probability of choosing i-variable and j-variable with the same dimensions, we can use the total number of usages of variables with the j-physical dimension and the total number of usages of all variables in the code:

$$D_{ij} = \frac{n_{ij}}{N_{VAR}} \frac{\left(\sum_{i=1}^{N_V} n_{ij}\right) - n_{ij}}{N_{VAR} - n_{ij}} \tag{7}$$

According to (7), the probability of choosing two random variables that have the same physical dimension is given by the following equation:

$$P_{dim} = \sum_{i=1}^{N_V} \sum_{j=1}^{N_D} \left(\frac{n_{ij}}{N_{VAR}} \frac{\left(\sum_{k=1}^{N_V} n_{kj}\right) - n_{ij}}{N_{VAR} - n_{ij}} \right) \tag{8}$$

Here n_{ij} represents the element of the N matrix representing the quantity of usage for the i-variable with the j-physical dimension; N_{VAR} represents the total number of variable usages in the code; N_D represents the total number of different dimensions of variables in the code; and N_V represents the total number of variables in the code.

For increasing the conditional probability detection of incorrect use of software variables we need to use other independent properties of variables. Using additional independent properties of variables can help increase the conditional probability detection of incorrect use of software variables. This is because using multiple properties helps to reduce the chance of false positives and increase the reliability of the detection model.

2.4.3. Simple Model for Detection of Incorrect Use of Variables Based on Orientational Analysis

In many cases, variables in CPS or the IoT have not only physical dimensions but also orientation information, which can be utilized to enhance the software quality of these systems. Therefore, we introduce a simple model for the detection of incorrect variable use based on orientational analysis (see Figure 5).

According to Figure 5, the initial state of the model is 'Variable'. This state has two transitions to states 'OK' and 'Check orientation' with probabilities $1 - P_{def}$ and P_{def}, respectively. In the state 'Check orientation', we can evaluate the required physical orientation of the variable using orientation analysis, such as l_0, l_x, l_y, and l_z. If the physical orientation of the variable matches the required orientation, we cannot detect a software defect. However, if the physical orientation is different from the required orientation, we can detect a software defect. These cases have probabilities P_{orient} and $1 - P_{orient}$, where P_{orient} is the probability that two randomly selected variables have the same physical orientation.

Now we need to define P_{orient}.

This defect model is similar to the dimension defect model of variables. However, in this case, we have four different orientations and N_V different variables. We can describe the relationship between variables and their orientations using an M matrix:

$$
\begin{array}{c c c c c c}
 & l_0 & l_x & l_y & l_z & \\
\text{var}_1 & m_{11} & 0 & 0 & 0 & \\
\text{var}_2 & m_{11} & 0 & 0 & 0 & \\
\text{var}_3 & 0 & m_{11} & 0 & 0 & \\
\text{var}_4 & 0 & m_{11} & 0 & 0 & \\
\ldots & \ldots & \ldots & \ldots & \ldots & \\
\text{var}_{N_V} & 0 & 0 & 0 & m_{N_V,A} &
\end{array}
\tag{9}
$$

where l_k is a direction of orientation and $k = 0, x, y, z$.

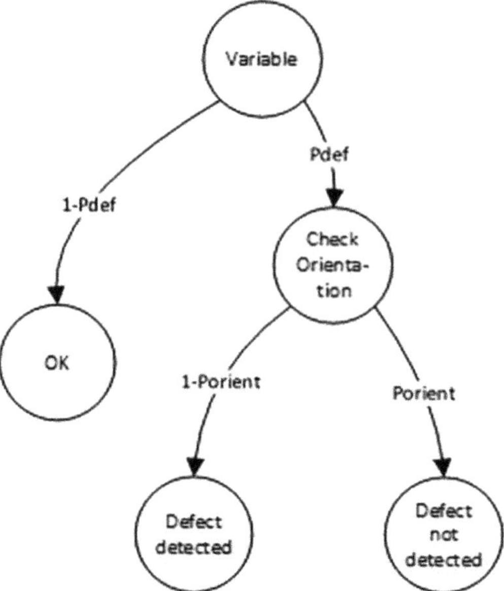

Figure 5. Simple model for the detection of incorrect use of variables based on orientational analysis.

Because, every software variable (as a host of a physical value) has only one orientation, every i-row of the M matrix has only one non-zero number m_{ik}—the item of M matrix—the number of using of i-variable which has k-orientation.

We can use Expression (8) for defining the probability of choosing two random variables that have the same physical orientation; this is given by the following Equation (10):

$$
P_{orient} = \sum_{i=1}^{N_V} \sum_{k=0,x,y,z} \left(\frac{n_{ik}}{N_{VAR}} \frac{\left(\sum\limits_{n=1}^{N_V} m_{nj}\right) - m_{ik}}{N_{VAR} - m_{ik}} \right)
\tag{10}
$$

Here m_{ik} represents the element of the M matrix representing the quantity of usage for the i-variable with the j-physical orientation. N_{VAR} represents the total number of variable usages in the code; N_V represents the total number of variables in the code.

We can increase the conditional probability of software defect detection by concurrently using both dimensional and orientational analysis. By combining these two methods, we can improve the accuracy of defect detection and reduce the likelihood of undetected defects.

2.4.4. Complex Model for Detection of Incorrect Use of Variables Based on Dimensional and Orientational Analysis

The complex model of the detection of incorrect use of variables based on both dimensional and orientational analysis is described in Figure 6.

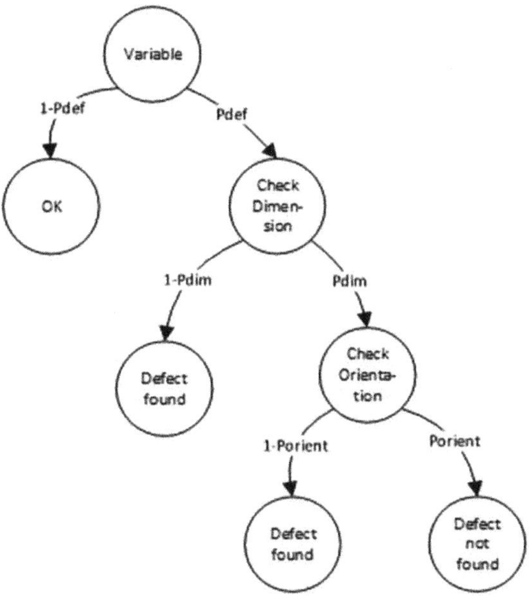

Figure 6. Complex model for the detection of incorrect use of variables based on dimensional and orientational analysis.

According to Figure 6, the initial state of the model is 'Variable'. This state has two transitions to the states 'OK' and 'Check Dimension' with probabilities $1 - P_{def}$ and P_{def}, respectively. In the 'Check Dimension' state, we evaluate the required physical dimension of the variable using dimensional analysis, such as length, mass, time, thermodynamic temperature, etc.

If the actual physical dimension matches the required dimension, we cannot detect the software defect with a probability of P_{dim}. However, if they differ, we can identify the software defect with a probability of $1 - P_{dim}$, where P_{dim} represents the probability that two randomly selected variables have the same physical dimension.

When we cannot detect the software defect, in the state 'Check orientation', we evaluate the required physical orientation of the variable using orientational analysis. If the physical orientation matches the required orientation, we cannot detect a software defect. However, if the physical orientation differs from the required orientation, we can detect a software defect. These cases have probabilities P_{orient} and $1 - P_{orient}$, where P_{orient} represents the probability that two randomly selected variables have the same physical orientation.

$$P_{VD} = \frac{P_{def}(1 - P_{\dim}) + P_{def}P_{\dim}(1 - P_{orient})}{P_{def}(1 - P_{\dim}) + P_{def}P_{\dim}(1 - P_{orient}) + P_{def}P_{\dim}P_{orient}}$$

$$P_{VD} = 1 - P_{\dim}P_{orient} \tag{11}$$

After the substitution of Expressions (8) and (10) in (11) we have

$$P_{VD} = 1 - \sum_{i=1}^{N_V}\sum_{j=1}^{4}\left(\frac{m_{ij}}{N_{VAR}}\frac{\left(\sum_{k=1}^{N_V} m_{kj}\right) - m_{ij}}{N_{VAR} - m_{ij}}\right)\sum_{i=1}^{N_V}\sum_{j=1}^{N_D}\left(\frac{n_{ij}}{N_{VAR}}\frac{\left(\sum_{k=1}^{N_V} n_{kj}\right) - n_{ij}}{N_{VAR} - n_{ij}}\right), \tag{12}$$

Here m_{kj} represents the element of the M matrix, which represents the quantity of usage for the k-variable with the j-physical orientation, n_{ij} represents the element of the N matrix representing the quantity of usage for the i-variable with the j-physical dimension, N_D represents the total number of different dimensions of variables in the code, N_V represents the total number of variables in the code, and N_{VAR} represents the total number of variable usages in the code. Furthermore, there are four different orientations (l_0, l_x, l_y, l_z) in the code.

According to Equation (12), P_{VD} denotes the conditional probability of detecting the incorrect use of software variables. The probability depends on the distribution of the software variables according to different dimensions and orientations.

To evaluate the correctness of the conditional probability of software defect detection, we need a model for detecting the incorrect use of operations. This model should take into account the types of operations that are commonly used in CPS and IoT software, as well as their potential incorrect use.

2.4.5. Model for Detection of Incorrect Use of Operations Based on Dimensional and Orientational Analysis

Let us consider three subsets of C++ operations: "additive" (A), "multiplicative" (M), and "other" (O) operations.

$$\begin{aligned}
A &= \{"+", "-", "=", "==", ">=", "<=", "!=", "<", ">", "++", "--", ".*", "->*", ",", ".", "->", "+=", "-=", "**"\}, \\
M &= \{"*", "/", "\%", "*=", "/=", "\%="\}, \\
O &= \{"||", "\&\&", "\&", "|", "\^", "~", "<<", ">>", "::", "?", "<<=", ">>=", "\&=", "|=", "\^="\}
\end{aligned} \tag{13}$$

In addition, we are given three probabilities associated with the utilization of this operation in the source code, namely, P_A, P_M, and P_O. Let us define the sum of these probabilities as the full group probability:

$$P_A + P_M + P_O = 1. \tag{14}$$

Let us define P_A, P_M, and P_O as follows:

$$P_A = \frac{N_A}{N_A + N_M + N_O}, \quad P_M = \frac{N_M}{N_A + N_M + N_O}, \quad P_O = \frac{N_O}{N_A + N_M + N_O}, \tag{15}$$

Here, N_A represents the total number of "additive" operations in a file, N_M represents the total number of "multiplicative" operations in a file, and N_O represents the total number of "other" operations in the file.

In this case, we can build a decision tree for the detection of incorrect use of operations based on dimensional and orientational analysis. The model allows us to define the conditional probability of operation defect detection (see Figure 7).

According to Figure 7, the initial state of the model is 'Operation'. This state has three transitions to the states 'Additive Operation', 'Multiplicative Operation', and 'Other Operation' with probabilities P_A, P_M, and P_O, respectively.

In the state 'Additive Operation', any 'additive' operation can be replaced by another operation. This event has the probability P_{def}. We then transition to the 'Using incorrect operation 1' state with this probability. In the other case, with a probability of $1 - P_{def}$, we transition to the 'Using correct operation 1' state.

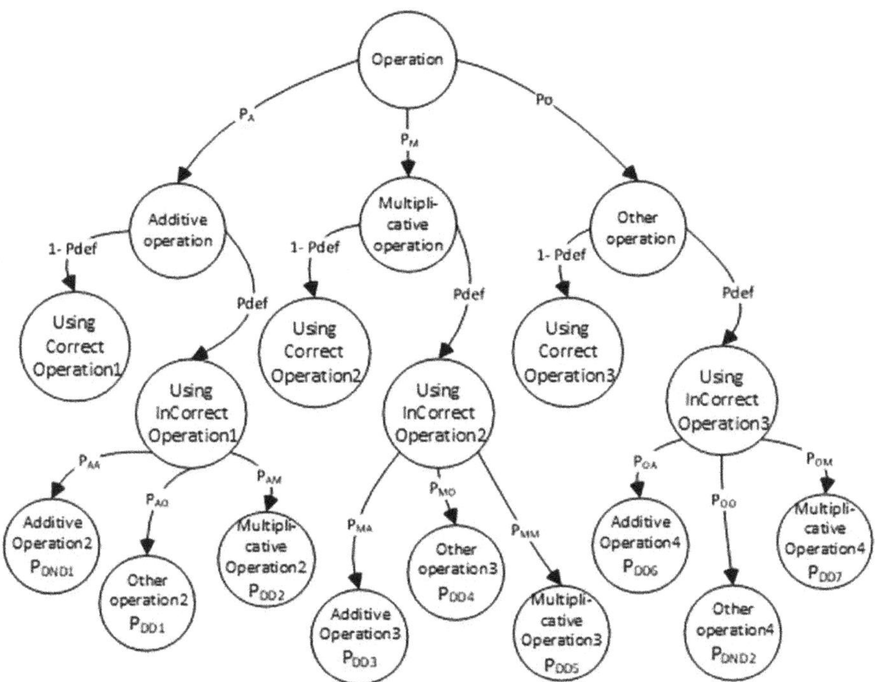

Figure 7. Model for the detection of incorrect use of operations based on dimensional and orientational analysis.

In the state 'Using incorrect operation 1', for a mutation where an additive operation mutates to another additive operation, we can transition to the 'Additive Operation 2' state with a probability of P_{AA}. In this case, we cannot detect a software defect, and the full probability for this scenario is P_{DND1}.

For a mutation where an additive operation mutates to other operations, we can transition to the 'Other Operation 2' state with a probability of P_{AO}. In this case, we can detect a software defect, and the full probability for this scenario is P_{DD1}.

For a mutation where an additive operation mutates to a multiplicative operation, we can transition to the 'Multiplicative Operation 2' state with a probability of P_{AM}. In this case, we can detect a software defect, and the full probability for this scenario is P_{DD2}.

In the state 'Multiplicative Operation', any 'multiplicative' operation can be replaced by another operation. This event has the probability P_{def}, and we transition to the 'Using incorrect operation 2' state. In the other case, with a probability of $1 - P_{def}$, we transition to the 'Using correct operation 2' state.

In the state 'Using incorrect operation 2', for a mutation where a multiplicative operation mutates to other additive operations, we can transition to the 'Additive Operation 3' state with a probability of P_{MA}. In this case, we can detect a software defect. The full probability for this scenario is P_{DD3}.

For a mutation where a multiplicative operation mutates to 'other' operations, we can transition to the 'Other Operation 3' state with a probability of P_{MO}. In this case, we can detect a software defect. The full probability for this scenario is P_{DD4}.

For a mutation where a multiplicative operation mutates to other multiplicative operations, we can transition to the 'Multiplicative Operation 3' state with a probability of

P_{MM}. In this case, we can detect a software defect. The full probability for this scenario is P_{DD5}.

In the state 'Other Operation', any 'other' operation can be replaced by another operation. This event has the probability P_{def}, and we transition to the 'Using incorrect operation 3' state. In the other case, with a probability of $1 - P_{def}$, we transition to the 'Using correct operation 3' state.

In the state 'Using incorrect operation 3', for a mutation where an 'other' operation mutates to additive operations, we can transition to the 'Additive Operation 4' state with a probability of P_{OA}. In this case, we can detect a software defect. The full probability for this scenario is P_{DD6}.

For a mutation where an 'other' operation mutates to other 'other' operations, we can transition to the 'Other Operation 4' state with a probability of P_{MO}. In this case, we cannot detect a software defect. The full probability for this scenario is P_{DND2}.

For a mutation where an 'other' operation mutates to multiplicative operations, we can transition to the 'Multiplicative Operation 4' state with a probability of P_{OM}. In this case, we can detect a software defect. The full probability for this scenario is P_{DD7}.

Additionally, we can conclude, that $P_{AA} + P_{AM} + P_{AO} = 1$, $P_{MA} + P_{MM} + P_{MO} = 1$, $P_{OA} + P_{OM} + P_{OO} = 1$. We can define the conditional probability of detecting incorrect use of software operations as follows:

$$P_{OD} = \frac{\sum\limits_{i=1}^{7} P_{DDi}}{\sum\limits_{i=1}^{7} P_{DDi} + P_{DND1} + P_{DND2}}.$$

Here, $P_{DD1} = P_A P_{def} P_{AO}$, $P_{DD2} = P_A P_{def} P_{AM}$, $P_{DND1} = P_A P_{def} P_{AA}$, $P_{DD3} = P_M P_{def} P_{MA}$, $P_{DD4} = P_M P_{def} P_{MO}$, $P_{DD5} = P_M P_{def} P_{MM}$, $P_{DD6} = P_O P_{def} P_{OA}$, $P_{DD7} = P_O P_{def} P_{OM}$, $P_{DND2} = P_O P_{def} P_{OO}$.

After substitution

$$P_{OD} = \frac{P_A P_{AO} + P_A P_{AM} + P_M + P_O P_{OA} + P_O P_{OM}}{P_A + P_M + P_O}$$

According to $P_A + P_M + P_O = 1$, $P_{OD} = P_A(1 - P_{AA}) + P_M + P_O(1 - P_{OO})$. Because $P_{AA} \approx P_A$ and $P_{OO} \approx P_O$, then $P_{OD} \approx P_A(1 - P_A) + P_M + P_O(1 - P_O)$

$$P_{OD} \approx 1 - P_A^2 - P_O^2 \tag{16}$$

Here, P_A represents the conditional probability of 'additive' operations in a file and P_O represents the conditional probability of 'other' operations in a file.

According to Equation (16), P_{OD} denotes the conditional probability of detecting the incorrect use of software operations. The value of P_{OD} depends on the square of the probabilities of additional operations (such as +, −, =, <, etc.) and other operations (such as = {" | | ", "&&", "&", " | ", "^", "~", "<<", ">>", "::", "?" etc.) in a source file. In order to evaluate P_{OD}, it is necessary to define the values of P_A and P_O. This evaluation requires analyzing the real source code.

Using Equation (2), which defines the conditional probability of software defect detection as a function of P_{var} (defined by the source code of the file) and the conditional probabilities of variable usage defect detection (P_{VD}, defined by Equation (12)) and operation usage defect detection (P_{OD}, defined by Equation (16)), allows us to evaluate the conditional probability of software defect detection.

3. Results

After analyzing the source code of Unmanned Aerial Vehicle Systems, which had a total volume of 2 GB and a total number of files of 20,000, saved on GitHub using our own statistical analyzer, we were able to determine the necessary statistical characteristics of the C++ source code.

Distribution N_{var}—total number of different variables per file (Figures 8 and 9).

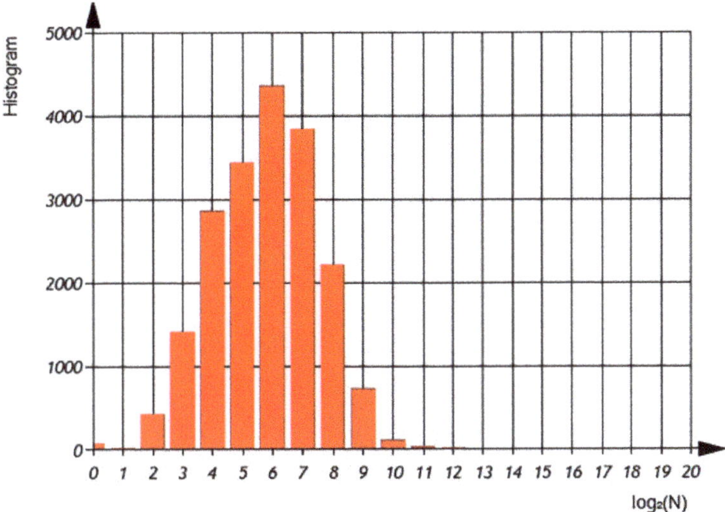

Figure 8. Histogram of the total number of variables per file (semi-log scale).

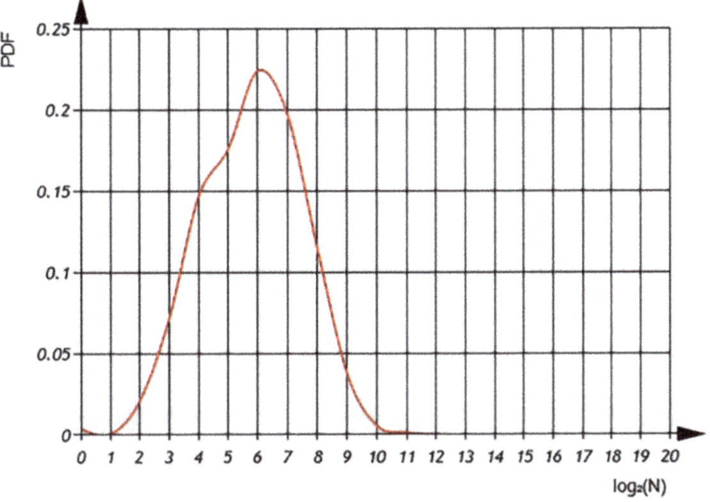

Figure 9. Probability density function of total variables per file (semi-log scale).

According to Figures 8 and 9 we can observe the distribution of different variables per file in semi-logarithmic coordinates. The histogram (Figure 8) reveals that the average number of different variables is 2^6, with a maximum of 64 variables observed in 4500 files. However, there are files that contain only one variable, as well as files with 1024 variables. The sum of the histogram columns corresponds to the total number of files, which is 20,000.

In Figure 9, presented subsequently, we can examine the probability density function of variable distributions. It is important to note that the integral of the probability density function should always equal one, ensuring a proper probability distribution.

According to Figures 10 and 11 we can observe the distribution of N_{VF}, which represents the average number of variables uses per file, in semi-logarithmic coordinates. The histogram (Figure 10) shows that the average number of variable usages is 5. The sum of the histogram columns corresponds to the total number of files, which is 20,000.

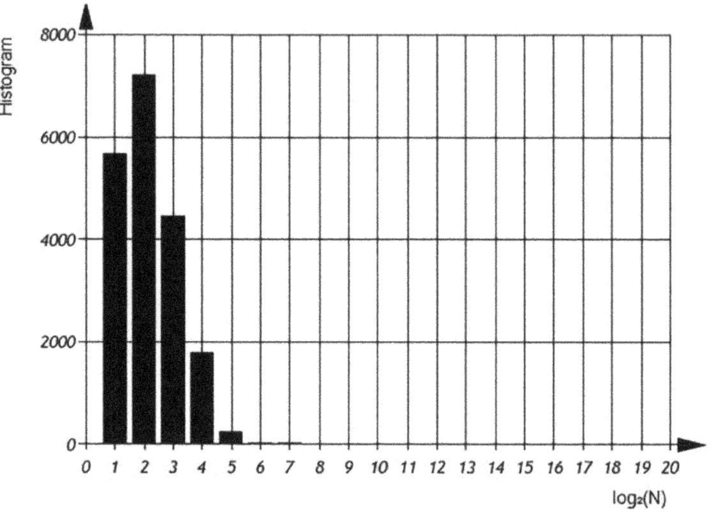

Figure 10. Histogram of variable usage per file using a semi-logarithmic scale.

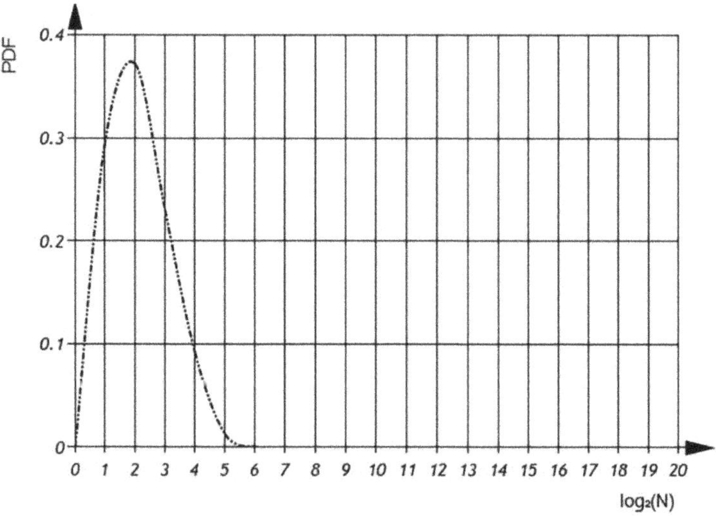

Figure 11. Probability density function of variable usage per file (semi-log scale).

In the subsequently presented Figure 11, we can analyze the probability density function of the distributions for the average usage of variables. According to the Probability Density Function, we can see that there are files with an average usage of 64 variables. It is crucial to note that the integral of the probability density function should always equal one, ensuring a proper probability distribution.

According to Equation (12) and the distributions described in Figures 8 and 10, as well as the uniform distribution for both n_{ij} and m_{ij}, we can obtain the Probability Density

Functions of conditional probability of detecting incorrect use of software variables P_{VD}, shown in Figure 12.

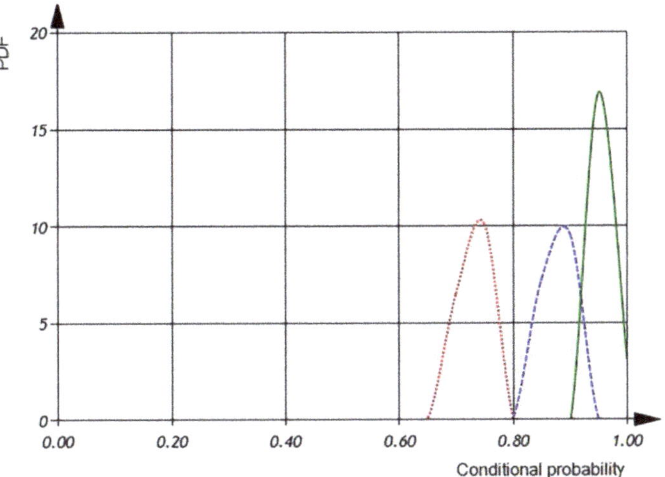

Figure 12. Probability density functions of conditional probability for detecting the incorrect use of software variables: (1) dimensional analysis (blue dash line); (2) orientational analysis (red dot line); (3) orientational and dimensional analysis (green solid line).

According to Figure 12, when combining dimensional and orientational analysis, the conditional probability for detecting incorrect use of software variables is greater than 0.9.

Additional statistical characters of C++ source code was evaluated: N_A—total numbers of additive operations per file; N_O—total numbers of 'other' operations per file; N_M—total numbers of multiplicative operations per file. Corresponded distributions shown on the Figures 13 and 14.

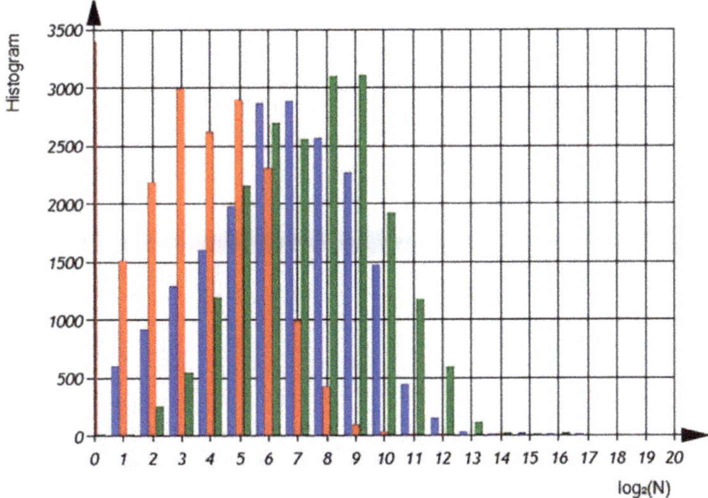

Figure 13. Histogram of software operations for the total number of additive operations (blue lines), multiplicative operations (red lines), and other operations (green lines) per file on semi-logarithmic coordinates.

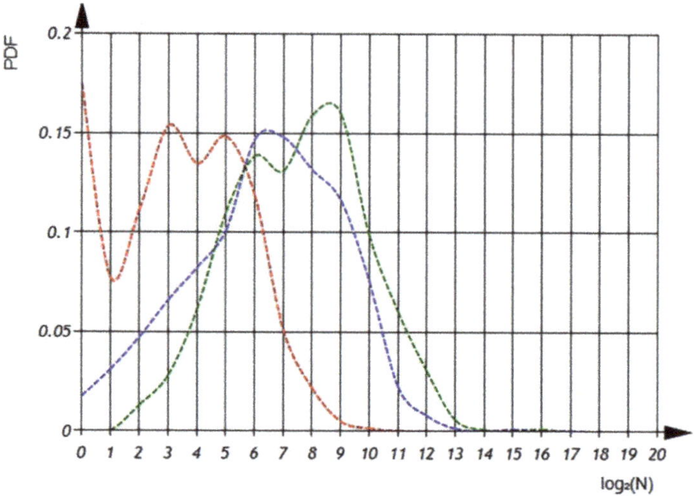

Figure 14. Probability density functions of software operations for the total number of additive operations (blue line), multiplicative operations (red line), and other operations (green line) per file on semi-logarithmic coordinates.

In Figure 14 we can observe the distributions of N_A, N_M, and N_O, which represent the average number of operations per file in semi-logarithmic coordinates. The histograms show that the average number of 'additive' operations is 128, observed in 2500 files. However, there are files that only contain 2 additive operations, as well as files with an average usage of 8192 additive operations. The sum of the histogram columns corresponds to the total number of files, which is 20,000. The histogram shows that the average number of 'multiplicative' operations is 16, observed in 2600 files. However, there are files that only contain 2 multiplicative operations, as well as files with an average usage of 512 multiplicative operations. The histogram shows that the average number of 'other' operations is 256, observed in 3000 files. However, there are files that only contain 4 'other' operations, as well as files with an average usage of 4096 'other' operations. The sum of the histogram columns corresponds to the total number of files, which is 20,000.

In the subsequently presented, we can examine the probability density function of the distributions for different operations per file. It is crucial to note that the integral of the probability density function should always equal one, ensuring a proper probability distribution.

By referring to Equation (15) and the distributions of N_A (total number of additive operations), N_M (total number of multiplicative operations), and N_O (total number of other operations) (as shown in Figure 15), we can calculate the distributions of P_A, P_M, and P_O.

In Figures 15 and 16, we can observe the distribution of conditional probabilities of operations per file in semi-logarithmic coordinates. The histogram shows that the average of conditional probabilities of operations per file. The value of conditional probability of 'additive' operations is $P_A = 0.309 \pm 0.161$ [0.000 ... 0.75]. The value of conditional probability of 'multiplicative' operations is $P_M = 0.056 \pm 0.056$ [0.000 ... 0.636]. The value of conditional probability of 'other' operations is $P_O = 0.635 \pm 0.155$ [0.2 ... 0.992].

In the subsequently presented, we can examine the probability density functions of the conditional probability of different operations. It is crucial to note that the integral of the probability density function should always equal one, ensuring a proper probability distribution.

Based on the distributions of P_A, P_M, and P_O we can calculate the distribution of the conditional probability of operation defect detection P_{OD} (as depicted in Figure 17 The histogram reveals that the average conditional probability of operation defect detection per file is $P_{OD} = 0.45 \pm 0.161$ [0.000 ... 0.8].

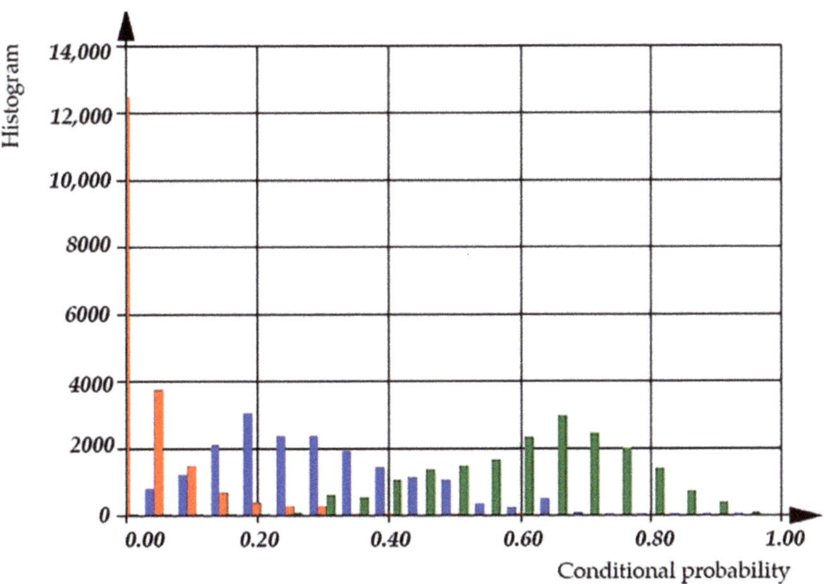

Figure 15. Histogram of the conditional probability of additive operations (blue lines), multiplicative operations (red line), and other operations (green line) per file based on dimensional and orientational analysis.

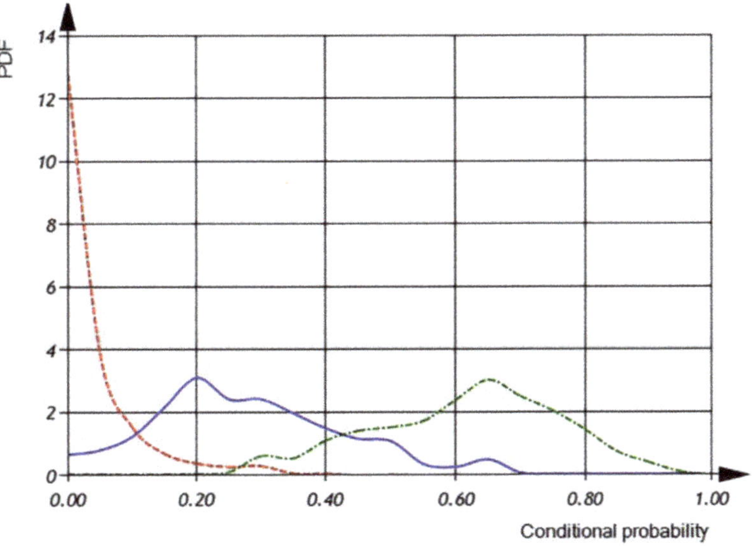

Figure 16. Probability density functions of the conditional probability of additive operations (blue solid line), multiplicative operations (red dash line), and other operations (green dash-dot line) per file based on dimensional and orientational analysis.

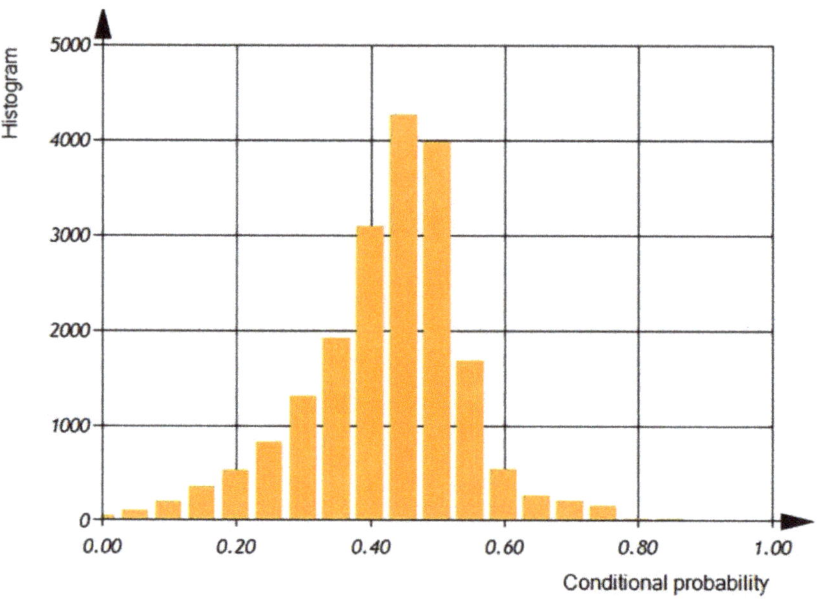

Figure 17. Histogram of the conditional probability of software operation defect detection based on dimensional and orientational analysis.

According to Figure 18, we can see that the mean value of the conditional probability of software operation defect detection is 0.5.

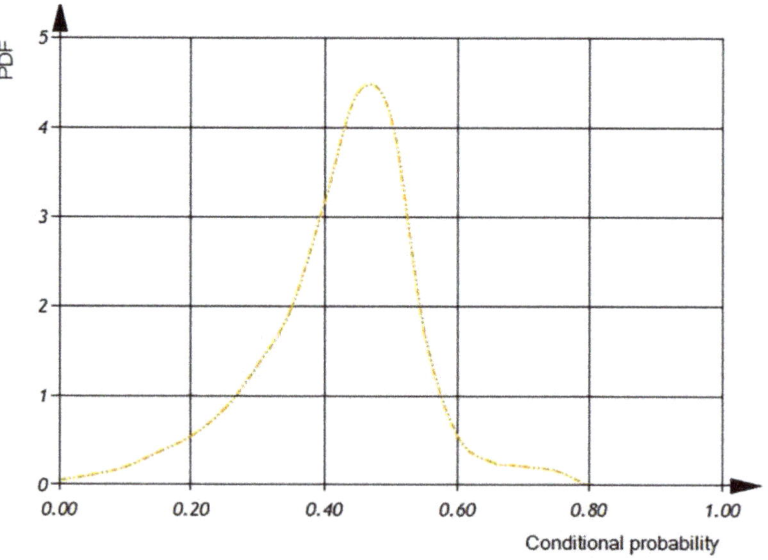

Figure 18. Probability density function of the conditional probability of software operation defect detection based on dimensional and orientational analysis.

After analyzing real C++ code statistically, we built the distributions of P_O and P_{var} (as shown in Figures 19 and 20).

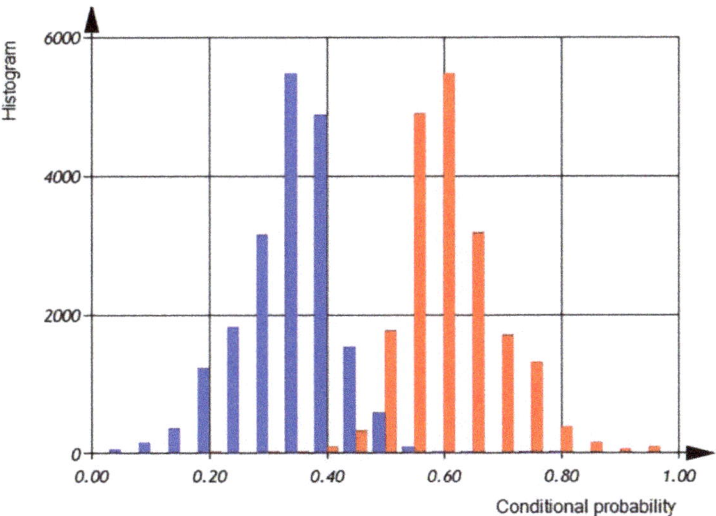

Figure 19. Histogram of conditional probabilities of variables (P_{var}—blue lines) and operations (P_O—red lines) per file.

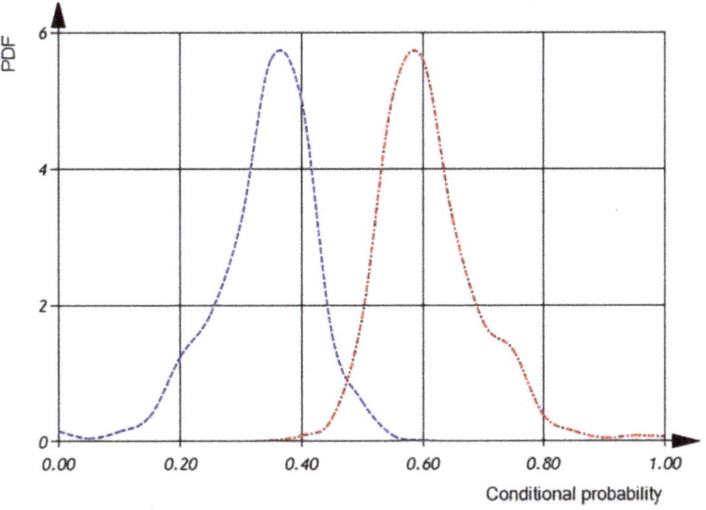

Figure 20. Probability density functions of conditional probabilities of variables (P_{var}—blue dash line) and Operations (P_O—red dash-dot line) per file.

The histogram shows that the average conditional probability of operations is $P_O = 0.65 \pm 0.12\ [0.4 \ldots 0.9]$. Similarly, the histogram reveals that the average conditional probability of variables is $P_{var} = 0.35 \pm 0.12\ [0.000 \ldots 0.55]$.

Figure 20, displays two distributions of the conditional probability for software operations and software variables, with peaks at 0.35 and 0.65, satisfying the equation $P_{var} + P_O = 1$.

Now, we can calculate the conditional probability of software defect detection based on the embedded source code and the proposed defect models (see Expression (2)). Let η be the conditional probability, then we have $\eta = P_{var}P_{VD} + (1 - P_{var})P_{OD}$, where P_{var} is the conditional probability of variables in the source code, P_{VD}—is the conditional probability of the defects detection of variables using the defect in the source code, and P_{OD} is the conditional probability the defects detection of operation using defect in the source code.

In the upcoming figures, Figure 21 presents a histogram of the conditional probabilities of defect detection, while Figure 22 shows probability density functions of software defect detection.

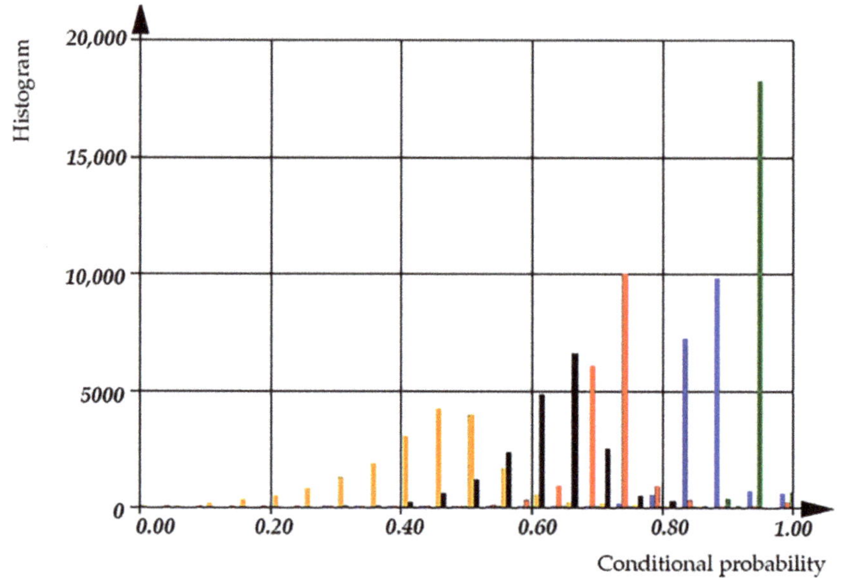

Figure 21. Histogram of conditional probabilities for the detection of (1) incorrect use of software operations and variables based on dimensional and orientational analysis (black solid lines); (2) incorrect use of software operations (orange lines); (3) incorrect use of software variables based on orientational analysis (red lines); (4) incorrect use of software variables based on dimensional analysis (blue lines); (5) incorrect use of software variables based on dimensional and orientational analysis (green lines).

According to the histogram (see Figure 21), the conditional probability for the detection of incorrect usage of variables is 0.95, while the conditional probability for the detection of incorrect usage of operations has a mean of 0.5. We can observe that the conditional probability of the incorrect usage of variables increases after incorporating both dimensional and orientational analysis. Using dimensional analysis alone yields a conditional probability of a defect detection of 0.9, while orientational analysis alone provides a conditional probability of a defect detection of 0.73. However, the overall conditional probability for the detection of the incorrect usage of operations or variables has a mean value of 0.60.

According to Figure 22, the conditional probability for the detection of the incorrect usage of variables has a mean value of 0.95 and is distributed in the interval of 0.4 to 0.8, while the conditional probability for the detection of the incorrect usage of operations has a mean of 0.5 and is distributed in the interval of 0 to 0.65. It is evident that the conditional probability of the incorrect usage of variables increases after incorporating both dimensional and orientational analysis. The incorporation of both analysis methods narrows the interval of distribution. When using dimensional analysis alone, the conditional probability of a defect detection is 0.9 within the interval of 0.8 to 0.95, while orientational analysis alone provides a conditional probability of a defect detection of 0.73 within the interval of 0.6 to 0.83. However, the overall conditional probability for the detection of the incorrect usage of operations or variables has a mean value of 0.60 within the interval of 0.4 to 0.75.

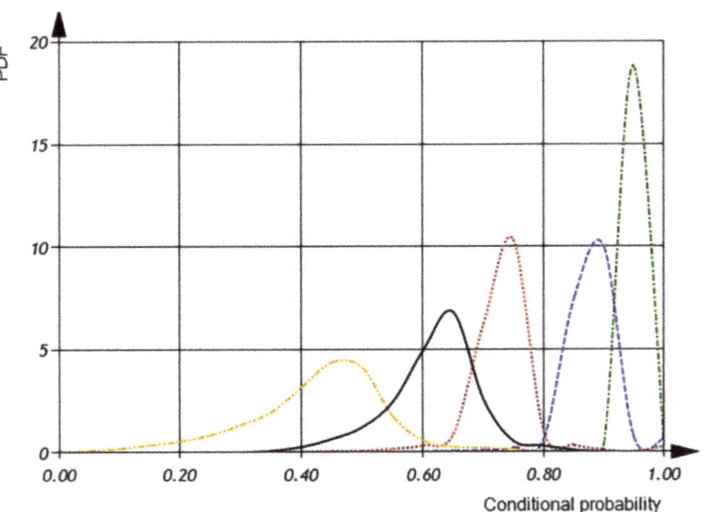

Figure 22. Probability density function of conditional probabilities for the detection of (1) incorrect use of software operations and variables based on dimensional and orientational analysis (black solid line); (2) incorrect use of software operations (orange dash-dot line); (3) incorrect use of software variables based on orientational analysis (red dot line); (4) incorrect use of software variables based on dimensional analysis (blue dash line); and (5) incorrect use of software variables based on dimensional and orientational analysis (green dash-dot line).

4. Discussion

The proposed method of formal software verification based on dimensional analysis and orientational analysis appears to be an effective approach to detecting software defects. The fact that it can detect over 60% of software defects, including those related to the incorrect usage of variables, operations, and functions, is noteworthy and suggests that it could be a valuable tool in software development.

However, it is important to note that no single method can detect all types of software defects, and different methods may be better suited for different types of defects. Thus, while the proposed method shows promise, it should be evaluated and compared to other approaches to determine its overall efficacy and limitations.

To fully assess its overall effectiveness and limitations, it is crucial to develop a concrete tool that can evaluate and compare the proposed method with other existing tools.

In the future, the proposed method will demonstrate a high level of effectiveness, detecting 90% of incorrect uses of software variables and more than 50% of incorrect uses of operations. One of the imminent tasks is to create a type of library for the formal verification of CPS and IoT software during compile-time.

5. Conclusions

This article focuses on a formal software verification method based on software invariants derived from both dimensional and orientational analysis.

The advantages of the proposed method are as follows:

1. Early detection of software defects in compile-time.
2. Reduced testing time via formal verification in compile-time and run-time
3. By catching a large number of defects early on, the method can help minimize the need for extensive debugging, maintenance, or post-release updates, resulting in overall cost reduction.
4. The ability to detect over 60% of latent defects suggests that the method contributes to improving software quality.

5. The method can be seen as a complementary approach that focuses on detecting latent defects based on software characteristics.

6. Continuous improvement efforts can contribute to even higher detection rates, leading to further advancements in software quality assurance.

Overall, the high detection rate of the proposed method and its potential benefits in reducing testing time and improving both reliability and software quality demonstrate its value as an efficient and effective approach to software defect detection.

However, the method has certain limitations, such as the need to know the physical dimensions and orientations of source variables at compile-time. Nonetheless, it offers several advantages, including improved programmer productivity, as programmers no longer need to spend time tracking down dimensional and orientational errors during development and run-time. Additionally, the method enables a comprehensive analysis of dimensional and orientational correctness during compile-time and run-time, including the correct use of software variables, operations, functions, and procedures through added argument checking.

Although the proposed method has the potential to enhance software reliability, it requires further research and development of specialized analysis tools to realize its full effectiveness.

Author Contributions: Conceptualization, Y.M.; methodology, Y.M. and Y.S.; software, Y.M. and Y.S.; validation, Y.M. and Y.S.; formal analysis, Y.M. and Y.S.; resources, Y.M. and Y.S.; data curation, Y.M.; writing—original draft preparation, Y.M. and Y.S.; writing—review and editing, Y.M. and Y.S.; visualization, Y.M. and Y.S.; supervision, Y.M.; project administration, Y.S. All authors have read and agreed to the published version of the manuscript.

Funding: This research received no external funding.

Data Availability Statement: Data are contained within this article.

Conflicts of Interest: The authors declare no conflict of interest.

References

1. Bai, L.S.; Dick, R.P.; Dinda, P.A. Archetype-based design: Sensor network programming for application experts, not just programming experts. In Proceedings of the 8th International Conference on Information Processing in Sensor Networks, IPSN 2009, San Francisco, CA, USA, 13–16 April 2009.

2. Internet of Things—Number of Connected Devices Worldwide 2015–2025. Available online: https://www.statista.com/statistics/471264/iot-number-of-connected-devices-worldwide/ (accessed on 27 November 2016).

3. Buffoni, L.; Ochel, L.; Pop, A.; Fritzson, P.; Fors, N.; Hedin, G.; Taha, W.; Sjölund, M. Open Source Languages and Methods for Cyber-Physical System Development: Overview and Case Studies. *Electronics* **2021**, *10*, 902. [CrossRef]

4. Cyber-Physical Systems (CPS). Available online: https://www.nsf.gov/publications/pub_summ.jsp?ods_key=nsf21551&org=NSF (accessed on 11 January 2021).

5. Cyber-Physical Systems (CPS). Available online: https://ptolemy.berkeley.edu/projects/cps/ (accessed on 11 January 2021).

6. Forecast End-User Spending on IoT Solutions Worldwide from 2017 to 2025 (in Billion U.S. Dollars). Available online: https://www.statista.com/statistics/976313/global-iot-market-size/ (accessed on 11 January 2021).

7. Global IoT Market Size to Grow 19% in 2023—IoT Shows Resilience Despite Economic Downturn. Available online: https://iot-analytics.com/iot-market-size/ (accessed on 7 February 2023).

8. Top 3 Programming Languages for IoT Development in 2018. Available online: https://www.iotforall.com/2018-top-3-programming-languages-iot-development (accessed on 9 October 2018).

9. Functional Mock-Up Interface, Version 2.0. Interface Specification. Available online: https://fmi-standard.org/downloads/ (accessed on 25 July 2014).

10. System Structure and Parameterization, Version 1.0. Interface Specification. Available online: https://ssp-standard.org (accessed on 1 March 2019).

11. A Unified Object-Oriented Language for Physical Systems Modeling-Language Specification Version 3.4. Available online: https://www.modelica.org/documents/ModelicaSpec34.pdf (accessed on 10 April 2017).

12. Technical Report on C++ Performance. Information Technology—Programming languages, Their Environments and System Software Interfaces. Available online: https://www.open-std.org/jtc1/sc22/wg21/docs/TR18015.pdf (accessed on 15 February 2006).

3. Xie, H.; Wei, L.; Zhou, J.; Hua, X. Research of Conformance Testing of Low-Rate Wireless Sensor Networks Based on Remote Test Method. In Proceedings of the International Conference on Computational and Information Sciences (ICCIS 2013), Shiyang, China, 21–23 June 2013.
4. Parisot, A.; Bento, L.M.S.; Machado, R.C.S. Testing and selecting lightweight pseudo-random number generators for IoT devices. In Proceedings of the 2021 IEEE International Workshop on Metrology for Industry 4.0 & IoT (MetroInd4.0&IoT), Rome, Italy, 7–9 June 2021.
5. Bae, H.; Sim, S.-H.; Choi, Y.; Liu, L. Statistical Verification of Process Conformance Based on Log Equality Test. In Proceedings of the 2nd International Conference on Collaboration and Internet Computing (CIC), Pittsburgh, PA, USA, 1–3 November 2016.
6. Silva, D.S.; Resner, D.; de Souza, R.L.; Martina, J.E. Formal Verification of a Cross-Layer, Trustful Space-Time Protocol for Wireless Sensor Networks. In *Information Systems Security*; Ray, I., Gaur, M., Conti, M., Sanghi, D., Kamakoti, V., Eds.; ICISS 2016; Lecture Notes in Computer Science; Springer: Berlin/Heidelberg, Germany, 2016; Volume 10063, pp. 426–443.
7. Ahmad, A. Model-Based Testing for IoT Systems: Methods and Tools. Ph.D. Thesis, University Burgundy Franche-Comté, Besançon, France, 1 June 2018.
8. Clarke, E.M.; Grumberg, O.; Kroening, D.; Peled, D.; Veith, H. *Model Checking*, 2nd ed.; MIT Press: Cambridge, MA, USA, 2018; 424p.
9. Back, R.-J. Invariant based programming: Basic approach and teaching experiences. *Form. Asp. Comput.* **2009**, *21*, 227–244. [CrossRef]
10. Lloyd, R. Metric Mishap Caused Loss of NASA Orbiter. Available online: http://edition.cnn.com/TECH/space/9909/30/mars.metric.02/index.html (accessed on 30 September 1999).
11. Mars Climate Orbiter Mishap Investigation Board. Phase I. Report. Available online: https://llis.nasa.gov/llis_lib/pdf/100946 4main1_0641-mr.pdf (accessed on 10 November 1999).
12. Cowing, K. NASA Finds the Metric System too Hard to Implement for Constellation. Available online: https://nasawatch.com/cev-calv-lsam-eds/nasa-finds-the-metric-system-too-hard-to-implement-for-constellation/ (accessed on 1 June 2009).
13. Disneyland Roller Coaster Derails. Available online: https://www.japantimes.co.jp/news/2003/12/06/national/disneyland-roller-coaster-derails/ (accessed on 6 December 2003).
14. Fuel Trouble Blamed for Forcing Jet Down on Car-Racing Strip. *The Leader-Post*, 25 July 1983; p. A1.
15. Schabel, M.; Watanabe, S. The Boost Units Library: A C++ Library for Zero-Overhead Dimensional Analysis and Unit/Quantity Manipulation and Conversion. Available online: https://www.researchgate.net/publication/287106084_The_Boost_Units_Library_A_C_library_for_zero-overhead_dimensional_analysis_and_unitquantity_manipulation_and_conversion (accessed on 1 December 2015).
16. Pebesma, E.; Mailund, T.; Hiebert, J. Measurement Units in R. *R J.* **2016**, *8*, 486–494. [CrossRef]
17. McKeever, S.; Bennich-Björkman, O.; Salah, O.-A. Unit of measurement libraries, their popularity and suitability. *Softw. Pract. Exp.* **2020**, *51*, 711–734. [CrossRef]
18. Syme, D. The early history of F#. *Proc. ACM Program. Lang.* **2020**, *4*, 1–58. [CrossRef]
19. SI Units. Available online: https://www.nist.gov/pml/owm/metric-si/si-units (accessed on 7 March 2023).
20. The International System of Units (SI) 9th Edition 2019 V2.01. Available online: https://www.bipm.org/documents/20126/4148 3022/SI-Brochure-9-EN.pdf (accessed on 1 December 2022).
21. Siano, D.B. Orientational Analysis—A Supplement to Dimensional Analysis. *J. Frankl. Inst.* **1985**, *320*, 267–283. [CrossRef]
22. Siano, D.B. Orientational analysis, tensor analysis and the group properties of the SI supplementary units. *J. Frankl. Inst.* **1985**, *320*, 285–302. [CrossRef]
23. Dos Santos, L.F.; De Freitas, A.C. Orientational Analysis of the Vesic's Bearing Capacity of Shallow Foundations. *Soils Rocks* **2020**, *43*, 3–9. [CrossRef]
24. Martínez-Rojas, J.A.; Fernández-Sánchez, J.L. Combining dimensional analysis with model based systems engineering. *Syst. Eng.* **2022**, *26*, 71–87. [CrossRef]
25. Euler's Equations of Motion for Rigid-Body Rotation. Available online: https://phys.libretexts.org/Bookshelves/Classical_Mechanics/Variational_Principles_in_Classical_Mechanics_(Cline)/13%3A_Rigid-body_Rotation/13.17%3A_Eulers_equations_of_motion_for_rigid-body_rotation (accessed on 14 March 2021).
26. Yari, M.; Armaghani, D.J.; Maraveas, C.; Ejlali, A.N.; Mohamad, E.T.; Asteris, P.G. Several Tree-Based Solutions for Predicting Flyrock Distance Due to Mine Blasting. *Appl. Sci.* **2023**, *13*, 1345. [CrossRef]

Article

Simulation of Multi-Phase Flow to Test the Effectiveness of the Casting Yard Aspiration System

Serghii Lobov [1,*], Yevhen Pylypko [2], Viktoriya Kruchyna [1] and Ihor Bereshko [1]

[1] Department of Ecology and Technogenic Safety, National Aerospace University "Kharkiv Aviation Institute", 61070 Kharkiv, Ukraine; v.kruchyna@khai.edu (V.K.)

[2] Ukrainian Scientific and Technical Center of the Metallurgical Industry "Energostal", 61070 Kharkiv, Ukraine; pilipkoev@gmail.com

* Correspondence: s.lobov@khai.edu

Abstract: The metallurgical industry is in second place among all other industries in terms of emissions into the atmosphere, and air pollution is the main cause of environmental problems arising from the activities of metallurgical enterprises. In some existing systems for localization, in the trapping and removal of dust emissions from tapholes and cast-iron gutters of foundries, air flow parameters may differ from the optimal ones for solving aspiration problems. The largest emissions are observed in the area of the taphole (40–60%) and from the ladles during their filling (35–50%). In this paper, it is proposed to consider a variant of the blast furnace aspiration system with the simultaneous supply of a dust–gas–air mixture from two-side smoke exhausters and two upper hoods with two simultaneously operating tapholes, that is, when the blast furnace operates in the maximum emissions mode. This article proposes an assessment of the effectiveness of the modernized blast furnace aspiration system using computer CFD modeling, where its main parameters are given. It is shown that the efficiency of dust collection in the proposed system is more than 90%, and the speed of the gas–dust mixture is no lower than 20 m/s, which prevents gravitational settling on the walls. The distribution fields of temperatures and velocities are obtained for further engineering analysis and the possible improvement of aspiration systems.

Keywords: aspiration system; blast furnace; casting yards; multi-phase flow processes; dust pollution

Citation: Lobov, S.; Pylypko, Y.; Kruchyna, V.; Bereshko, I. Simulation of Multi-Phase Flow to Test the Effectiveness of the Casting Yard Aspiration System. *Computation* **2023**, *11*, 121. https://doi.org/10.3390/computation11060121

Academic Editors: Mykola Nechyporuk, Vladimir Pavlikov and Dmitriy Kritskiy

Received: 13 May 2023
Revised: 12 June 2023
Accepted: 15 June 2023
Published: 20 June 2023

1. Introduction

It is air pollution that is the main cause of environmental problems arising from the activities of metallurgical enterprises. Emissions from pipes lead to soil pollution, destruction of vegetation and the formation of man-made wastelands around large factories. In addition, the environmental problems of domestic metallurgy are exacerbated due to high wear and tear of equipment and outdated technologies. In many countries, including Ukraine, about 70% of all capacities of the metallurgical industry are worn out, outdated and unprofitable, while metallurgy (ferrous and non-ferrous) accounts for about a third of all industrial emissions into the atmosphere [1–3].

Similar problems are observed in many countries where ferrous metallurgy is developed [4–13]. As a rule, carbon oxides (67.5%), suspended dust particles (15.5%), sulfur dioxide (10.8%) and other compounds are predominant in atmospheric emissions. The sum of the emissions of pollutants into the atmosphere can be 83.6% of the total volume of volatile pollutants. In Ukraine, for example, the population of large industrial cities and adjacent areas suffers primarily from the concentrated location of metallurgical industry enterprises. It is in these cities that there is a trend towards an increase in the incidence of environmentally dependent diseases (especially in children) [3–5].

When carrying out technological processes of casting iron in closed volumes, as is the case in various furnaces, the main part of dust and gas emissions is systematically re-moved through gas exhaust paths and chimneys. In conditions when one or another technological process is open, an important place in the fight against air pollution is occupied by the prevention of dust and gas emissions by suppressing them in places of formation. Depending on the specific conditions of the foundry process, dust and gas emissions can be suppressed in various ways [6–8], but the most common is the method of capturing dust and gas emissions using aspiration systems with exhaust hoods, side exhausts, etc. To purify gases from chemical gaseous impurities, the following three methods can be used: absorption, adsorption and the transformation of gaseous impurities into a solid or liquid state [9].

The proposed task of computing the aspiration system was carried out as part of a project to improve the aspiration system of a blast furnace with an increase in its design capacity in terms of the amount of air being cleaned. The project was carried out for three years, including a technological audit of the blast furnace operation modes, a technological audit of the aspiration system, and a construction survey of gas duct structures. The parameters of the aspirated air before and after the treatment equipment were also measured, which included the following: productivity, velocity, dustiness, temperature, rarefaction in the gas duct, and chemical composition.

This information largely determined the initial data for calculating and modifying the geometry of the system.

Preparations for the reconstruction of the aspiration system began in 2022. At the moment, a part of the existing gas duct has been dismantled, exhaust hoods have been installed and a part of the designed branches of the gas duct has been laid in the places where they go through new air ducts. Unfortunately, at the moment it is not possible to continue work on the construction of the aspiration system.

The relevance of this work is due to the fact that, at the moment, at a metallurgical plant with the foundry yard under research, the existing aspiration system has a number of significant imperfections:

- Initially, the required amount of aspiration air was incorrectly calculated to capture the dust–gas–air mixture released at the points of release.
- The cross-sectional area of the gas ducts was chosen incorrectly, and therefore the velocity of the dust–gas–air mixture in some areas has a velocity of less than 20 m/s, while dust particles are accumulated in the gas ducts, and in some cases it is more than 35 m/s, and as a result there is abrasion of the structural elements [14–16].
- The main line of gas ducts was laid with a large number of turns, which are not curved along the radius, but at an angle of $90°$, forming many stagnant zones in which dust accumulates.
- All existing gas ducts and local exhausters were laid below the marks at which the dust–gas–air mixture was released, which reduced the efficiency of the system as a whole, since the gases had a high temperature and raised sharply from the point of release.
- Local aspiration hoods have a small cross-sectional area, which causes emissions to escape past the hoods.
- The temperature regime of the system was incorrectly determined, which led to errors in the selection of structural materials. As a result, the gas duct walls were deformed and the connections lost their tightness.
- The existing system was designed only for the standard mode of operation—the discharge of cast iron on one taphole. In this case, the maximum release of the dust–gas–air mixture occurs sequentially—at the beginning of the process, when the taphole is opened, and then after some time when the taphole is blown. The non-standard mode of operation with simultaneous discharge of cast iron from two tapholes and parallel separation of the dust–gas–air mixture for the existing system was not considered during its development.

The proposed modernized aspiration system eliminates these imperfections. The scheme of the modernized aspiration system is shown in Figures 1 and 2. The total size of the foundry yard is 100×100 m.

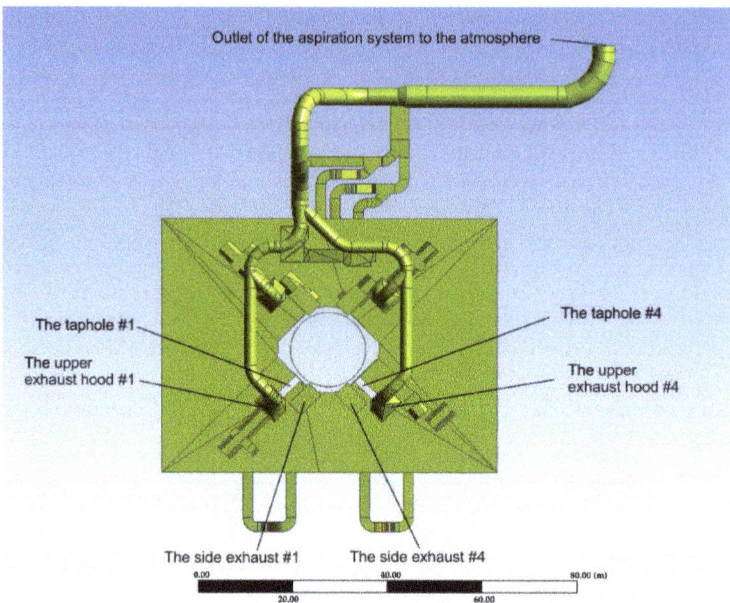

Figure 1. Scheme of the proposed aspiration system and part of the casting yard with places of emissions (top view).

It is obvious that the gas duct line also has a fairly large number of turns, but with smooth bends with a large bend radius, which eliminates the formation of stagnant zones. Local exhausters are located above the emission zones, directly along the trajectory of the dust–gas–air mixture. A new geometry has been developed and the cross-sectional area of aspiration hoods has been increased to completely remove the dust–gas–air mixture.

Mainly, the proposed system was designed for the simultaneous parallel operation of two tapholes (#1 and #4 on Figure 1). When computing, it was taken into account that on one taphole, cast iron is tapped, and on the other taphole, it is completed and the taphole is blown. At this moment, the maximum amount of the dust–gas–air mixture enters the aspiration.

Based on accumulated practical experience, specialists involved in the development of aspiration systems chose the most unfavorable operating mode with the simultaneous operation of two tapholes—tapholes #1 and #4. These entrances were chosen as the most distant from the aspiration unit (bag filter in the outlet of the system) and having the longest and geometrically heterogeneous route.

To check the geometry calculations and the effectiveness of the aspiration system, it was necessary to perform mathematical CFD modeling of the entire system. In other words, the main goals and tasks of modeling were as follows:

- Creation of a 3D mathematical model of the movement of a dust–gas–air mixture when ejected from two zones during the parallel operation of two tapholes when draining cast iron on one and blowing on the other.
- Checking the aerodynamic mode of operation of the aspiration system to identify zones in the gas duct with a reduced flow velocity and structural elements with a large coefficient of local resistance.
- Determination of the concentration of coarse aerosols in the flow at the inlet to the gas duct and the gas duct itself.

- Visual modeling of the aerodynamic field of suction plumes of local exhausters, calculated dimensions and geometry to determine their efficiency in terms of the amount of dust–gas–air mixture captured.
- Creation of a flow map of temperature and velocity at the inlet to the gas duct and in the gas duct itself for the selection of structural materials.

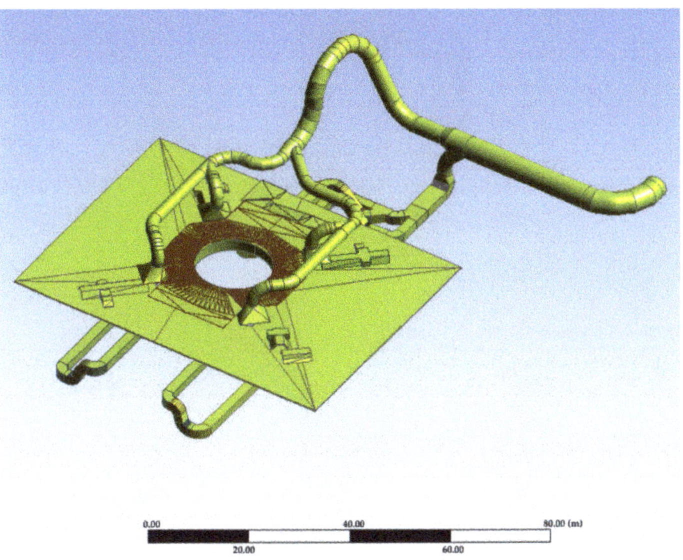

Figure 2. Scheme of the proposed aspiration system and part of the casting yard (isometric view).

2. Initial Data and Method of Modelling and Computing

The cast-iron taphole is a rectangular channel 250–300 mm wide and 450–500 mm high. The channel is made in chimney refractory masonry at a height of 600–1700 mm from the surface of the hearth. Channels for slag holes are laid out at a height of 2000–3600 mm. The channel of the cast-iron taphole is usually closed with a refractory mass. A cast-iron taphole is opened by drilling a hole with a diameter of 50–60 mm with a drilling machine. After the release of cast-iron and slag (on modern large blast furnaces, the release of cast iron and slag is carried out through cast-iron tapholes), the holes are clogged with an electric gun. The tip of the electric gun is inserted into the taphole and a taphole refractory mass is fed into it from the gun under pressure. The slag taphole at the blast furnace is protected by water-cooled elements, which are collectively called slag stoppers and a lever design with a remotely controlled pneumatic actuator.

Large volume blast furnaces (3200–5500 m^3) are equipped with four cast-iron tapholes operating alternately and one slag taphole (Figures 1 and 3).

When performing technological operations for the production of cast iron, a significant number of mechanical particles (dust), various gases and thermal energy are released into the air. The gases escaping from the taphole and partly above the main gutter form a fairly powerful and stable thermal jet adhering to the furnace casing and rising upwards to the ceiling of the casting yard.

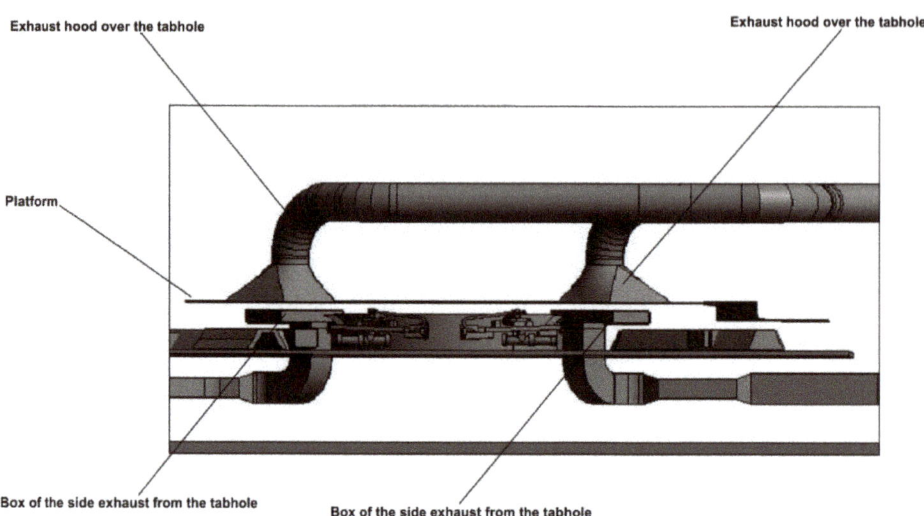

Figure 3. Aspiration system (side elevation) in casting yard.

Basic data for modeling are as follows:

1. Productivity at the outlet of the aspiration system: 1,400,000 m^3/h.
2. Dust content of aspirated air at the starting point: 5 . . . 6 g/m^3.
3. The temperature of the dust and gas flow at the starting zone is 350 °C.
4. Ambient temperature: +30 °C.
5. Air temperature of the working area: +30 °C.
6. Figure 4 shows the chemical composition of dust emitted from the taphole by weight. Since dust contains (by mass) most of all Fe$_2$O$_3$, it was decided to carry out the main calculation of the solid phase for the most significant (and heaviest) part of it (Fe$_2$O$_3$).

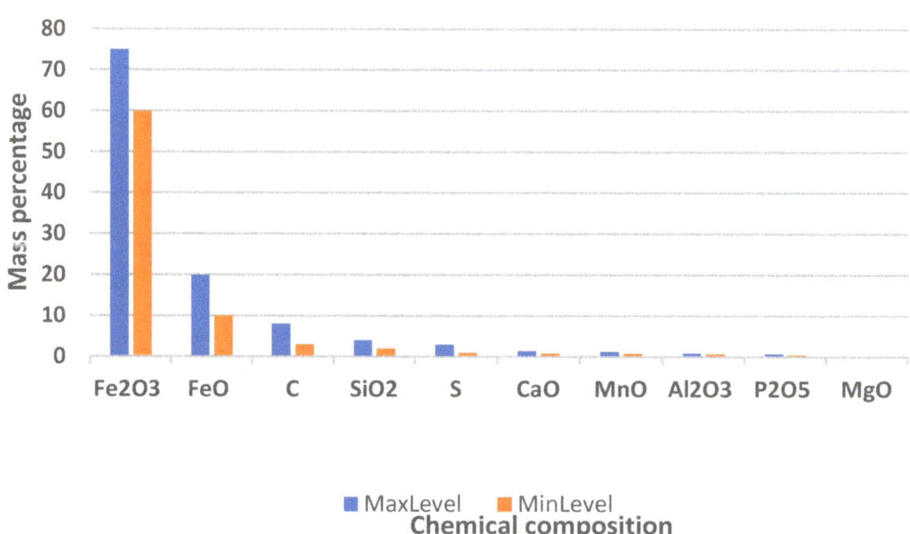

Figure 4. The distribution of the chemical composition of dust by weight in percent.

7. The particles size distribution of the selected dust component is shown in Figure 5.

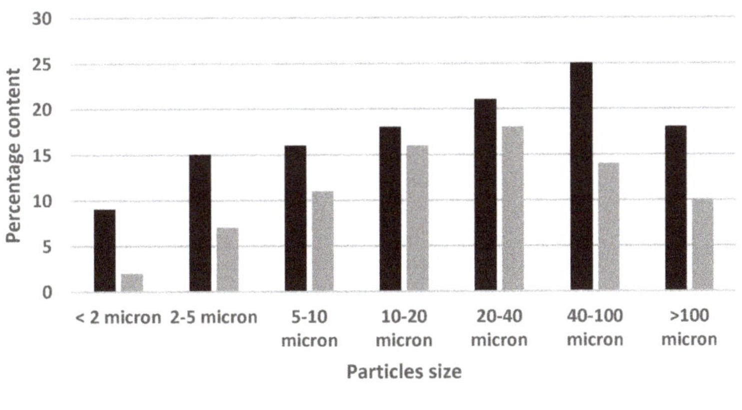

Figure 5. Dependence of percentage on Fe_2O_3 dust particle size.

As described above, the most unfavorable operating mode was taken for mathematical analysis; tapholes #1 and #4 operate simultaneously, since the gas duct route from tapholes #1 and #4 is the longest and has a greater number of local resistances (bends, transitions, T-joins, etc.) as is shown in Figures 1 and 6.

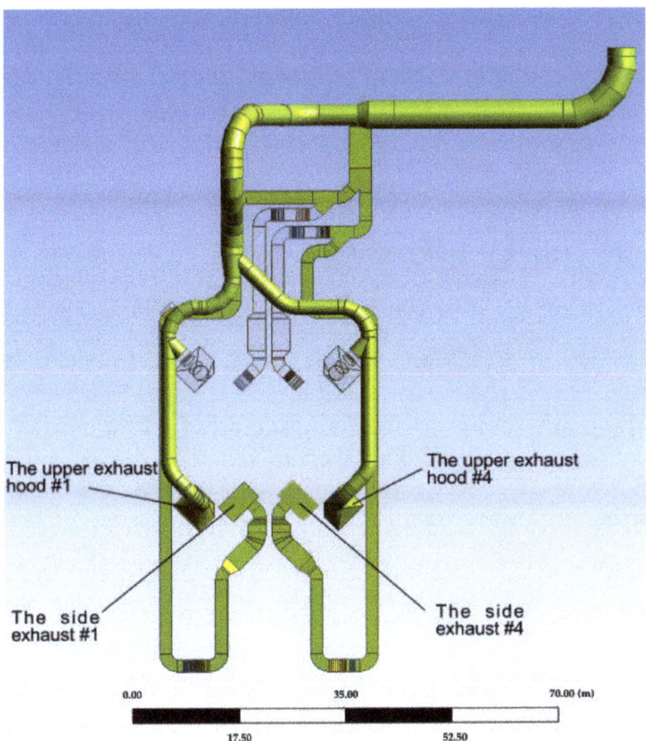

Figure 6. Air ducts that are involved in the considered mode of operation.

Through the tapholes under hoods #1 and #4, the flow rate of the dust–gas mixture with a full cross section is set at a velocity of 7 m/s and a value of 280,000 m^3/h according to the "Recommendations for the design of aspiration systems for capturing and cleaning dust and gas emissions from casting yards of blast furnaces with a volume of 2700–5500 m³" [17].

Three-dimensional problem. The most general theoretical description of the flow and heat transfer in the flow path of the system can be performed on the basis of the direct application of the laws of conservation of mass, momentum, and energy to a continuous gaseous medium [18]. The mathematical form of writing the laws of conservation of a viscous gas is the complete system of Navier–Stokes equations, for a three-dimensional region of an arbitrary shape with given boundary conditions with the introduction of simplifying assumptions, can be solved by numerical methods. This approach makes it possible to calculate the characteristics of the system that cannot be determined within the framework of a one-dimensional formulation, in particular, the flow structure, the uneven distribution of the velocity, the maximum temperatures of the structural elements, the distribution of the solid particles and dust of the gas flow, etc.

When performing a three-dimensional aero–gas–dynamic calculation of the aspiration system, the following main assumptions were made:

- The flow of air, blast furnace gas and gas–dust mixture in the flow part of the system is compressible, turbulent, quasi-stationary.
- The difference in specific heat capacities and molecular weights of blast-furnace gas and air is neglected (kg ≈ kv = 1.4; Rg ≈ Rv = 287 J/kg K).

For the simulation of the processes of the flow of a multiphase mixture carrier, in this case, air, a model of gas ducts "with capture" of space in the immediate vicinity of the source of dusting (tapholes #1 and #4) was built. It is shown on Figure 7. This space is limited above and below by platforms. On the other sides, the "walls" of this space were set as transparent—that is, open to the surrounding air.

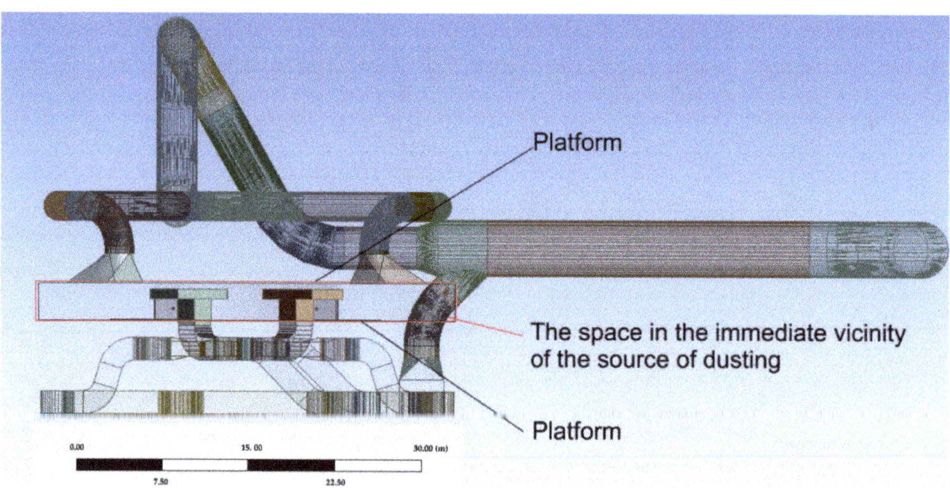

Figure 7. Solid 3D model of the flow path of the aspiration system (front view).

For mathematical modeling of a three-dimensional turbulent flow of air, blast furnace gas and the solid phase of the mixture in the flow path of the aspiration system, the ANSYS Fluent computer program was used, which implements the numerical solution of the complete system of Reynolds-averaged Navier–Stokes equations [19–22]. The numerical solution was carried out in the subdomain of space shown in Figure 7, which was covered with a hybrid calculation mesh.

When solving the problem, naturally, the most difficult thing is the simulation of the computational mesh. In the end, the authors came to the following model parameters:

mesh refinement depending on the curvature of the surface, cell volumetric centering tolerance was chosen as fine, and mesh smoothing was chosen as medium. The maximum element size was 0.24 m. Since the aspiration system includes the connection of pipes of various diameters and gas ducts of various rectangular cross sections, it was decided not to set it manually, but to use the software "control" of boundary layers. So, the "automatic inflation" option was used with the following values: the maximum number of layers 5, the Growth Rate 1.2 and, accordingly, mesh refinement for the boundary layers (as shown in Figure 8). With this mesh configuration, the total number of elements was 2,757,970 and the number of nodes (nodes) was 659,305.

The calculations were carried out under steady conditions. The convergence of the solution was controlled using the convergence criterion, which was adopted by a Residual Target = 10^{-6}. With the described computational mesh, it was necessary to perform about 350 iterations before the convergence of the solution was achieved. The calculation time in this case reached one and a half hours when using a personal computer based on Intel(R) Core(TM) i5-10600KF CPU 4.10 GHz with RAM 16 Gb.

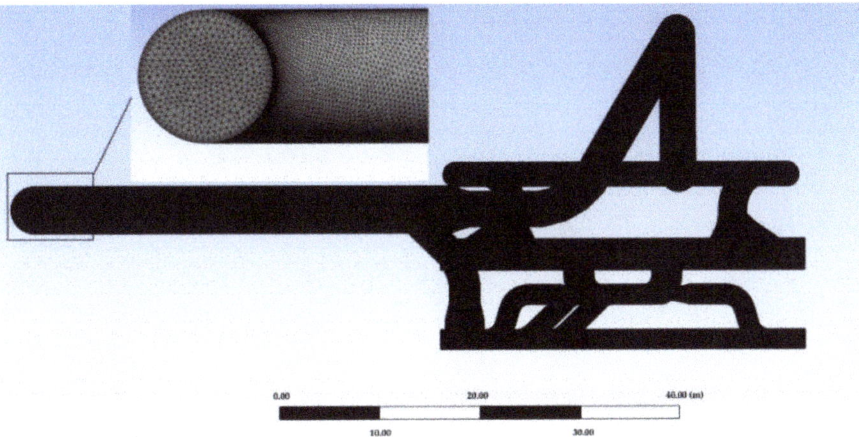

Figure 8. The hybrid calculation grid.

Turbulence modeling is carried out using the standard k–ω turbulence model. Modeling of a multiphase flow is carried out using the Euler equations for multiphase media in a dispersed formulation. The flow simulation is based on the Reynolds-averaged Navier–Stokes equation. The gas phase is modeled according to the ideal gas equations. The mass flow rate of the resulting dust is calculated based on the given dust content of the aspiration flow [23–26].

When modeling the dusty flow, the following restrictions and assumptions were taken:

1. Near-wall functions were used to take into account the boundary layers not resolved by the grid.
2. In the areas of the computational domain corresponding to the outlet from the aspiration system, zero excess static pressure, total temperature and turbulence parameters of the gas–air mixture were set.
3. In the sections of the computational domain corresponding to the inlet of the aspiration system, the total pressure, total temperature, and parameters of air turbulence were set.
4. The material of the air duct of the suction system was specified as steel, 10 mm thick. Density = 8030 kg/m^3, Specific heat Cp = 502.48 J/(kg·K) and Thermal conductivity = 16.27 W/(m·K). Since the solution of the gas–dynamic problem is impossible without taking into account thermodynamic conditions, the wall material is a participant in the heat exchange processes. Since the temperature of the gas–dust

mixture initially reaches a temperature of 350 °C, and the ambient temperature is assumed to be 30 °C, then, in addition to thermal conductivity, there will be heat exchange by radiation from the wall surface, which was also taken into account when solving the problem.

The discretization was performed by the control volume method using the COUPLED algorithm which solves the momentum and pressure-based continuity equations together. Full implicit coupling is achieved through an implicit discretization of pressure gradient terms in the momentum equations, and an implicit discretization of the face mass flux including the Rhie–Chow pressure dissipation terms.

3. Main Results

Figures 9 and 10 show the streamlines of the solid phase in the area of emission from tapholes and at the entrance to the exhaust elements of the aspiration system. Obviously, part of the solid phase, unfortunately, is not captured by the proposed system.

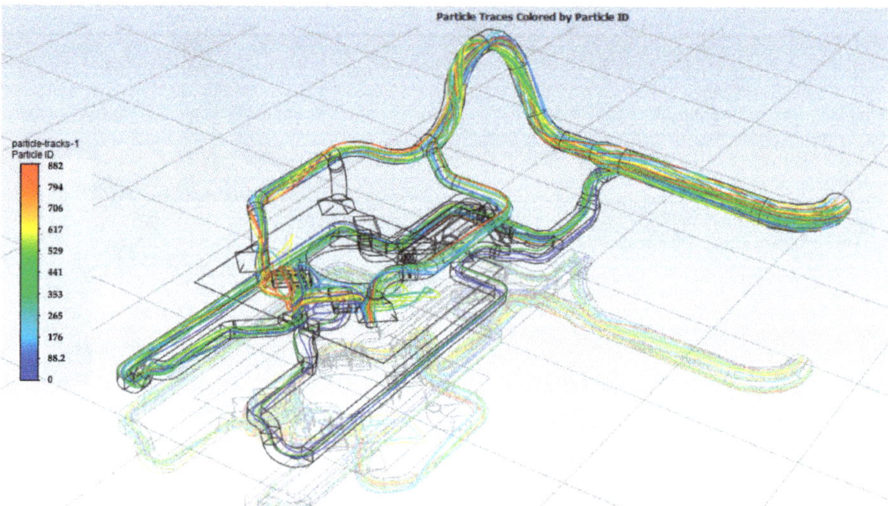

Figure 9. Solid phase (dust only) streamlines.

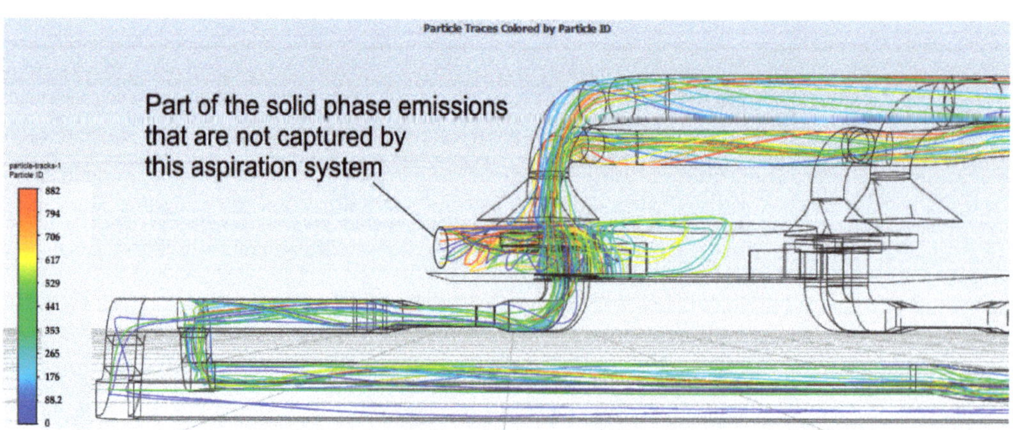

Figure 10. Solid phase (dust only) streamlines, focused on the place of "ejection".

The efficiency of the aspiration system is determined on the basis of the integral indicators of the mass flow rates of the solid phase of gas and dust flows coming from sources of pollution and dust flows leaving the aspiration system.

1. The average integral mass flow rate of the solid phase (dust) at the outlet from pollution sources is 0.233 kg/s + 0.466 kg/s = 0.699 kg/s.
2. The average integral mass flow rate of the solid phase (dust) at the outlet from the aspiration system pollution is 0.638 kg/s.

The efficiency of dust collection by the aspiration system can thus be estimated as follows:

$$\eta = (0.638/0.699) \cdot 100\% = 91.27\%$$

It can be seen that in the worst case (the most dangerous from the point of view of the operation of the aspiration system for the operating mode in accordance with Figure 6), the efficiency of collecting the solid phase of the gas–dust mixture is more than 90%.

It is clear that the next question for the authors of the work was the following: is it possible to increase the efficiency of dust collection while maintaining the geometry of the proposed aspiration system? Obviously, by varying the productivity of the system, you can significantly affect the efficiency. So, with an increase in productivity of only 10%, the cleaning efficiency of dust collection would be almost 93%, with a further increase—for example, up to 30%, the efficiency will already be at the level of 95%, as shown in Figure 11. However, everything has two faces. It can be seen that with the productivity and efficiency of dust collection of the system, the area-weighted average velocity in the outlet of the aspiration system of the gas–dust mixture also increases and, starting from a 5–10% increase in productivity, this velocity already exceeds 30 m/s and further increases as productivity increases up to 42 m/s. Such a level of velocity of the gas–dust mixture for aspiration systems is considered dangerous from the point of view of the abrasive wear of wall materials [14–16]. Therefore, the authors of the work believe that increasing the efficiency of dust collection by increasing productivity by more than 10% for the proposed geometry is inappropriate and can even be dangerous for materials walls.

Figure 11. Dependence of efficiency of dust collection and velocity of the gas–dust mixture on productivity of the aspiration system.

Figure 12 shows a graphical representation of the volume gradient of the distribution of gas–dust flow velocities through the aspiration system. One-dimensional estimation calculations cannot give a volumetric distribution of velocities over the cross section of the air duct. Thanks to the proposed modeling, it was shown that the velocity of the dust and gas flow along the length of the aspiration system is in the range of about $(20 \ldots 35)$ m/s at the inlet to 29.05 m/s at the outlet of the system.

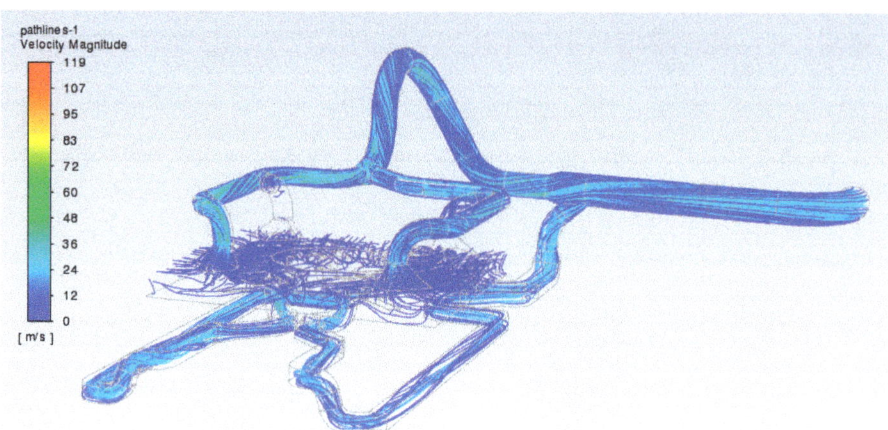

Figure 12. The field of velocities streamlines of the gas–dust-air mixture.

Figure 13 shows the average values of velocities in the sections of the gas ducts along the length for all four exhaust lines. Thus, the velocity range excludes the possibility of settling of the solid phase on the surfaces of the air channels of the aspiration system. On the other hand, the velocity of movement of the dust–gas flow is not large enough to cause significant abrasive wear of the structure.

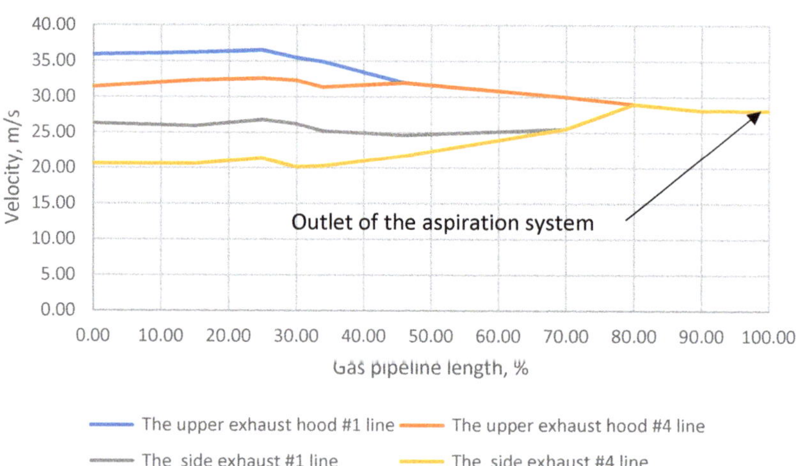

Figure 13. The value of the area-weighted average velocity along the length of the gas ducts of the aspiration system.

In the 3D modeling of the design of the aspiration system (walls of gas ducts), an ideal model of a solid body (with steel as the material) was used without taking into account the following: the physical wear of gas ducts during operation, the possible adhesion of dust to the walls and, accordingly, a decrease in the cross-sectional area of the gas duct, which in turn leads to an increase in the resistance of the route and the resistance of stop valves. This means that the actual experimental values of the gas–dynamic flow velocity may be somewhat lower than the calculated ones. Unfortunately, the proposed model of the aspiration system has not yet been manufactured and, as a result, the calculated values obtained have not been verified, especially with a long operating time, so minor discrepancies can be predicted.

The change in the temperature of the gas–dust mixture in the aspiration system deserves special attention. The fact is that the temperature from the source of the release—the taphole—reaches 350 °C (Figures 14 and 15), and the temperature at the outlet of the system before the proposed simulation could only be estimated (or measured immediately after manufacture and installation). Obviously, the temperature in the space above the entrances and on the platforms is also significant. However, the outlet temperature does not exceed 80 degrees Celsius. This air temperature is acceptable for the gas–dust mixture filtration system, which should be located immediately after the aspiration system. The temperature field of the gas–dust mixture (Figure 16) allows us to additionally select heat-insulating materials or make a decision on heat removal for secondary heat recovery both in industrial and domestic tasks.

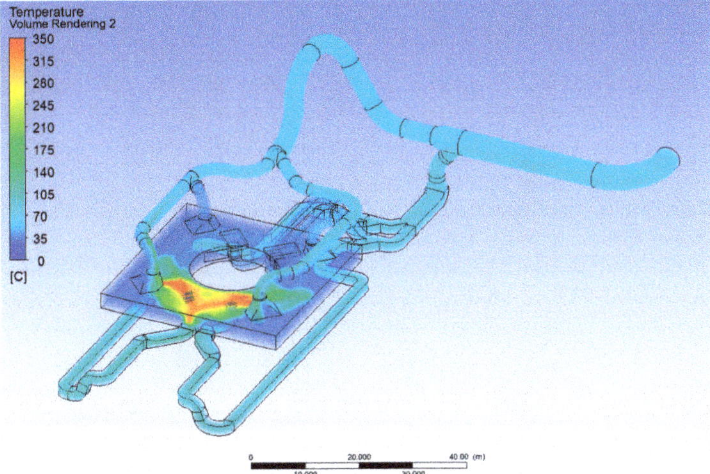

Figure 14. Temperature field of the dust–gas mixture.

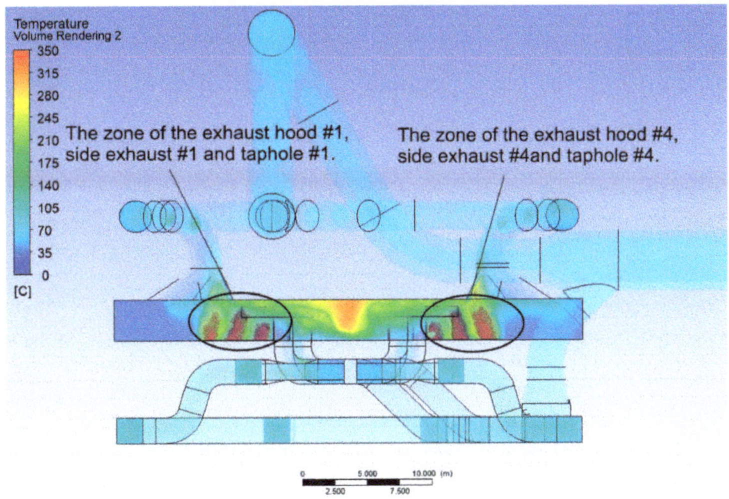

Figure 15. Temperature field of the dust–gas mixture, with focus at the "ejection" site.

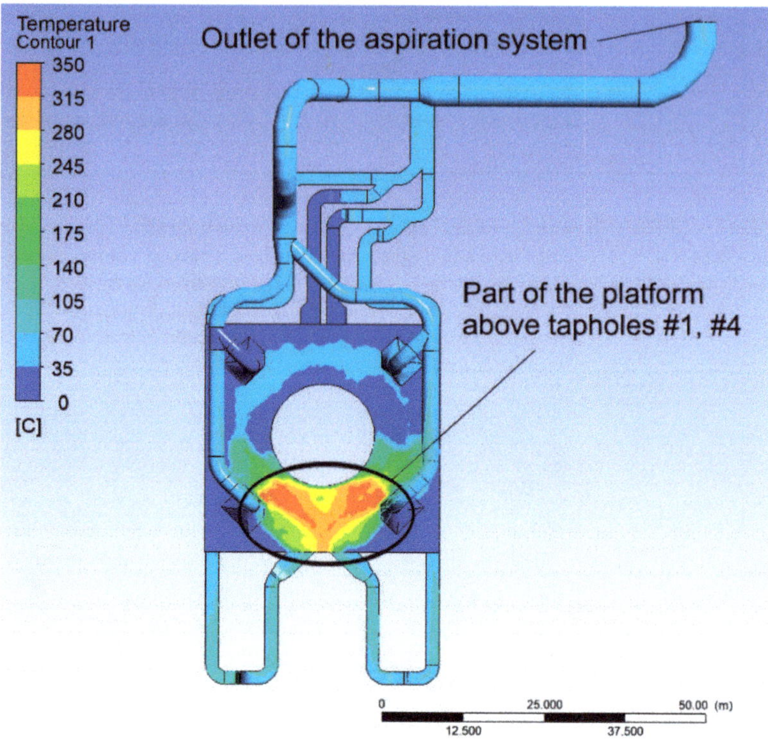

Figure 16. Temperature of solid surfaces (walls of the aspiration system).

In accordance with the 3D calculation, the main parameters of the gas and dust flow at the outlet of the aspiration system of the foundry were obtained. They are presented in Table 1.

Table 1. The main parameters of the gas–air mixture at the outlet from the aspiration system.

Parameter	Computed Value
Mass flow, kg/s	361.1
Flow velocity, m/s	29.5
Flow temperature, 0 °C	80.5
Mass flow rate of the dust, kg/s	0.638
Density, kg/m^3	0.986

Unfortunately, as mentioned above, it is not possible to verify the data obtained because, for objective reasons, the construction of the aspiration system at the object of study has not yet been carried out.

4. Conclusions

The calculation of the amount of aspiration air necessary for the almost complete removal of the evolved dust–gas–air mixture was carried out.

The designed gas ducts have a precisely calculated cross-sectional area, which makes it possible to keep the velocity of the movement of the dust–gas–air mixture uniform throughout the entire route of the gas duct within the permissible level.

Data on the required aspiration volumes were obtained from the developed design of local suctions.

The study of the dispersed composition and concentration of aspirated dust from the developed design of local suctions was carried out, and the law of distribution of the dispersed composition of aspirated dust was obtained.

The method for calculating the concentration of dust in the air of the working area has been refined when using a complex of dust removal systems, including aspiration systems, in relation to the foundry industry.

It is shown that the efficiency of dust collection of the aspiration system is 91.27%.

It is calculated that the weighted average flow velocity of the gas–dust mixture along the ducts of the aspiration system at the most loaded operating mode is not lower than 20 m/s, which should prevent the gravitational settling of the solid phase on the walls of the gas ducts. On the other hand, this speed does not exceed 35 m/s in a small section of the gas duct, which slightly exceeds the speed range recommended in the literature: from 20 to 29 m/s.

The obtained value of the temperature of the gas and dust flow at the outlet of the aspiration system is 80.5 degrees Celsius, which is below the allowable value for the filter system.

As a recommendation, it can be proposed to change the area of the flow section of the gas duct from exhaust hood #1 to the point of connection of this line with the line of exhaust hood #4. This will reduce the average velocity to more acceptable values. To improve the efficiency of the system, it is recommended to increase the productivity of the proposed aspiration system by no more than 10 percent, which will increase the efficiency to 93.5 percent without significantly increasing the average velocity of the gas–dust flow. However, it is necessary to additionally evaluate the feasibility of increasing productivity, because the cost of increasing productivity can be much higher than the cost of the filter system.

Author Contributions: Conceptualization, Y.P. and S.L.; formal analysis, S.L.; investigation, S.L.; methodology, Y.P. and V.K.; project administration, I.B. and V.K.; resources, I.B.; software, S.L.; supervision, S.L. and V.K.; validation, I.B.; visualization, S.L. and I.B.; writing—original draft, S.L.; writing—review and editing, Y.P. and V.K. All authors have read and agreed to the published version of the manuscript.

Funding: This research received no external funding.

Institutional Review Board Statement: Not applicable.

Informed Consent Statement: Not applicable.

Data Availability Statement: Not applicable.

Conflicts of Interest: The authors declare no conflict of interest.

References

1. Masikevich Yu, G.; Gryn, G.Y.; Sladky, V.D. *Technoecology*; Chernivtsi: Zelena Bukovyna, Ukraine, 2006; Volume 192.
2. Apostoliuk, S.O.; Dzhigirei, V.S.; Sokolovsky. *Industrial Ecology*; Znannia: Kyiv, Ukraine, 2012; Volume 430.
3. Simonyan, L.M.; Kochetov, A.I. *Environmentally Clean Metallurgy*; MISiS: Moscow, Russia, 2004; Volume 140.
4. Christof, L.; Wilfried, P.; Robert, N.; Christoph, F. Emissions and Removal of Gaseous Pollutants from the Top-gas of a Blast Furnace. *ISIJ Int.* **2019**, *59*, 590–595. [CrossRef]
5. Nowak, M.; Paul, A.; Srivastava, R.; Radziwon, A. Coal Conversion. *Encycl. Energy* **2004**, *1*, 425–434. [CrossRef]
6. Hemeon, W.C.L. Air Pollution Problems of the Steel Industry Technical Coordinating Committee T-6 Steel Report. *J. Air Pollut. Control. Assoc.* **2012**, *7*, 62–67. [CrossRef]
7. Boscolo, M.; Padoano, E.; Parussini, L.; Petronelli, N.; Dimastromatteo, V.; Nazzarri, S. CFD Analysis of Low-Cost Solutions to Minimize Gas and Dust Emissions during the Emergency Opening of Blast Furnace Bleeders. *Appl. Sci.* **2022**, *12*, 2266. [CrossRef]
8. Collection of Methods for Calculating Air Emissions of Pollutants by Various Industries. Available online: http://eco.com.ua/content/sbornik-metodik-po-raschetu-vybrosov-v-atmosferu-zagryaznyayushchih-veshchestv-razlichnymi (accessed on 15 September 2021).

9.   Wu, Q.; Gao, W.; Wang, S.; Hao, J. Updated atmospheric speciated mercury emissions from iron and steel production in China during 2000–2015. *Atmos. Chem. Phys.* **2017**, *17*, 10423–10433. [CrossRef]

10. Arens, M.; Worrell, E.; Eichhammer, W.; Hasanbeigi, A.; Zhang, Q. Pathways to a low-carbon iron and steel industry in the medium-term—The case of Germany. *J. Clean. Prod.* **2017**, *163*, 84–98. [CrossRef]

11. Adrian, T. Reducing Blast Furnace Use to Save the Planet. AZoM. 2021. Available online: https://www.azom.com/news.aspx?newsID=57209 (accessed on 11 June 2023).

12. Suer, J.; Ahrenhold, F.; Traverso, M. Carbon Footprint and Energy Transformation Analysis of Steel Produced via a Direct Reduction Plant with an Integrated Electric Melting Unit. *J. Sustain. Metall.* **2022**, *8*, 1532–1545. [CrossRef]

13. Stefanenko, V.T.; Stefanenko, A.V.; Popova, N.P. Technical solutions for efficiency increasing of blast furnace casting yards aspiration. Ferrous Metallurgy. *Bull. Sci. Tech. Econ. Inf.* **2018**, *10*, 82–87. [CrossRef]

14. SHushlyakov, D.A.; Kubyshkin, A.V.; Soldatov, D.A. The examination of the aspiration system of the casting yard, bunker overpass, receiving device with transshipment units of the blast furnace «Rossiyanka». *Ekol. I Stroit.* **2019**, *3*, 20–27. [CrossRef]

15. Matarneh, M.I.; Al Quran, F.M.; Andilakhay, A.A.; Haddad, J.S. Speed of abrasive particles in the process of abrasive blasting by submerged jets. *Int. J. Acad. Res. Part A* **2014**, *6*, 240–245. [CrossRef]

16. *Recommendations on the Design of Aspiration Systems for the Capture and Purification of Dust and Gas Emissions from Foundries of Blast Furnaces with a Volume of 2700–5500 m³*; Ministry of Ferrous Metallurgy CHERMETENERGOOCISTKA: Kharkiv, Ukraine, 1977; Volume 7.

17. Romanenko, I.; Martseniuk, Y.; Bilohub, O. Modeling the Meshing Procedure of the External Gear Fuel Pump Using a CFD Tool. *Computation* **2022**, *10*, 114. [CrossRef]

18. Kim, J.-G.; Lee, I.-B.; Lee, S.-Y.; Park, S.-J.; Jeong, D.-Y.; Choi, Y.-B.; Decano-Valentin, C.; Yeo, U.-H. Development of an Air-Recirculated Ventilation System for a Piglet House, Part 1: Analysis of Representative Problems through Field Experiment and Aerodynamic Analysis Using CFD Simulation for Evaluating Applicability of System. *Agriculture* **2022**, *12*, 1139.

19. Fletcher, C.A.J. *Fletcher Computational Techniques for Fluid Dynamics*; Springer: Sydney, Australia, 1991.

20. Shih, T.-H. A New Eddy-Viscosity Model for High Reynolds Number Turbulent Flows Model Development and Validation. *Comput. Fluids* **1995**, *24*, 227–238. [CrossRef]

21. Prikhodko, A.A. *Computer Technologies in Aerohydrodynamic and Heat-Mass Transfer*; Naukova dumka: Kyiv, Ukraine, 2003; Volume 379.

22. Nikushchenko, D.V. *Investigation of Viscous Incompressible Fluid Flows based on the FLUENT Computational Complex*; St. Petersburg State Marine Technical University: St. Petersburg, Russia, 2006; Volume 94.

23. Shin, D.; Chergui, J.; Juric, D. *Direct Simulation of Multiphase Flows with Modeling of Dynamic Interface Contact Angle. Theoretical and Computational Fluid Dynamics*; Springer: Berlin/Heidelberg, Germany, 2018; Volume 32, pp. 655–687.

24. Comtet, J.; Keshavarz, B.; Bush, J.W.M. Drop impact and capture on a thin flexible fiber. *Soft. Matter.* **2016**, *12*, 149–156. [PubMed]

25. Yamamoto, Y.; Tokeida, K.; Wakimoto, T.; Ito, T.; Katoh, K. Modeling of the dynamic wetting behavior in a capillary tube considering the macroscopic-microscopic contact angle relation and generalized Navier boundary condition. *Int. J. Multiph. Flow* **2014**, *59*, 106–112.

26. Tryggvason, G.; Scardovelli, R.; Zaleski, S. *Direct Numerical Simulations of Gas-Liquid Multiphase Flows*; Cambridge University Press: Cambridge, UK, 2011.

 *computation*

Article

Stress-Strained State of the Thrust Bearing Disc of Hydrogenerator-Motor

Oleksii Tretiak [1,*], Dmitriy Kritskiy [2,*], Igor Kobzar [3,*], Mariia Arefieva [1,*], Volodymyr Selevko [4,*], Dmytro Brega [1,*], Kateryna Maiorova [5,*] and Iryna Tretiak [6,*]

[1] Department of Aerohydrodynamics, National Aerospace University Kharkiv Aviation Institute, 61070 Kharkiv, Ukraine
[2] Department of Information Technology Design, National Aerospace University Kharkiv Aviation Institute, 61070 Kharkiv, Ukraine
[3] Special Design Office for Turbogenerators and Hydrogenerators, Join Stock Company "Ukrainian Energy Machines", 61037 Kharkiv, Ukraine
[4] Section of Postgraduate and Doctoral Studies, National Aerospace University Kharkiv Aviation Institute, 61070 Kharkiv, Ukraine
[5] Aircraft Manufacturing Department, National Aerospace University, Kharkiv Aviation Institute, 61070 Kharkiv, Ukraine
[6] Aerospace Thermal Engineering Department, National Aerospace University, Kharkiv Aviation Institute, 61070 Kharkiv, Ukraine
* Correspondence: alex3tretjak@ukr.net (O.T.); d.krickiy@khai.edu (D.K.); ivkobzar@ukr.net (I.K.); marii.arefieva@gmail.com (M.A.); v.selevko@khai.edu (V.S.); brega10.04@gmail.com (D.B.); k.majorova@khai.edu (K.M.); irina.ii3t@gmail.com (I.T.)

Abstract: In this article, the main causes of vibration in the thrust bearing of hydrogenerator motors rated 320 MW are considered. The main types of internal and surface defects that appear on the working surface of the thrust bearing disc during long-term operation are considered. A method of three-dimensional modeling of such defects is presented, and an assessment of the stress-strain state of the heel disc is proposed, taking into account the main forces acting on the working surface using the finite element method. An analysis of the possible further operation of discs with similar defects, in accordance with the technical requirements, is carried out, and we consider ways to eliminate them.

Keywords: hydrogenerator; thrust bearing; strength; rigidity

Citation: Tretiak, O.; Kritskiy, D.; Kobzar, I.; Arefieva, M.; Selevko, V.; Brega, D.; Maiorova, K.; Tretiak, I. Stress-Strained State of the Thrust Bearing Disc of Hydrogenerator-Motor. *Computation* **2023**, *11*, 60. https://doi.org/10.3390/computation11030060

Academic Editor: Demos T. Tsahalis

Received: 15 February 2023
Revised: 6 March 2023
Accepted: 13 March 2023
Published: 16 March 2023

1. Introduction

Nowadays, in Ukraine, the main part of electrical power is produced at thermal, nuclear, and hydroelectric power stations, where turbo- and hydrogenerators work, respectively. Vertical-type hydrogenerators stand out among them. At the same time, an important element of the design of the generator, which perceives vertical loads from the weight of the entire unit, is the thrust bearing.

High-power generators have quite significant external geometric dimensions and consist of structural elements of various scales, which complicate (and often makes impossible) strength analysis of the generator design as a whole. At the same time, the elements of the generator structures work under complex load conditions caused by the joint action of inertial forces from the rotation of the rotor, gravitational forces, component loads arising from the seating of parts with tension, as well as temperature loads that arise, first of all, as a result of the release of heat in the active circuit and are determined by the operating parameters of their forced ventilation system. In complex generator design, this leads to the need to consider a whole complex of problems related to the determination of the thermally stressed state of structures, complicated by previous stresses, the influence of temperature fields that depend on the operating parameters of ventilation systems, and many other factors.

Figure 1 shows the design of the thrust bearing on adjustable screw supports [1]. The thrust bearing is located in an oil bath with water-oil coolers. The moving part of the support structure consists of a hub, which is mounted on the upper end of the shaft, and a ring into which the segments are inserted. The lower part of the hub is equipped with a disc with a polished surface, namely a mirror of the thrust bearing. All the vertical load is transmitted through the mirror of the bracket to the segments. The segments of the thrust bearing are arranged in one or two rows on the rest seat of the thrust bearing that accepts the load. The steel segment consists of the body of the segment and a cushion covered with a layer of anti-friction material, namely babbit or fluoroplastic. The segment rests on a plate (segment support), which is supported by spherical heads of bolts, which ensure their self-installation during operation. To prevent rotation (turning), the segments are held by special radial rests.

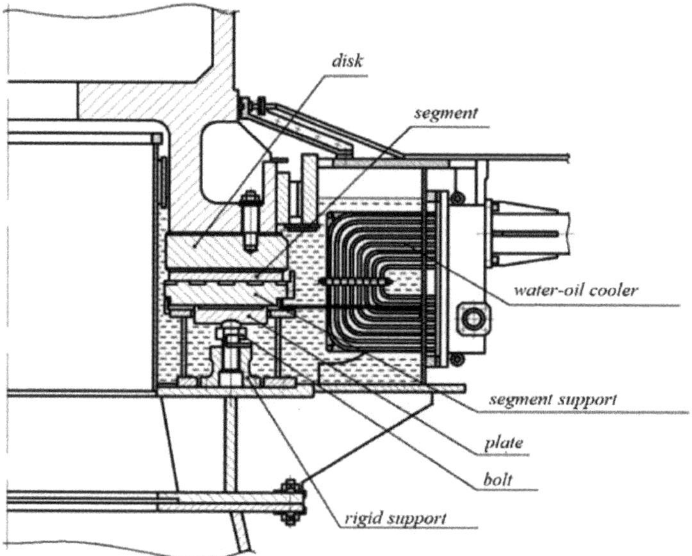

Figure 1. The design of the thrust bearing rest is on adjustable screw supports.

During the operation of hydro-aggregates, it is necessary to monitor the vibration of the bearings, and in the event of a vibration that is significantly greater than usual, it is necessary to make sure that the magnitude of the vibration does not exceed the permissible level before connecting the generator to the network. Otherwise, start-up is stopped, and the generator is stopped until the causes of the increased vibrations are identified. After reaching the rotation speed, which is close to the rated one, the generator is switched on for parallel operation in the network. Activation can be carried out by exact synchronization or self synchronization. Accurate synchronization assumes that the generator is connected to the network at the moment when the voltage of the generator and the network are the same in magnitude, and the voltage vectors of the generator and the network coincide in phase. This occurs when the frequency of the generator coincides with the frequency of the network. However, due to the fact that the frequency of the network constantly changes within small limits due to fluctuations in the load in the system, there is practically no long-term coincidence of the frequencies of the generator and the network. Therefore, switching on is carried out with a large or small deviation from the moment of complete coincidence of the vectors.

In work [2], the author's team arranged a detailed review of the structures of geometric forms of thrust bearings and evaluated the effectiveness of the structures. The work highlights the main problems of modeling this complex node and offers methods that take into account the hydrodynamics and incompressibility of substances, considering the influence of temperatures. However, this work is appropriate for the simulation of new bearings; the possibility of reusing discs after long-term operation is not considered.

In work [3], a detailed review of the design of thrust bearing pads with self-adjusting pads for inclination, which is similar to the thrust bearing studied in our work, is submitted. A detailed technique of three-dimensional modeling of hydraulic and thermal processes on the surface of the segments is presented, which makes it possible to estimate the temperature effect and forces on the surface of the disc, as a result of which, defects may appear and develop on the mirror surface.

It is necessary to pay attention to the fact that the damage that occurs on the mirror surface of the thrust bearing disc as a result of operation causes an increase in the vibration of the structure.

The work [4] provides a detailed review of damaged bearing elements as a result of overheating and established cause-and-effect relationships between vibration, overheating, and the subsequent failure of the unit. The work [5] presents in detail the types of failures of hydro-aggregate units and analyzes the economic losses caused by the failure of at least one unit. The presented work is an overview and really reveals many factors affecting the appearance of additional vibrations; however, its use is limited by the lack of a three-dimensional formulation of the task under consideration.

2. Research Task

The purpose of this work is the analysis of the structure, the determination of typical defects, and the development of a methodology for calculating heel pads with various types of defects on the mirror surface.

The work considers a synchronous three-phase reversible hydrogenerator motor of the vertical type. Its main elements are the stator and the rotor. The stator consists of a welded body, active steel with a winding, a set of foundation plates and studs. The rotor includes a shaft, a core, a stacked-up rim, a pole with a winding, and a current supply with slip rings. In addition, the hydrogenerator motor includes: a spider consisting of a central part with an oil bath of a guide bearing, bearing segments, and oil coolers; feet of spacer jacks, overlapping of the upper cross-piece of the hydrogenerator motor; an oil bath with a thrust bearing, consisting of a bath with a seal, a thrust bearing housing with supports and segments, a thrust bearing disc, and oil coolers; main and neutral terminals of the stator winding; stand with traverse, excitation bus-bars; covering of the pump-turbine shaft consisting of beams and segments; water and oil pipelines with fittings and control and measuring devices; the ventilation system, consisting of air coolers with branch-pipes, upper and lower air distribution shields, as well as the braking system, consisting of brake jacks with stands, brake piping, a high-pressure pump, and a braking cubicle.

At the same time, it should be noted that the mass of the structural elements is, namely, the stator is 540 t, the rotor is 780 t, and the spider is 93.4 t.

3. Types of Surface Defects

As mentioned above, the main cause of surface and internal defects are the vibration of the structure and the influence of temperature. The review work [6] shows the most common types of surface defects that occur on the working surface of thrust bearings. This work provides comprehensive knowledge about the cause-and-effect relationships of the occurrence of defects of a certain type. The work [7] not only provides information on the causes of defects, but also contains recommendations for their elimination.

However, in this work, only the main types of defects that are most often met in the practice of disc operation are considered. The main surface defects considered are torn surfaces, hairs, hot cracks, chains (belts), bubbles, and others (see Table 1).

Table 1. Basic types of defects studied in this work.

No.	1	2	3	4
Description	surface tear	hot cracks	gas bubbles	hair
Drawing				

4. Feature of Long-Term Operation of the Hydro-Aggregates

During the operation of the hydrogenerator, there are vibrations of the rotor caused by the load of the thrust bearing disc, and its damage leads to a significant increase in vibration of the structure in general. Figure 2 presents the results of the study of the beating of the friction surface of the thrust bearing disc with decomposition into harmonic components by the Fourier method [8].

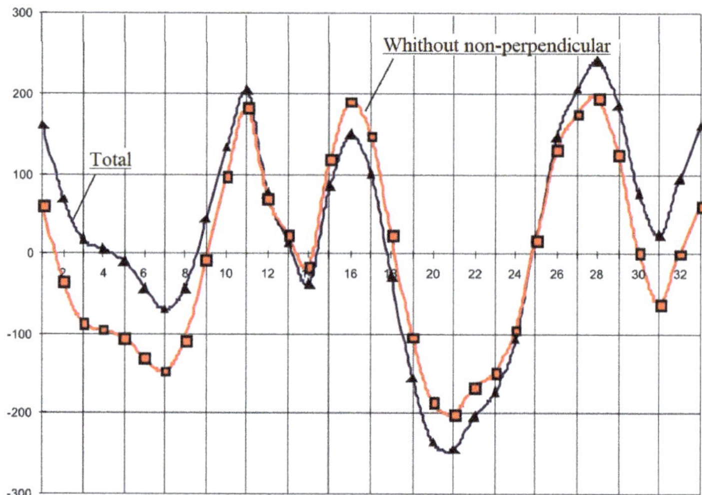

Figure 2. Harmonic components of vibrations of the thrust bearing disc due to beating.

However, the permissible vibration (for oscillations with a frequency of up to 100 Hz in the established symmetrical mode), which is regulated by the vibration of the frontal parts of the stator winding in the tangential and radial directions, should not exceed 100 μm [9].

5. The Main Causes of Vibration

The presence or absence of vibration of the hydro-aggregate determines the possibility of long-term reliable operation of the unit and is one of the main quality indicators of its design, manufacturing technology, and installation work. The increased vibration of the hydro-aggregate can lead to an emergency state, a decrease in efficiency (coefficient of useful action), and additional power losses. Therefore, when the vibration of the unit exceeds the permissible values, the causes of the increased vibration shall be established and eliminated.

The causes of increased vibration of the hydro-aggregate, depending on the source of the disturbing force, can be divided into three types: mechanical, hydraulic, and electrical.

The mechanical causes include: unbalance of the generator rotor and turbine impeller; incorrect condition and position of the shaft axis of the hydro-aggregate; problems in bearing units; weak fastening of the supporting parts of the unit or their insufficient rigidity; involvement of rotating parts of the unit with stationary parts [10].

Hydraulic causes are: hydraulic imbalance of the impeller; incorrect height position of the impeller of the radial-axis turbine in relation to the guide apparatus; incorrectly established combinatorial dependence in rotary vane turbines; turbine operation in cavitation modes [11].

The electrical causes of vibration of the unit are usually unevenness of the attraction of the rotor to the stator (electromagnetic imbalance), which is mainly caused by unevenness of the air gap of the generator, exciter, and subexciter; the oval shape of the generator rotor; and short-circuits of the windings of the rotor poles [12,13].

In the above-mentioned works, the main structural shortcomings of heel pads are presented, while the problem of vibrations in the framework of cause-and-effect relationships is not covered. However, in [14], this problem is considered taking into account modern modeling methods.

6. Conditions for Calculating the Stress-Strained State of the Thrust Bearing Disc

In order to calculate the stress-strained state, the mechanical loads of the mirror surface of the thrust bearing disc with long-term service defects were determined [15]. Taking into account the above, all the denominations and designations of the defect are characterized according to DSTU 2658-94 [16].

The technological requirements for the mirror surface of the thrust bearing disc during its manufacture [17] are indicated below:

1. The roughness of the mirror surface of the disc should be no more than 0.32 μm (class 9) and not less than 0.16 μm (class 10). In some places, which make up no more than 10% of the mirror surface of the disc, the permissible purity is 0.63 μm (8-th grade).
2. Measurements of the roughness of the mirror surface of the thrust bearing disc shall be carried out during overhauls of the unit, as well as when signs of deterioration of the mirror surface cleanliness appear (increase in the temperature of all segments at a constant oil temperature in the thrust bearing bath, etc.).
3. The roughness of the mirror surface of the thrust bearing disc under operating conditions can be checked by removing casts on plastic material (for example, oil-gutta-percha mass) and then examining them under a microscope or using a profilometer.
4. If the roughness of the mirror surface of the disc is worse than specified in point 1, and if there are a large number of scratches, the mirror surface of the disc shall be processed (superfinishing and subsequent polishing) and brought to 0.32 μm.
5. The mirror surface of the thrust bearing discs of umbrella-type hydrogenerators can be processed under operating conditions with the help of a special self-propelled machine installed in the thrust bearing, from which the segments have been removed. The machine can be made for every size of thrust bearing.
6. The mirror surface of the disc of the hanger of the suspended hydrogenerators is processed in the conditions of hydroelectric power plants with the help of simpler devices after removing the bush with the disc from the shaft. If possible, the disc is sent to the factory for machining.

Summarizing the above, during long-term operation of the thrust bearing disc, there may be cases when, after mechanical processing during major repairs, as well as friction during work, internal defects shall be revealed.

The basic document on the production of forgings is presented in [18], where it is stated that these defects may appear as a result of mechanical effects on the part.

On the machined surfaces of forgings, individual defects are allowed without removal, if their depth, determined by a control cut or scraping, does not exceed 75% of the actual one-sided machining allowance for forgings produced by forging and 50% for forgings produced by dyeing.

7. Mechanical Calculation of the Disc

Conditions for mechanical calculation:

- Vertical load on to the thrust bearing disc is 1600 tf;
- Speed is 53.6 rpm.

Since the surface of the disc with clearly expressed defects interacts with the fluoro-plastic coating of the heel segments, the friction force Ftr = 16000 N, the friction coefficient $f = 0.0001$ (actual 0.05) appears accordingly. In the calculation, it is assumed that the defects are located at a uniform distance from each other and do not have mutual influence (see Figure 3).

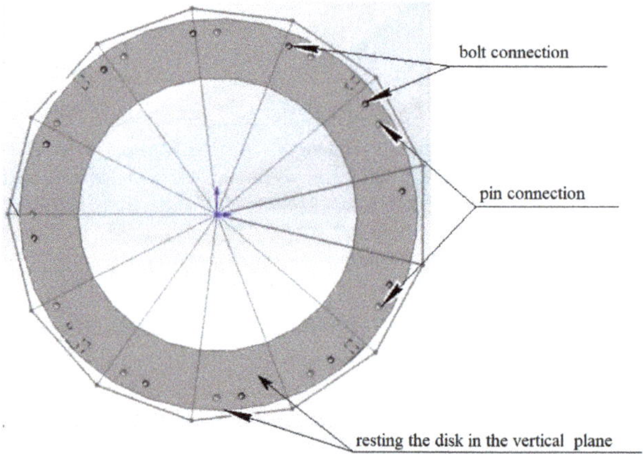

Figure 3. Calculation model of the thrust bearing disc.

One of the defining moments in solving finite elements method problems is the choice of a finite element. Two types of tetrahedrons (Figure 4) with different approximations of movements inside the element are used as basic finite elements. The first tetrahedron has units at the vertices (Figure 4a) and is based on a linear approximation of movements inside the element, and the second is an oblique tetrahedron that has units at the vertices of the element and in the middle of its ribs; it is based on a quadratic approximation of movements inside the element (Figure 4b).

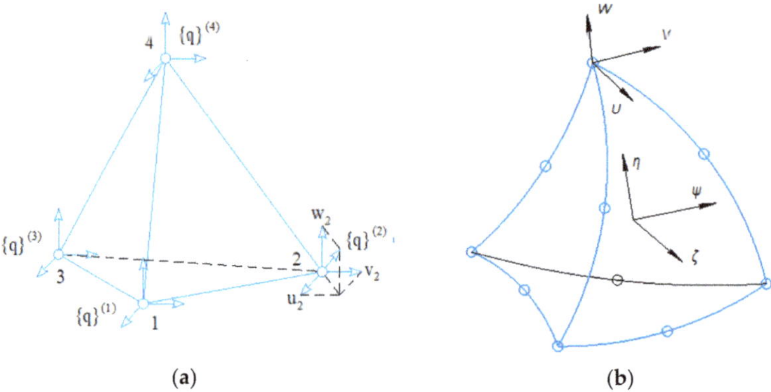

Figure 4. A finite element in the form of a tetrahedron.

An oblique tetrahedron allows for a more accurate description of the geometry and deformation process of the research object; however, it has 10 internal units and contains 30 unknown values, which is almost three times the number of unknowns for an ordinary tetrahedron with four units and, accordingly, with 12 sought values.

Therefore, a classical tetrahedron shall be used for a qualitative study of stress-strained state parameters, and an oblique one shall be used for more accurate, final calculations. In SolidsWork, which is used to solve all problems, these are TETRA4 and TETRA10 finite elements, respectively [19].

The calculation grid is built for each individual calculation of the disc section. In places where there are defects, a grid control element is used.

At the same time, there shall be at least three grid elements for the minimum geometric element. Convergence of the task is carried out by reducing the grid in such a way that the results do not differ by more than 0.5%. The accepted thickness of the oil film is 0.05 mm. Due to the fact that the depth of defects exceeds this value in the defect zones, additional forces will occur on the end surfaces.

Symmetry conditions are set on the face end surfaces of the disc segments to simplify modeling. For all defects, forces are set on the contact surface of the disc in the area of defect development. The disc-surface contact of the segment is specified as forces. At the same time, the calculation coefficient of friction in a pair (disc-fluoroplastic through oil film) according to the methodology of JSC "Ukrainian Energy Machines" is 0.05, and defects make up a very small part of the total mirror surface; then the accepted coefficient of friction for end forces is taken equal to 0.0001.

The material of the disc is forging, strength group KP245 with a yield strength of 245 MPa. In accordance with DSTU 9182:2022, defects, flakes, and cracks on the surface of the forging are permissible no more than the depth of machining. The thermal loads on the disc are set from the condition of contact heat exchange from the end surfaces of the disc and the bushing, provided by the pretension of the M48 bolts. The temperature of the disc in the summer is 30 °C, and the maximum drop is no more than 150 °C (boundary conditions of the first kind for thermal calculation). For the calculation, the disc material is set as isotropic. The change in the modulus of elasticity with temperature changes is not taken into account.

7.1. Defect No. 1 Surface Tear

Calculation results of stress for defect No. 1 are shown in Figures 5–9.

According to the obtained results, the average stress along the mirror surface of the disc comprises 50 MPa. In the defect location zone, the maximum stress is 625 MPa and the average stress is 520 MPa. These values exceed the strength limit (470 MPa, according to DSTU 9182:2022), yield strength (245 MPa, according to DSTU 9182:2022), and permissible stresses for rotating parts (233 MPa, according to IEC 60034-33:2022) [18,20].

7.2. Defect No. 2 Hot Cracks

Calculation results of stress for defect No. 2 are shown in Figures 10–13.

According to the obtained results, the average stress along the mirror surface of the disc comprises 50 MPa. In the defect location zone, the maximum stress is 901 MPa and the average stress is 750 MPa. These values exceed the strength limit (470 MPa, according to DSTU 9182:2022), yield strength (245 MPa, according to DSTU 9182:2022), and permissible stresses for rotating parts (233 MPa, according to IEC 60034-33:2022) [18,20].

According to the obtained results, the average stress along the mirror surface of the disc comprises 50 MPa. In the defect location zone, the maximum stress is 728 MPa and the average stress is 500 MPa. These values exceed the strength limit (470 MPa, according to DSTU 9182:2022), yield strength (245 MPa, according to DSTU 9182:2022), and permissible stresses for rotating parts (233 MPa, according to IEC 60034-33:2022) [18,20].

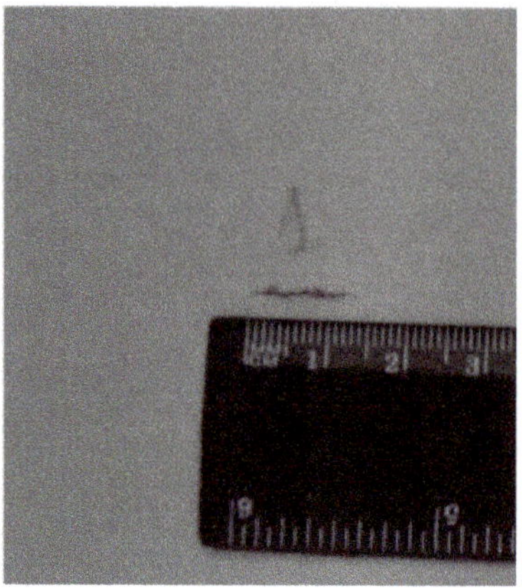

Figure 5. Actual defect and calculation grid for a defect on a disc segment.

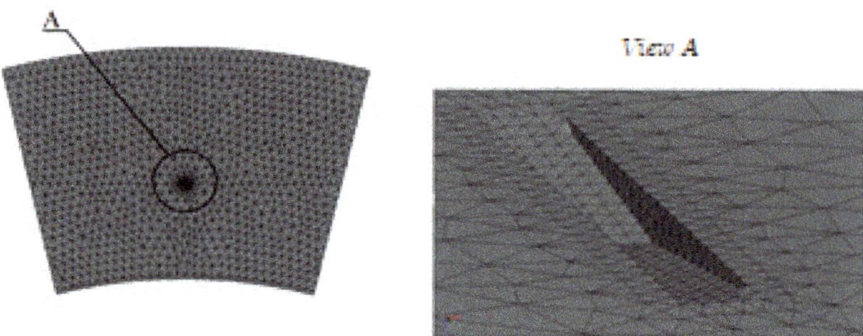

Figure 6. Conditions of the disc fastening.

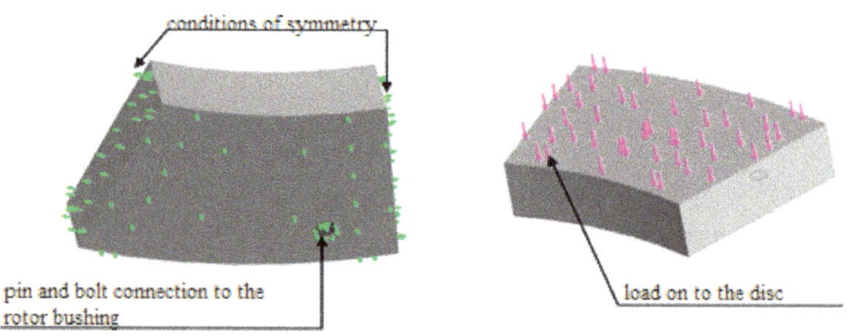

Figure 7. Acting load on the mirror surface of the disc.

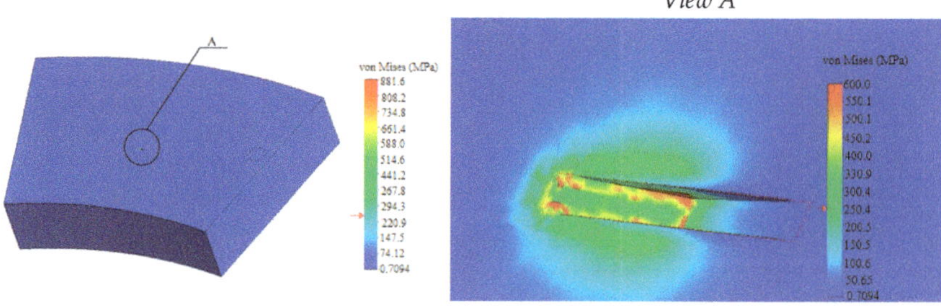

Figure 8. The stress field in the part of the thrust bearing disc and defect No. 1.

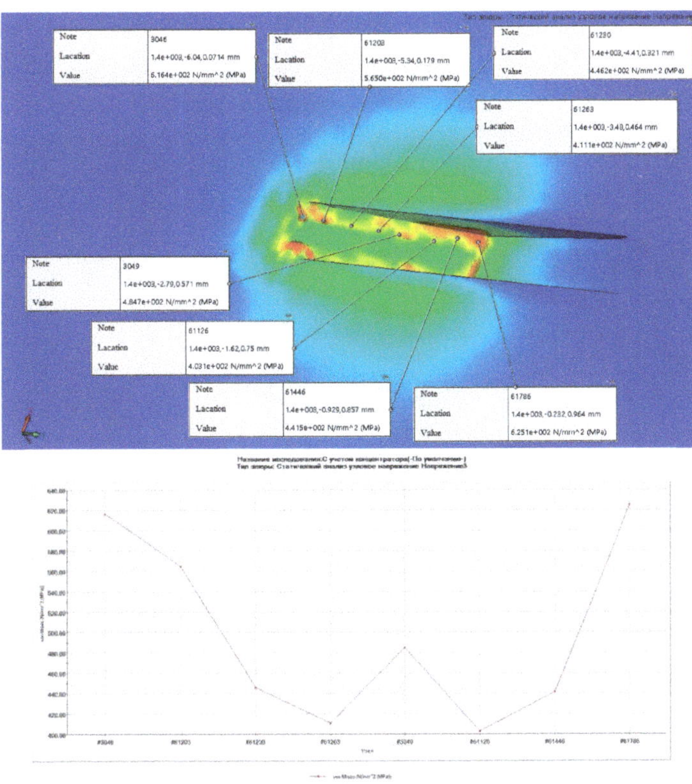

Figure 9. Change in stresses along the length of defect No. 1.

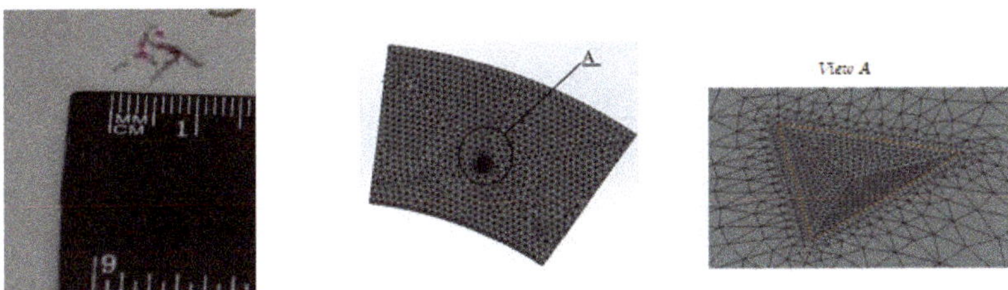

Figure 10. Actual defect and calculation grid for a defect on a disc segment.

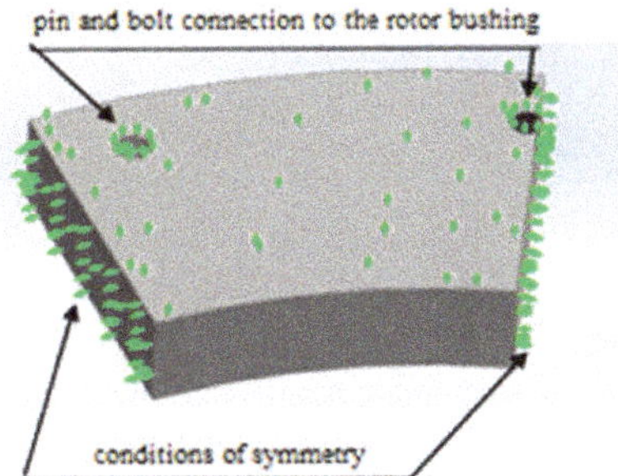

Figure 11. Conditions of the disc fastening.

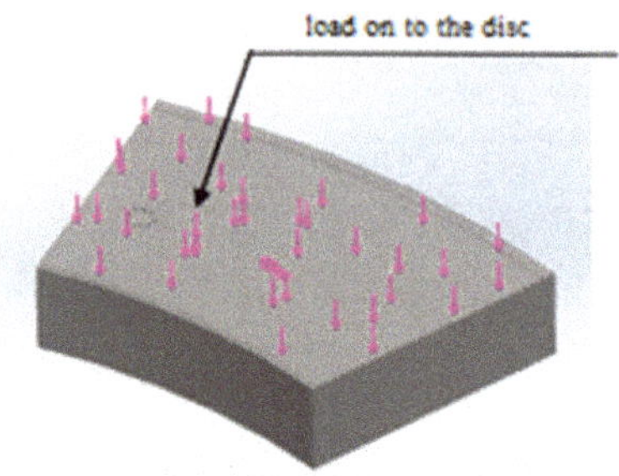

Figure 12. Acting load on the mirror surface of the disc.

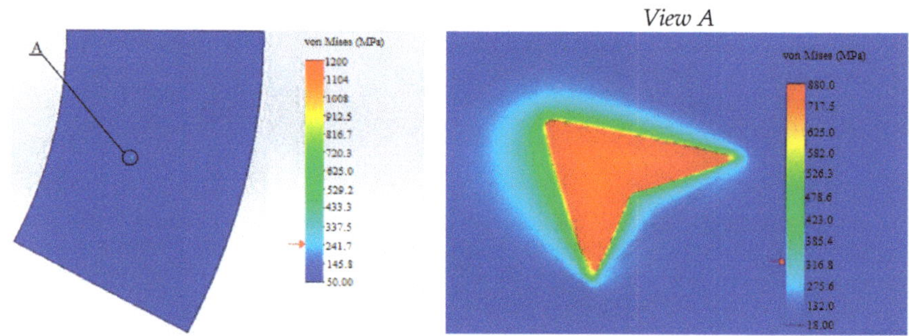

Figure 13. The stress field in the part of the thrust bearing disc and defect No. 2.

7.3. Defect No. 3 Gas Bubbles

Calculation results of stress for defect No. 3 are shown in Figures 14–18.

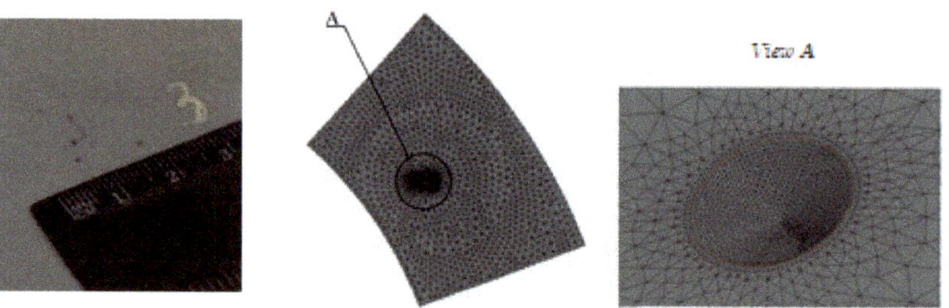

Figure 14. Actual defect and calculation grid for a defect on a disc segment.

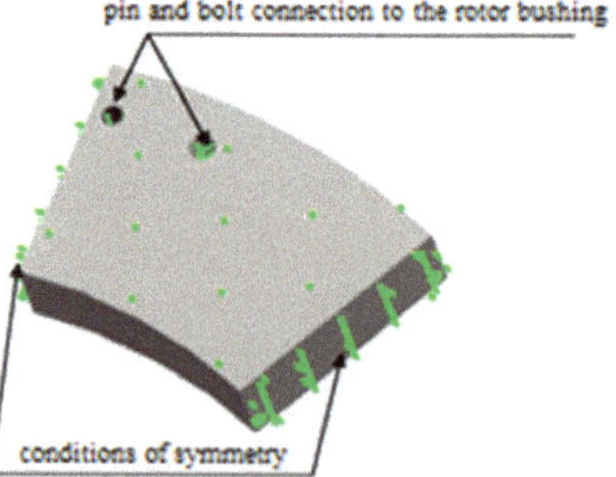

Figure 15. Conditions of the disc fastening.

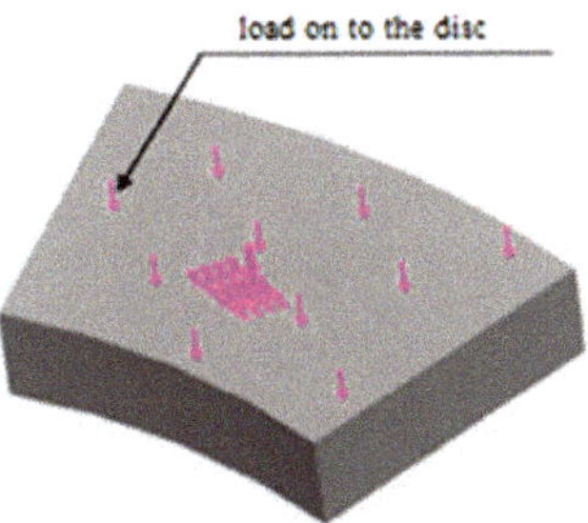

Figure 16. Acting load on the mirror surface of the disc.

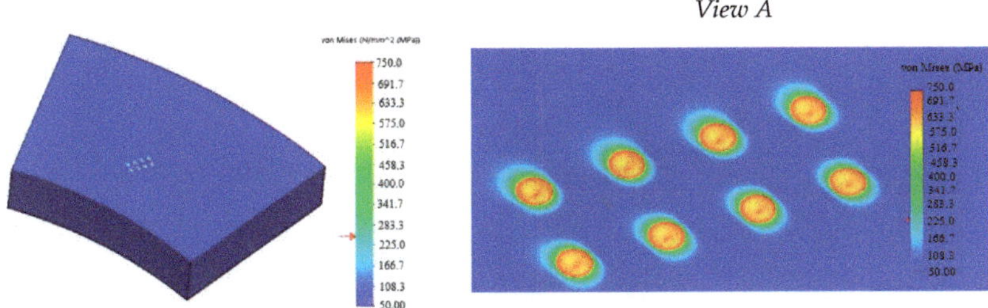

Figure 17. The stress field in the part of the thrust bearing disc and defect No. 3.

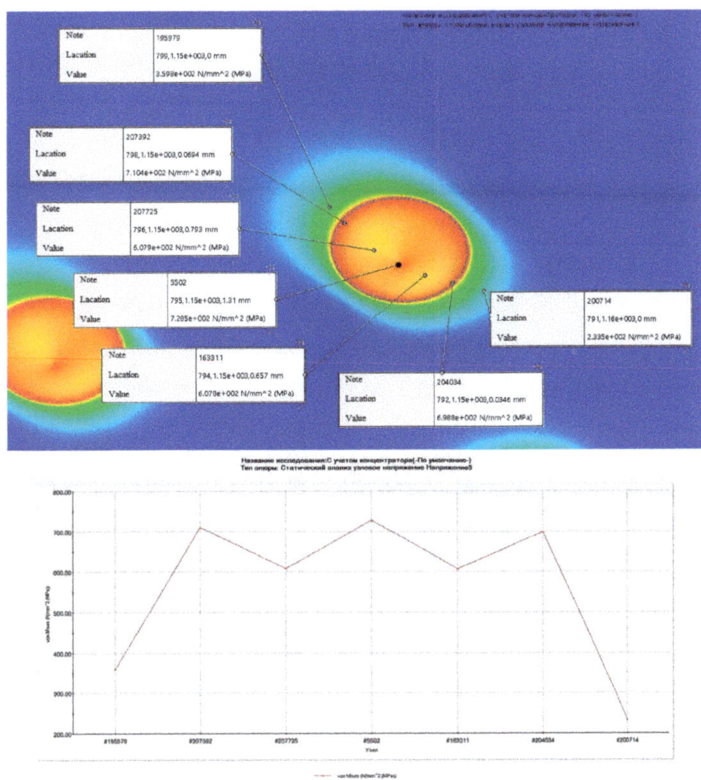

Figure 18. Change in stresses along the length of defect No. 3.

7.4. Defect No. 4 Hair

Calculation results of stress for defect No. 4 are shown in Figures 19–22.

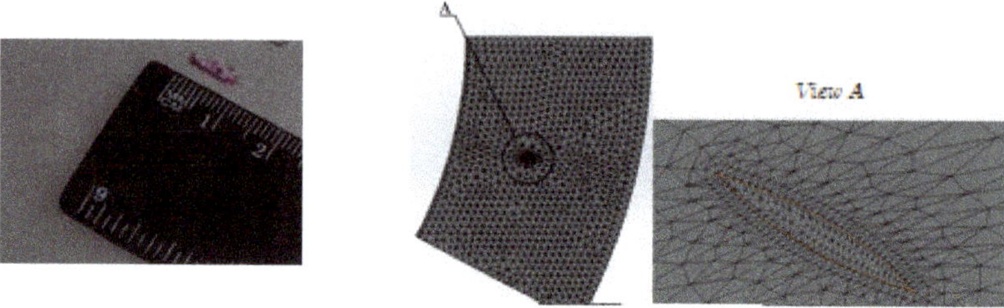

Figure 19. Actual defect and calculation grid for a defect on a disc segment.

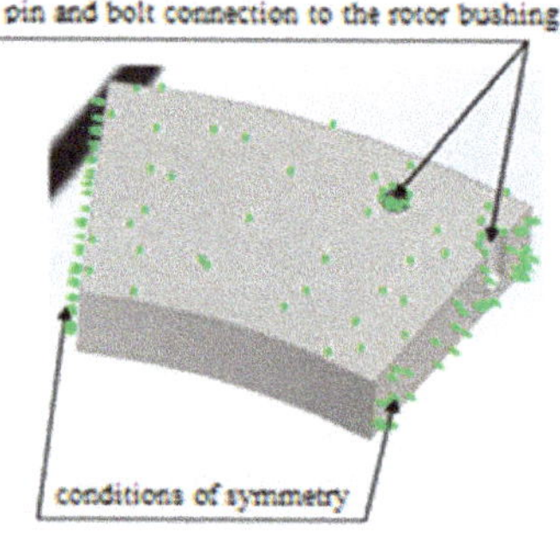

Figure 20. Conditions of the disc fastening.

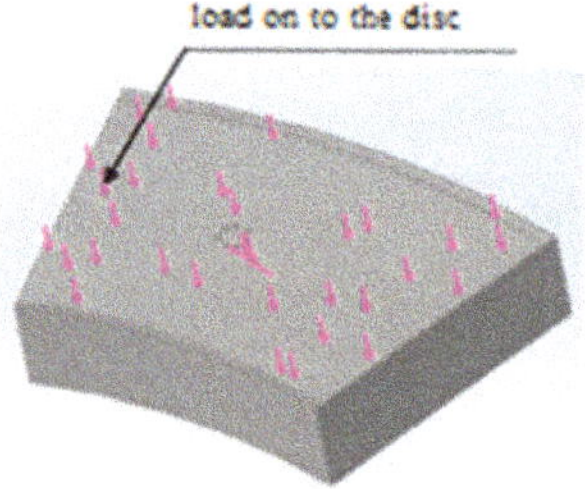

Figure 21. Acting load on the mirror surface of the disc.

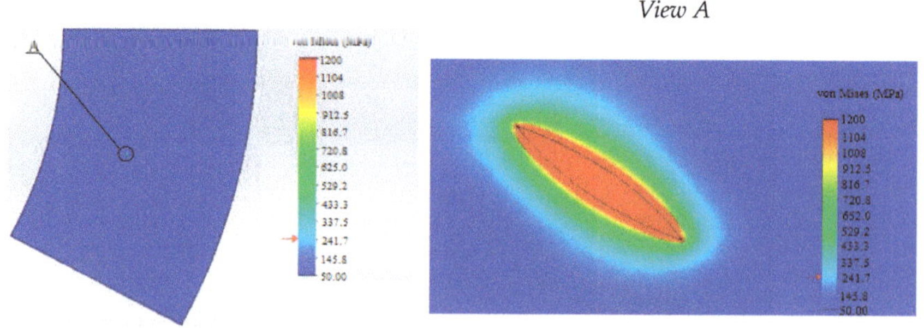

Figure 22. The stress field in the part of the thrust bearing disc and defect No. 4.

According to the obtained results, the average stress along the mirror surface of the disc comprises 50 MPa. In the defect location zone, the maximum stress is 1873 MPa and the average stress is 1150 MPa. These values exceed the strength limit (470 MPa, according to DSTU 9182:2022), yield strength (245 MPa, according to DSTU 9182:2022), and permissible stresses for rotating parts (233 MPa, according to IEC 60034-33:2022) [18,20].

8. Discussion

As a result of the carried-out calculations, for all defects, the stress values in the defect zone exceed the permissible yield strength $\sigma t = 245$ MPa for the material from which the disc is made (forging KP 245). Namely, they exceed the permissible $0.95\sigma t$ from the yield point for rotating parts of hydrogenerators (IEC 60034-33:2022) [20], the permissible $2/3\sigma t$ from the yield point for parts of the rotor of the hydrogenerator in the rated operational mode, and $0.9\sigma t$ from the yield point at runaway speed. Due to the fact that the calculated stresses exceeded the strength limit, the issue of fatigue was not considered.

The obtained stress values in the defect zones indicate the possibility of their further development, while the operating mode of 700 cycles per year (no more than four times per day) cannot be definitely ensured.

Removal of such defects will require a significant reduction in the height of the thrust bearing disc, which in the process of operation shall lead to an increase in the effect of temperature deformations and the appearance of gaps between the rotor bush and the thrust bearing disc. In these gaps, conditions shall arise for the initiation and development of cavitation of the contact surfaces, namely, for the appearance of microcracks and microexplosions. Therefore, it is not recommended to use a thrust bearing disc with such defects.

Author Contributions: Writing—review and editing O.T., D.K., I.K., M.A., V.S., D.B., K.M. and I.T.; All authors have read and agreed to the published version of the manuscript.

Funding: This research received no external funding.

Institutional Review Board Statement: Not applicable.

Data Availability Statement: Data not available due to confidentiality.

Conflicts of Interest: The authors declare no conflict of interest.

References

1. Generator Cradle. TPK Cheboksary-Electra. Available online: https://chebelektra.ru/dlya-ges/podpyatnik-generatora (accessed on 10 January 2023).
2. Fouflias, D.G.; Charitopoulos, A.G.; Papadopoulos, C.I.; Kaiktsis, L.; Fillon, M. Performance comparison between textured, pocket, and tapered-land sector-pad thrust bearings using computational fluid dynamics thermohydrodynamic analysis. *Proc. Inst. Mech. Eng. Part I J. Eng. Tribol.* **2015**, *229*, 376–397. [CrossRef]
3. XL_ThrustBearing®: A Computational Physics Analysis Tool for Tilting Pad Thrust Bearings (Both Regular and Self-Equalizing Types). Available online: https://rotorlab.tamu.edu/TRIBGROUP/TPTB%20Koosha.html (accessed on 6 March 2023).
4. Iliev, H. Failure Analysis of Hydro-Generator Thrust Bearing. *Wear* **1999**, *225–229*, 913–917. [CrossRef]
5. Liu, X.; Luo, Y.; Wang, Z. A review on fatigue damage mechanism in hydro turbines. *Renew. Sustain. Energy Rev.* **2016**, *54*, 1–14. [CrossRef]
6. Branagan, L.A. Survey of Damage Investigation of Babbitted Industrial Bearings. *Lubricants* **2015**, *3*, 91–112. [CrossRef]
7. Strecker, W. Troubleshooting Tilting Pad Thrust Bearings. In *Machinary Lubrication*; Noria Corporation: Tulsa, OK, USA, 2004; Volume 3.
8. Savchenko, Y.V. About the supports of hydrounits of Dniprovskaya HPP-1. *Hydroenergetics Ukr.* **2010**, *4*, 20–24.
9. *SOU-N EE 20.302:2020.* Testing Standards Electrical Equipment Approved by Order of the Ministry of Energy and Environmental Protection of Ukraine. Available online: https://docs.dtkt.ua/download/pdf/1225.943.1 (accessed on 11 February 2023).
10. Zhang, W.; He, Y.; Xu, M.; Zheng, W.; Sun, K.; Wang, H.; Gerada, D. A comprehensive study on stator vibrations in synchronous generators considering both single and combined SAGE cases. *Int. J. Electr. Power Energy Syst.* **2022**, *143*, 108490. [CrossRef]
11. Li, J.; Chen, D.; Liu, G.; Gao, X.; Miao, K.; Li, Y.; Xu, B. Analysis of the gyroscopic effect on the hydro-turbine generator unit. *Mech. Syst. Signal Process* **2019**, *132*, 138–152. [CrossRef]
12. Valavi, M.; Nysveen, A.; Nilsen, R.; Le Besnerais, J.; Devillers, E. Analysis of magnetic forces and vibration in a converter-fed synchronous hydrogenator. In Proceedings of the 2017 IEEE Energy Conversion Congress and Exposition (ECCE), Cincinnati, OH, USA, 1–5 October 2017; IEEE: Piscataway, NJ, USA, 2017; pp. 1838–1844. [CrossRef]

13. Tétreault, A. Rotor shape vs. rotor field pole pole shorted turns: Impact on rotor induced vibrations on hydrogenerators. In Proceedings of the IEEE International Conference on Condition Monitoring and Diagnosis, Bali, Indonesia, 23–27 September 2012; IEEE: Piscataway, NJ, USA, 2012; pp. 133–136. [CrossRef]
14. Liu, J.; Wang, L.; Shi, Z. Dynamic modelling of the defect extension and appearance in a cylindrical roller bearing. *Mech. Syst. Signal Process* **2022**, *173*, 109040. [CrossRef]
15. Gurvich, A.K.; Ermolov, I.N.; Sazhin, S.G. Non-Destructive Testing. In *Book. I. General Questions. Control by Penetrating Substances*; Sukhorukova, V.V., Ed.; Higher School: Moscow, Russia, 1992; 241p.
16. *DSTU 2658-94*; Black Metal Rolling. Terms and Definition of Surface Defects. Ukrainian Scientific and Advanced Institute of Metals "UkrNDIMet": Kiev, Ukraine, 1995; 70p. Available online: http://online.budstandart.com/ua/catalog/doc-page.html?id_doc=93138 (accessed on 6 March 2023).
17. *RD 34.31.501-88*; Methodical instructions for the Operation of Vertical Hydro Turbine Units (Hydro Units). Virobniche Association for Improvement, Improvement of Technology and Operation of Power Plants and Measures of "Soyuztechenergo": Kiev, Ukraine, 1998. Available online: http://online.budstandart.com/ua/catalog/doc-page?id_doc=69527 (accessed on 6 March 2023).
18. *DSTU 9182:2022*; Carbon and Alloy steel Forgings Made by Forging on Presses. Allowances and Tolerances. [25/05/2022] Ukrainian Scientific and Advanced Institute of Metals "UkrNDIMet": Kiev, Ukraine, 2022. Available online: http://online.budstandart.com/ru/catalog/doc-page.html?id_doc=97897 (accessed on 6 March 2023).
19. Baiborodov, Y.I. Development of Methods and Means of Increasing the Strength, Operational Characteristics and Durability of Heavily Loaded Sliding Bearings of Power Plant Rotors. Ph.D. Thesis, Tech. Sciences, Kiev, Ukraine, 2008; 406p. Available online: https://fizmathim.com/razrabotka-metodov-i-sredstv-povysheniya-prochnosti-rabotosposobnosti-i-dolgovechnosti-tyazhelonagruzhennyh-opor-skolzhen (accessed on 12 March 2023).
20. *IEC 60034-33:2022*; Rotating Electrical Machines—Part 33: Synchronous Hydrogenerators Including Motor-Generators—Specific Requirements. International Standard: Geneva, Switzerland, 2022. Available online: https://webstore.iec.ch/publication/30487 (accessed on 12 March 2023).

 *computation*

Article

Research on Scientific Directions for Flying Cars at the Preliminary Design Stage

Andrii Humennyi, Liliia Buival * and Zeyan Zheng

Department of Airplanes and Helicopters Design, Faculty of Aircraft Engineering, National Aerospace University "Kharkiv Aviation Institute", 61070 Kharkiv, Ukraine; a.gumennyy@khai.edu (A.H.); zhengzw585@foxmail.com (Z.Z.)
* Correspondence: l.buival@khai.edu

Abstract: This article was written to investigate the research on the scientific directions for flying cars at the preliminary design stage to provide a rationale for the choice of scientific research in the area of flying cars. At present, the population of the Earth is gradually increasing, and traffic congestion will become a common phenomenon in cities in the future. This work used the methods of theoretical and statistical analysis to form an overall picture of this area of research. We researched the statistical data analysis conducted by scientists who dealt with flying cars and the associated issues. We gave a rationale for the choice of the object of scientific research, which is flying cars. People can read this information to have a starting point in their understanding of flying car design. This analysis of famous scientific works provides possible scientific directions that the research can take with respect to designing a flying car that combines the advantages of an airplane and a car and can take off and land on a normal highway for a short distance, as well as help people reach their destination quickly and easily.

Keywords: flying car; scientific direction; analysis

Citation: Humennyi, A.; Buival, L.; Zheng, Z. Research on Scientific Directions for Flying Cars at the Preliminary Design Stage. *Computation* **2023**, *11*, 58. https://doi.org/10.3390/computation 11030058

Academic Editor: Xiaoqiang Hua

Received: 17 November 2022
Revised: 28 February 2023
Accepted: 8 March 2023
Published: 10 March 2023

1. Introduction

The current ground transportation system is suffering from various challenges, including the high cost of infrastructure development, limited land space, and a growing urban population. When driving in a city, especially in major cities, drivers who encounter traffic jams still cannot fly with their cars. A hundred years ago, this was a fantasy, but today, it is close to being a reality. Air mobility is a service that will improve existing transportation opportunities by bringing traffic into the air. Due to the increasing population, road traffic is growing at an alarming rate and many urban areas are experiencing traffic congestion. Flying cars will provide improved shortcuts for individuals to move between urban areas while reducing traffic congestion on land. The flying car is a vehicle that will become popular in future fast-paced urban life due to important characteristics such as short takeoff and landing capabilities, the ability to pass quickly regardless of terrain obstacles, and low emissions compared to large passenger aircraft.

Flying cars can solve traffic problems in the future, promote environmentally friendly urban ecosystems, and provide faster travel for people. Flying cars can be used for many different purposes, as they offer autonomous driving with vertical landing and take-off capabilities. The vehicle can be used in emergency operations, cargo transportation, air taxi operations, and security situations [1].

In recent years, the number of companies involved in the development of flying cars has increased rapidly, and some large companies have joined: Boeing has acquired the aviation technology company Aurora Flight Sciences and will join forces with Porsche to develop electric vertical takeoff and landing vehicles (eVTOL); Toyota has invested in Joby Aviation and SkyDrive; Geely Automobile has acquired the U.S. Terrafugia and invested in Germany's Volocopoter; and Chinese online technology giant Tencent has invested in

Lilium several times. Mohamed-Slim Alouini et al. describe several potential innovations that make communication between eVTOLs and the ground feasible. These innovations include three-dimensional cellular networks on the ground, tethered balloons, high-altitude platforms, and satellites [1].

Therefore, to develop a preliminary design for a flying car, a deep analysis of research on the relevant scientific directions should be performed. The result will provide significant direction for investigations on flying cars, and it should facilitate the rapidly increasing development in the field of flying car design since it will allow this field to not repeat the path already taken by other scientists but to instead take into account the gained experience.

The purpose of the work is to research the scientific directions of flying cars at the preliminary design stage and to provide a rationale for the choice of scientific research in the area of flying cars.

2. Collection and Analysis of Information on the Direction of Scientific Investigation on Flying Cars

This section is devoted to information gathered about the names and countries of scientists investigating flying cars, along with their works, that are the most influential in the preliminary stage of flying car research. To conduct this analysis more effectively, a decision about a wide area of science was made.

To collect relevant information from scientific journals, scientific books, articles, authors (scientists), and publications from official websites of flying car companies, the Scopus database, the Research Gate social network, the Web of Science, Science Direct, etc., were used. In particular, their features for the study were taken into account.

The findings and the analyses of results were based on the highest impact factor, the number of possible combinations of inner limits, the research impact, the categories of research directions on flying cars, range of years, number of citations, number of times read, downloads, the place of publication, and indexing of web resources.

The three sections below correspond to three key points on which the current research was based:

- Scientists, publications, and flying cars companies;
- Scientists' claims;
- Flying cars' features.

2.1. Scientists, Publications, and Flying Car Companies

The list below represents *scientists* studying flying cars and *their publications,* which were selected based on the considerations described above.

- Haktan Yağmur (Turkey). "Conceptual Design of a Novel Roadable Flying Car" [1].
- Nasir Saeed (Saudi Arabia). "Wireless Communication for Flying Cars" [2].
- Larisa Ivascu (Romania). "The Flying Car–A Solution for Green Transportation" [3].
- Steven C. Crow Starcar Development Co. "A Practical Flying Car" [4]
- Mohammad Adhitya (Indonesia). "Center of gravity analysis of a flying car" [5]; "Folded wing mechanism for flying car" [6]; "Flying-cars body manufacturing using spraying elastic waterproof and water-absorbing frame fabric method" [7]; "Design and analysis of tubular space-frame chassis on flying car with impact absorbers material" [8]; "Drag polar analysis for a flying car model using wind tunnel test method [9]"; "Take off and landing performance analysis for a flying car model using wind tunnel test method" [10]; "Wheel retraction mechanism design of flying vehicle project" [11];
- James R. McBride (Ford Motor Company, MI, USA). "Role of flying cars in sustainable mobility" [12];
- Wolfgang Ott (San Jose State Univ, CA, USA). "HELIos, a VTOL flying car" [13];
- Gaofeng Pan (Beijing Institute of Technology, Beijing, China). "Flying Car Transportation System: Advances, Techniques, and Challenges" [14];

- Kaushik Rajashekara (University of Houston, TX, USA). "Flying Cars: Challenges and Propulsion Strategies" [15].

The following list shows the flying car *companies* selected for consideration.

- Personal Air and Land Vehicle (PAL-V) Europe;
- AeroMobil;
- Klein Vision.

2.2. Scientists' Claims

Each publication is reviewed below, while the intermediate results of the analysis are discussed in terms of possible modern scientific directions for each one.

- In [6], the authors discussed involving aerospace engineering in the automotive industry, which presents a major gap and has many limitations, but it does not rule out the possibility of making a flying car. They claim that wings are needed on this project to handle the air while the object is flying. The authors considered a few types of hinges, and linkage was found to be a good design to achieve a folding mechanism that fits into the structure of the wing, allowing a folded wing mechanism to be used when it fully expands for flying and that can fold when driving on regular streets [6]. **Scientific direction**: A folded wing mechanism for flying cars that can be used when driving on a normal street.
- In ref. [3], the authors briefly analyze the urban European context based on the available data from the 2019 report. Then, they present an inventory of existing flying car models as innovative solutions developed based on electric cars for green cities' transportation systems [3]. **Scientific direction**: Flying cars can offer new solutions for green urban mobility.
- The authors of [1] have represented current ground-based transportation systems, which are subjected to various challenges, including the high cost of infrastructure development, limited land space, and a growing urban population. Therefore, the automotive and aviation industries are collaborating to develop flying cars, also known as electric, vertical, takeoff, and landing aircraft (eVTOL). They believe that these eVTOLs will allow for rapid and reliable urban and suburban transportation. The safe operation of eVTOLs, which the authors discussed, will require well-developed wireless communication networks [1]. **Scientific direction**: Electric, vertical, takeoff, and landing aircraft (eVTOL) could be a rapid and reliable form of urban and suburban transportation.
- In ref. [5], the authors investigated one of the stages in designing a flying car, namely, determining the center of gravity. The center of gravity of the aircraft must be in the range of 15–25% of the mean aerodynamic wing chord so that the aircraft can fly stably. In a flying vehicle, the center of gravity is determined by arranging the components of the vehicle so that the center of gravity falls within that range [5]. **Scientific direction:** The results showed that the center of gravity when the fuel tank was in the middle of the vehicle was located at 444.7 mm in front of the forward center of gravity limits, and when the fuel tank was in the back of the vehicle, it was located at 366.05 mm in front of it. The second configuration tends to be more stable. Additionally, the canard is unable to balance the aircraft's lift force in a stalling condition.
- In ref. [13], the authors presented a single-seat, three-wheel, vertical takeoff and landing (VTOL) flying car concept of HELIos. It uses counter-rotating propellers enclosed in ducts. This technology eliminates the need for a tail rotor and makes the vehicle more compact. It needs no modification to switch between drive mode and flight mode [13]. **Scientific direction:** A vertical takeoff and landing flying car concept eliminates the need for a tail rotor and makes the vehicle more compact.
- Ref. [4] presents the theory of wings hanging on the sides, in which the driver plugs them into the fuselage when they want to fly, but not while the car is in use. The authors discuss most of the functions that this design serves in both road and sky

modes [4]. **Scientific direction:** A hanging-wing working mechanism which works in both road and sky modes.

- In ref. [12], the authors show that the interest and investment in electric vertical takeoff and landing aircraft (VTOL), commonly known as flying cars, have grown significantly. However, the authors note that the sustainability implications are unclear. They report a physics-based analysis of primary energy and greenhouse gas (GHG) emissions of VTOLs vs. ground-based cars that are efficient for tilt-rotor/duct/wing VTOLs when cruising but consume substantial energy for takeoff and climbing [12]; **Scientific direction:** For a vertical takeoff and landing aircraft, VTOL GHG emissions per passenger-kilometer are 52% lower than internal combustion engine vehicles (ICEVs) and 6% lower than battery electric vehicles (BEVs). VTOLs offer fast, predictable transportation and could have a role in sustainable mobility.

- In ref. [15], increasing interest in flying vehicles and the greater electrification of these vehicles with the advances in engines, electric motors, power converters, and communications is shown [15]. **Scientific direction:** The authors examine the challenges and requirements of developing a hybrid or a pure electric flying car, propulsion strategies for operations such as automobiles and airplanes, and vertical takeoff and landing (VTOL).

- The authors of [7] discuss the manufacturing method of flying cars' bodies. The focus of this research is on the technique of manufacturing flying cars' bodies by coating the body's frame with elastic fabric and spraying it. Two types of fabrics were used in this study, namely, water-absorbing fabric and waterproof fabric. A mold ring was used as the body frame, and the elastic fabric forming the surface was then sprayed with resin to make it harder. After the elastic fabric had hardened, fiberglass was added to strengthen the material. Then, a tensile test and a stress analysis were performed to determine the strength and suitability of the material [7]. **Scientific direction:** The method of manufacturing flying cars' bodies when using GFRP (waterproof fabric specimen) has better strength than when using GFRFP (water absorbing fabric specimen).

- In ref. [8], the flying car is technically considered an airplane with the added feature of being able to move properly on a highway. To fulfill its function as an aircraft, the chassis construction of the vehicle must be strong enough to withstand the loads while flying or while functioning as a vehicle. The vehicle chassis is able to withstand collisions as much as possible as a passive safety system in the event of an accident [8]. **Scientific direction:** Compared with the chassis of a space-frame type without impact absorbers and filled with impact absorbers by filling rigid polyurethane foam, the stiffness of the chassis can increase by 2.9% and can reduce the displacement UY by 2.9% (displacement in the Y-direction of the selected reference coordinate system by SolidWorks Simulation).

- In ref. [9], a model is created at one-seventh the real size and is tested with a wind tunnel. The maximum value of this comparison is crucial for the determination of the overall design. The values are collected based on wind tunnel testing. This research is quantitative with a descriptive design [9]. **Scientific direction:** Drag analysis of a flying car model was carried out.

- The authors of [11] investigate how to maximize the limited space of a vehicle in terms of fuel storage so that it can travel a long distance. Without increasing the size and weight of the vehicle, efficiency can be achieved by reducing drag [11]. **Scientific direction:** A wheel retraction system is designed that can reduce parasite drag to 24%.

- The authors of [16] present the idea of this VTOL propulsion system, which is to combine the fan propulsion system with the car wheel system attached to the suspension system. Therefore, a special design was needed to allow the suspension system to change the takeoff or landing direction of the fan propulsion system and to perform its function (car wheel support system for driving and steering), specifically, when the flying car moves on the road. The selection of material and the wishbone shape are important aspects of wishbone design to meet the design requirements. The wishbone shapes are made of steel tubes. After analysis through simulation, combined with material and variations on the wishbone shape, the combination design Upper Wishbone without Bracing and material AISI 1040 was found to be the best combination design. The shape design of an upper wishbone without bracing was chosen because it is lighter, easier to fabricate, and generated a smaller drag value than other designs. The material AISI 1040 was chosen because the price is cheaper than that of material AISI 4130, although it is a little more expensive than the material AISI 1020 [16]. **Scientific direction:** The takeoff or landing direction of the fan propulsion system can be changed, and it can perform its function (car wheel support system as driving and steering), specifically when the flying car moves on the road.

- The author of [17] presents the development of an aeroelastic analysis approach for the dynamic response of a Z-shaped folding wing. The structural model is established by the finite element method (FEM) and the component mode synthesis (CMS) method, accounting for the configuration-changing effects on inertial and stiffness characteristics. The aerodynamic model is directly built using the continuous-time state-space unsteady vortex lattice method (UVLM). The analysis results show that the folding and unfolding processes have opposite effects on both the aerodynamic load and the aeroelastic characteristics. Moreover, the effects become more significant with an increasing morphing rate [17]. **Scientific direction:** The folding and unfolding processes of the Z-shaped wing is presented in different configurations.
- In ref. [18], the authors provide information about the first documented manned all-electric VTOL flights, which occurred in 2011–2012. These flights were taken by a co-axial twin-rotor helicopter and a multi-copter. They were bare-bones aircraft with a solo pilot, enabled by lightweight permanent magnet synchronous motors and compact Li-ion batteries. They flew for only a few minutes (5–10 min) and lacked all attributes of a practical aircraft—payload, range, endurance, and safety—but proved the viability of electric trackless aircraft that, if realized on a practical scale, could open up new opportunities in aviation due to their many inherent strategic advantages [18]. **Scientific direction:** An all-electric VTOL structure and its influence on mass parameters and flight performance are presented.

2.3. Flying Cars' Features

The features of flying cars based on their inventors' scientific research interests are shown below.

The Curtiss Autoplane (Figure 1): in 1917, invented by Glenn Curtiss, who could be called the father of the flying car, unveiled the first attempt to build such a vehicle as roadable aircraft. It was shown at the Pan-American Aeronautic Exposition in New York City in February 1917. It made a few short hops before the entry of the United States into World War I in April 1917 ended the development of the Autoplane [16]. Although the vehicle was capable of lifting off the ground, it never achieved full flight. His aluminum Autoplane sported three wings that spanned 40 feet (12.2 m). The car's motor drove a four-bladed propeller at the rear of the car. The Autoplane never truly flew, but it did manage a few short hops [19].

Figure 1. Curtiss Autoplane [20].

Arrowbile (Figure 2): Developed by **Waldo Waterman** in 1937, the Arrowbile was a hybrid Studebaker aircraft. Like the Autoplane, it had a propeller attached to the rear of the vehicle. The three-wheeled car was powered by a typical 100-horsepower Studebaker engine. The wings detached for storage. A lack of funding killed the project [19].

Figure 2. Waterman Arrowbile [21].

Airphibian (Figure 3): **Robert Fulton**, who was a distant relative of the inventor of the steam engine, developed the Airphibian in 1946. Instead of adapting a car for flying, Fulton adapted a plane for the road. The wings and tail section of the plane could be removed to accommodate road travel, and the propeller could be stored inside the plane's fuselage. It took only five minutes to convert the plane into a car. The Airphibian was the first flying car to be certified by the Civil Aeronautics Administration, the predecessor of the Federal Aviation Administration (FAA). It had a 150-horsepower, six-cylinder engine and could fly 120 miles per hour and drive at 50 mph. Despite his success, Fulton could not find a reliable financial backer for the Airphibian [19].

Figure 3. Fulton Airphibian [22].

ConvAirCar (Figure 4): in the 1940s, **Consolidated-Vultee** developed a two-door sedan equipped with a detachable airplane unit. The ConvAirCar debuted in 1947 and offered one hour of flight and a gas mileage of 45 miles (72 km) per gallon. Plans to market the car ended when it crashed on its third flight [19].

Figure 4. Convair Model 118 ConvAirCar [23].

Avrocar (Figure 5): The first flying car designed for military use was the Avrocar, developed in a joint effort between the Canadian and British military. The flying-saucer-like vehicle was supposed to be a lightweight air carrier that would move troops to the battlefield [19].

Figure 5. Avro Canada VZ-9 Avrocar [24].

Aerocar (Figure 6): Inspired by the **Airphibian and Robert Fulton**, whom he had met years before, Moulton "Molt" Taylor created perhaps the most well-known and most successful flying car to date. The Aerocar was designed to drive, fly, and then drive again without interruption. Taylor covered his car with a fiberglass shell. A 10-foot (3-m) drive shaft connected the engine to a pusher propeller. It cruised at 120 mph (193 kph) in the air and was the second and last roadable aircraft to receive FAA approval. In 1970, Ford Motor Co. even considered marketing the vehicle, but the decade's oil crisis dashed those plans [19].

Figure 6. A 1949 Taylor Aerocar—N4994P [25].

3. Rationale for the Choice of the Object of Scientific Research

From the conception of the flying car until today, there has been little research progress, and the weight design requirements of cars and airplanes are so different that flying cars nowadays are gradually losing their car-driving function.

The design of flying cars can give full play to the advantages of both cars and airplanes, and the design of folding up the wings can be achieved to minimize the aerodynamic impact on the driving process, which will facilitate the pilot's driving and improve the comfort of the ride.

The internal frame of the fuselage can be made of a titanium alloy, which increases the cost to a certain extent, but due to the small size of the fuselage, the increase in cost is limited, whereas the strength of the key connection parts of the fuselage is greatly improved, as is the safety. The overall fuselage is streamlined, and due to the wide-body design, it will provide part of the lift to the whole flying car during the flight.

The engine can adopt a replaceable modular design, which currently uses gasoline power. Because the use of battery power cannot meet the speed and power needs, perhaps with current developments, the efficiency of the use of electrical energy can improve and the battery size can be greatly reduced, so the flying car can also use electrical energy to provide power.

Good airfoil and wing design are fundamental to ensuring flight and can be very useful from the point of view of additional equipment and design techniques. For example, flaps and slits are widely used on modern aircraft to increase lift at takeoff and increase drag upon landing. The torsional angle design applied to the wing also helps to improve wingtip stall speed. Wingtips are effective in reducing drag and fuel consumption during flight, but for some short ranges, the design requirements of a vehicle with added wingtips would increase the weight of the wing structure, and the additional weight would not compensate for the fuel savings from adding them.

The primary sources of loads on the wing are shear, bending moments, and torques caused by the lift generated by the wing. The load increases gradually from the wing tip to the root, supposing abrupt changes in the shear moment at the concentrated load and a linear relationship between the effect of the bending moment acting on the load at the wing root. The wing can be made of carbon fiber composite skin wrapped with an aluminum alloy beam, and the wing is divided into two parts, which can be folded in the lateral direction.

The fixing device between the engine and the fuselage frame and the connecting part of the wing, which are inevitably related to the structural strength and the flight life of the flying car, is also an important part of the flying car that should be investigated.

In addition, the car-driving process needs downforce to give the wheels and the ground better friction, in addition to an elevator to provide downforce, and the nose part of this flying car also needs to add a device to provide downforce.

4. Discussion

The last section discusses important scientific directions not fully covered or unexplored, along with possible future directions that were discovered in the research of the current article.

After analyzing scientific directions, well-known publications, and flying car concepts and catalogs, it can be observed that most of the research work is still at the stage of theoretical studies, and the cost and practicality of manufacturing flying cars have not yet been studied in depth.

Table 1 shows the comparison of scientific directions at the preliminary stage.

The difficulties faced by flying cars in terms of regulations, technology, and the market have also emerged. In terms of regulations, the current certification process of flying cars needs to pass the flight airworthiness certification and car driving certification, and temporarily borrowing from general aviation and ground car standards cannot fully represent the characteristics of flying cars, although low-altitude flight and ground driving have their own regulations, the effective linkage between them and the management of flying car operation mechanism has not been formed. In terms of technology, the current technology for flying cars in a low-altitude and intelligent driving environment perception, as well as decision-making and control technology, is not yet mature. On the market side, there is a lack of hardware and software facilities required for the operation of flying cars, and countries around the world have not yet achieved the normal operation of flying cars. In addition, research is only beginning on the topics of setting up a "route", avoiding and defining the responsibilities for accidents, air safety supervision, and methods of law enforcement. However, with the progress of technology and society, these problems will be solved.

Table 1. Comparison of scientific directions at the preliminary stage.

Scientific Directions	Results Have Been Achieved	Not Fully Covered Scientific Directions	Unexplored Scientific Directions
Folded wing mechanism for flying cars	yes	no	no
Determination of the position of the center of gravity of the flying car	no	yes	no
Vertical takeoff and landing flying car, eliminating the need for a tail rotor and making the vehicle more compact.	yes	no	no
Detachable wings	yes	no	no
Vertical takeoff and landing aircraft can reduce pollutant gas emissions and reduce the environmental impact of traffic	yes	no	no
Hybrid or electric power	no	yes	no
The manufacturing method of flying cars' bodies	no	yes	no
Wheel retraction system to reduce drag during flight	no	yes	no
Selection of materials for internal frame structure	yes	no	no
The need for the vehicle's undercarriage material selection and structural design to be robust enough to withstand the loads imposed during flight or when operating as a vehicle.	no	yes	no
Drag analysis of flying car models	yes	no	no
Airworthiness certification for flying cars	no	no	yes
Design of the body shape (to meet the aerodynamic requirements of both road travel and air flight)	no	no	yes
Traffic rules related to flying cars and flight routes	no	no	yes
Material selection for flying car glass	no	no	yes
Economic study of flying cars (e.g., economic comparison with cars and air travel, manufacturing costs)	no	no	yes
Comparison of the advantages and disadvantages of flying cars requiring runway takeoff and vertical takeoff and landing methods	no	yes	no

In fact, cars and airplanes belong to two different categories. The former need to undergo crash tests, need high-strength body rigidity, and, from the aerodynamic point of view, need to produce more downforce at high speed when the car is driving to improve the grip of the tires. Meanwhile, for airplanes, the lighter the weight, the better, the whole body is assembled with aluminum alloy and composite materials, and they do not need to undergo crash tests, whereas, from the aerodynamic point of view, the fuselage is designed to generate lift at high speed. However, one thing in common between cars and airplanes is that their fuselage must be streamlined to reduce drag.

In addition, the front windshield of a car must be made of glass, whereas the front windshield of a light aircraft is generally made of plastic to be less likely to break in a collision and to be lighter in weight.

The prospect of the development of flying cars is supported by the traction of objective demand and by technology. On the demand side, urban traffic congestion is the most common problem in major cities today, and traditional initiatives such as building viaducts and underground tunnels have not easily or effectively solved the problem of the traffic flow network effect of urban congestion, helicopters due to noise, safety limitations and limited application scenarios. In addition, urban traffic urgently needs to develop and use the three-dimensional low-altitude space of urban centers through flying cars in a way that is safe and environmentally friendly, so as to achieve three-dimensional intelligent transportation. In terms of technology, the great progress of electric vehicles and intelligent-networked vehicles has made good technical and industrial progress for the development of electric flying cars; the emergence of electric vertical take-off and landing (eVTOL), i.e., electric flying cars, makes high safety and low-noise urban air traffic possible.

5. Conclusions

After researching more than 20 scientific directions from results that have been achieved, scientific directions that are not fully covered, and unexplored scientific directions that are important for social life, a new scientific direction was formed, devoted to designing a flying car that combines the advantages of an airplane and a car, can take off and land on a normal highway for a short distance, and can help people reach their destination quickly and easily. Additionally, this will provide a significant direction for flying car investigations, and it should allow for rapidly increasing development in the field of flying car design since it will mean that scientists will not repeat the path already taken by others, but will take into account the gained experience. The flying car should be designed to provide people with a variety of faster and more convenient modes of transportation to choose from, and it can quickly change between car-driving mode and airplane-flying mode while looking over the ground during the flight to enjoy the scenery. When the road is not congested, it should be possible to use the car-driving mode, and when traffic congestion occurs or the route to the destination is more complicated due to the construction of the topography, then it should be possible to use the flight mode to quickly reach the destination, so that drivers can easily cross rivers and lakes and shorter mountain slopes, thus ultimately saving time and improving efficiency.

Author Contributions: Investigation, Z.Z.; data curation, L.B.; writing—original draft preparation, Z.Z.; writing—review and editing, A.H., L.B. and Z.Z.; supervision, A.H. All authors have read and agreed to the published version of the manuscript.

Funding: This research received no external funding.

Data Availability Statement: Data is contained within this article.

Conflicts of Interest: The authors declare no conflict of interest.

References

1. Yağmur, H.; Bayar, C.; Filiz, T.; Ertatligül, B.; Serbest, K. Conceptual Design of a Novel Roadable Flying Car. *J. Smart Syst. Res. (JOINSSR)* **2021**, *2*, 111–134.
2. Saeed, N.; Al-Naffouri, T.; Alouini, M.-S. Wireless Communication for Flying Cars. *Front. Commun. Netw.* **2021**, *2*, 1–9. [CrossRef]
3. Ivascu, L.; Mocan, A.; Robescu, D.; Draghici, A. The Flying Car—A Solution for Green Transportation. Advances in Smart Vehicular Technology, Transportation, Communication and Applications. In Proceedings of the Third International Conference on VTCA, Arad, Romania, 15–18 October 2019; pp. 145–158. [CrossRef]
4. Crow, S. A Practical Flying Car. In Proceedings of the 1997 World Aviation Congress, Anaheim, CA, USA, 13 October 1997. [CrossRef]
5. Mulyono, R.; Adhitya, M. Center of gravity analysis of a flying car. *AIP Conf. Proc.* **2020**, *2227*, 020032. [CrossRef]
6. Wartojo, B.; Adhitya, M. Folded wing mechanism for flying car. *AIP Conf. Proc.* **2020**, *2227*, 020034. [CrossRef]
7. Sudirja; Adhitya, M. Flying-cars body manufacturing using spraying elastic waterproof and water-absorbing frame fabric method. *AIP Conf. Proc.* **2018**, *2008*, 020007. [CrossRef]
8. Pamungkas, P.; Adhitya, M. Design and analysis of tubular space-frame chassis on flying car with impact absorbers material. *AIP Conf. Proc.* **2018**, *2008*, 020008. [CrossRef]
9. Lubyana, K.; Adhitya, M. Drag polar analysis for a flying car model using wind tunnel test method. *AIP Conf. Proc.* **2020**, *2227*, 020036. [CrossRef]
10. Pardede, W.; Adhitya, M. Take off and landing performance analysis for a flying car model using wind tunnel test method. *AIP Conf. Proc.* **2020**, *2227*, 020030. [CrossRef]
11. Mastiawan, M.; Adhitya, M. Wheel retraction mechanism design of flying vehicle project. *AIP Conf. Proc.* **2020**, *2227*, 020033. [CrossRef]

2. Kasliwal, A.; Furbush, N.; Gawron, J.; McBride, J.; Wallington, T.; De Kleine, R.; Kim, H.C.; Keoleian, G. Role of flying cars in sustainable mobility. *Nat. Commun.* **2019**, *10*, 1–10. [CrossRef] [PubMed]
3. Ott, W. HELIos, a VTOL flying car. AIAA and SAE. In Proceedings of the 1998 World Aviation Conference, Anaheim, CA, USA, 30 September 1998. [CrossRef]
4. Pan, G.; Alouini, M.-S. Flying Car Transportation System: Advances, Techniques, and Challenges. *IEEE Access* **2021**, *4*, 1–18. [CrossRef]
5. Ajashekara, K.; Wang, Q.; Matsuse, K. Flying Cars: Challenges and Propulsion Strategies. *IEEE Electrif. Mag.* **2016**, *4*, 46–57. [CrossRef]
6. Pratomo, W.; Adhitya, M.; Putra, P. Design and Analysis of Upper Wishbone for Suspension System on Vertical Take Off and Landing (VTOL) Propulsion System Flying Car. *AIP Conf. Proc.* **2018**, *2008*, 020009. [CrossRef]
7. Xie, C.; Chen, Z.; An, C. Study on the Aeroelastic Response of a Z-shaped Folding Wing During the Morphing Process. In Proceedings of the AIAA SCITECH 2022 Forum, San Diego, CA, USA, 3 January 2022. [CrossRef]
8. Mike Hirschberg, A.D. Current E-VTOL Concepts. Available online: https://vtol.org/files/dmfile/tvf.wg2.yr2017draft.pdf (accessed on 15 November 2022).
9. Bonsor, K. How Flying Cars Will Work. Available online: https://auto.howstuffworks.com/flying-car1.htm (accessed on 15 February 2023).
10. Curtiss Autoplane. Available online: https://patents.google.com/patent/US1294413 (accessed on 15 February 2023).
11. Aerofiles. Available online: http://www.aerofiles.com/_water.html (accessed on 17 February 2023).
12. Fulton Airphibian. Available online: http://www.nasm.si.edu/research/aero/aircraft/fulton.htm (accessed on 15 February 2023).
13. Convair Model 118. Available online: http://www.fiddlersgreen.net/models/aircraft/Aerocar.html (accessed on 15 February 2023).
14. Yenne, W. From Focke-Wulf to Avrocar. In *Secret Weapons of World War II: The Techno-Military Breakthroughs That Changed History*; Berkley Books: New York, NY, USA, 2003; 320 p.
15. 1949 Taylor Aerocar-N4994P. Available online: https://www.eaa.org/eaa-museum/museum-collection/aircraft-collection-folder/1949-taylor-aerocar---n4994p (accessed on 15 November 2022).

 *computation*

Article

Algorithm for Determining Three Components of the Velocity Vector of Highly Maneuverable Aircraft

Volodymyr Pavlikov [1,*], Eduard Tserne [1,*], Oleksii Odokiienko [1], Nataliia Sydorenko [1], Maksym Peretiatko [1], Olha Kosolapova [1], Ihor Prokofiiev [1], Andrii Humennyi [2] and Konstantin Belousov [3]

[1] Aerospace Radio-Electronic Systems Department, National Aerospace University "Kharkiv Aviation Institute", 61070 Kharkiv, Ukraine

[2] National Aerospace University "Kharkiv Aviation Institute", 61070 Kharkiv, Ukraine

[3] Spacecraft, Measuring Systems and Telecommunications Department, Yuzhnoye SDO, 49000 Dnipro, Ukraine

[*] Correspondence: v.pavlikov@khai.edu (V.P.); e.tserne@khai.edu (E.T.)

Abstract: We developed a signal processing algorithm to determine three components of the velocity vector of a highly maneuverable aircraft. We developed an equation of the distance from an aircraft to an underlying surface. This equation describes a general case of random spatial aircraft positions. Particularly, this equation considers distance changes according to an aircraft flight velocity variation. We also determined the relationship between radial velocity measured within the radiation pattern beam, the signal frequency Doppler shift, and the law of the range changing within the irradiated surface area. The models of the emitted and received signals were substantiated. The proposed equation of the received signal assumes that a reflection occurs not from a point object, but from a spatial area of an underlying surface. It fully corresponds to the real interaction process between an electromagnetic field and surface. The considered solution allowed us to synthesize the optimal algorithm to estimate the current range and three components $\{V_x, V_y, V_z\}$ of the aircraft's velocity vector $\vec{V}$. In accordance with the synthesized algorithm, we propose a radar structural diagram. The developed radar structural diagram consists of three channels for transmitting and receiving signals. This number of channels is necessary to estimate the full set of the velocity and altitude vector components. We studied several aircraft flight trajectories via simulations. We analyzed straight-line uniform flights; flights with changes in yaw, roll, and attack angles; vertical rises; and landings on a glide path and lining up with the correct yaw, pitch, and roll angles. The simulation results confirmed the correctness of the obtained solution.

Keywords: aircraft radio electronics; velocity measurement; height measurement; signal processing algorithm

Citation: Pavlikov, V.; Tserne, E.; Odokiienko, O.; Sydorenko, N.; Peretiatko, M.; Kosolapova, O.; Prokofiiev, I.; Humennyi, A.; Belousov, K. Algorithm for Determining Three Components of the Velocity Vector of Highly Maneuverable Aircraft. *Computation* **2023**, *11*, 35. https://doi.org/10.3390/computation11020035

Academic Editor: Ali Cemal Benim

Received: 9 November 2022
Revised: 11 February 2023
Accepted: 14 February 2023
Published: 15 February 2023

1. Introduction

Autonomy is one of the most important characteristics of aviation systems. It refers to the ability to receive all necessary information about both a flight (e.g., coordinates in space, velocity, and flight altitude) and the detected surrounding objects with aviation equipment. Autonomy also helps pilots make appropriate decisions regarding aircraft flight control and solve the tasks assigned to them. It allows pilots to considerably expand the area of effective aviation applications. Pilots can usually use different autonomy systems and levels to solve different problems. However, engineers are trying to design multifunctional systems [1,2]. An important feature of such systems is the wide range of measured parameters and characteristics of the studied objects with minimum on-board equipment [3]. However, the implementation of multifunctional systems often requires the development of new, more complex operation algorithms [4,5]. Implementing such algorithms by increasing the computational performance of programmable logic devices while reducing their power consumption is possible [6].

To implement the autonomy of the aircraft, constantly obtaining information about the current parameters of its movement is necessary. To determine this, scholars have synthesized various separate radio systems for navigation and traffic control [7–9]. Among these radio systems, the presence of gauges of three components of the aircraft velocity and flight altitude is fundamental for autonomous aircraft systems [10]. Traditionally, two different systems measure these parameters: a Doppler radar measures the full speed and angle of attack, and a radio altimeter measures the true altitude of the aircraft. At the same time, current radio electronic components and high-speed processing systems open new possibilities for the design of multifunctional systems. Such systems will allow researchers to minimize the radar volume and weight and simultaneously reduce energy consumption. Besides the technical aspects, creating a new structure of the single signal processing system and a new method to estimate aircraft movement parameters and positions in space is also relevant. Scholars have paid particular attention to helicopters, which are characterized by a higher degree of freedom in movement than airplane-type vehicles (hovering, vertical flight, backward flight, low-speed flight, and so on). This imposes remarkable limitations on problem solutions concerning signal processing algorithm synthesis for such a radar operation [10].

We conducted scientific research and synthesized a signal-processing algorithm for an advance-functional radar for measuring the full vector $\{V_x, V_y, V_z\}$ of velocity $\overrightarrow{V}$ and altitude with [10,11]. We were able to perform such a synthesis because of the achievement of the theory of the statistical optimization of radio engineering systems and the availability of modern computer systems to measure aircraft motion parameters in quasireal time [12–14].

2. Materials and Methods

2.1. Geometry of the Problem. The Equation of the Distance to the Underlying Surface for a Maneuvering Aircraft

To measure the velocity and altitude of a helicopter-type aircraft, developing a new radar with special orientation of the antennas and their radiation patterns is necessary. The radiation pattern beam must not be directed vertically down, but at some angle relative to the nadir direction. In this way, scholars will avoid zero or near zero values of the Doppler frequency shift [15]. Thus, when developing a new algorithm for the operation of a multifunctional radar, the position of the rays is assumed to already be fixed and rigidly related to the direction of the longitudinal axis of the aircraft. Figure 1 shows the primary geometry of the stated problem and describes the key parameters of a radiation pattern beam position in space.

In Figure 1, the transmitting A_1 and receiving A_2 antennas are assumed to be located in the center of the coordinate system $x'\,y'\,z'$, which is related to some point of the aircraft. $\Delta\theta_{pa}$ is the beam width of the radiation pattern. The parameters D'_1 and D'_2 describe the shapes of the antenna apertures. The radius vectors $\overrightarrow{r}'_1$ and $\overrightarrow{r}'_2$ characterize the distance from the phase center of the antenna to any point within its aperture. The angles μ, φ, ρ, η are used to determine the position of the radiation pattern in space. The radiation pattern irradiates a certain area of the underlying surface located in the coordinate system $x\,y\,z'$. Several main parameters characterize the range from the antenna system to the irradiated area. The first parameter is the distance R between the phase center of the transmitting antenna and the center of the underlying surface of the irradiated area. The range to the nearest R_{min} and farthest R_{max} point of the irradiated area and the current range R_{fl} to an arbitrary point P within this area are also separately determined. Separately distinguishing a point within the area of equal distance, which is conventionally shown within the irradiated area of the surface, is not possible.

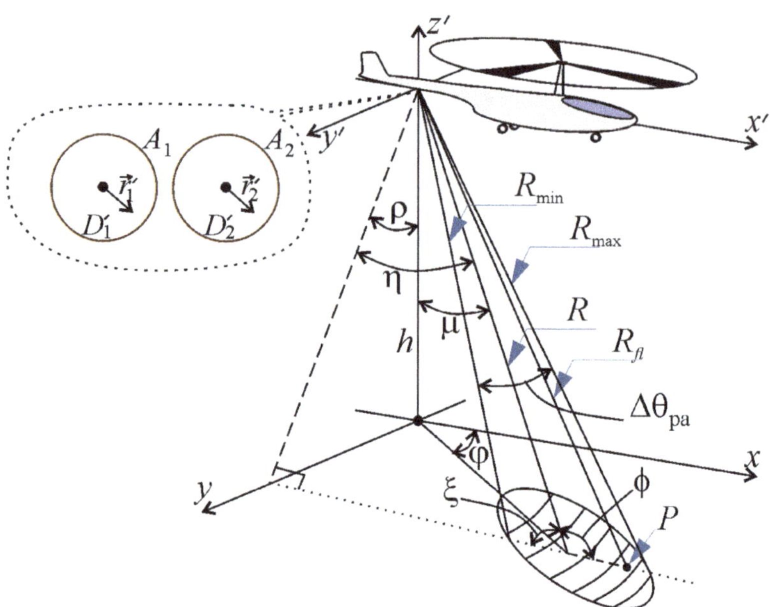

Figure 1. Primary geometry of the problem with marked physical parameters and geometric relationships for one radiation pattern beam.

A helicopter is a highly maneuverable aircraft [16]. Rapid changes in the helicopter position lead to essential variations in the antenna's radiation pattern direction in space. Considering this, the radiation pattern spot usually "slides" on the underlying surface along very complex trajectories. We used the geometry shown in Figure 2 to calculate the range with any changes in the yaw, pitch, and roll angles.

In Figure 2, the coordinate system 0xyz is related to the underlying surface. In this coordinate system, a helicopter is at a certain altitude h' in the center of the $0'x'y'z'$ coordinate system. This coordinate system is the initial one and corresponds to the case when the aircraft does not perform any maneuvers, i.e., the yaw, roll, and pitch angles are equal to zero. In this case, the radiation pattern direction coincides with the line R', which determines the current distance to the central irradiated point M' on the underlying surface x0y. The angle η to the axis $0'x'$ and the angle ρ to the axis $0'y'$ characterize the line R' position in space. The angle μ' between R' and the nadir direction and the angle φ' between the projection R' onto the underlying surface and the axis 0x are also used for calculations. When a yaw angle α' appears in an aircraft movement, the transition from the $0'x'y'z'$ coordinate system to $0x''y''z''$ occurs. At the same time, direction R'' focuses the radiation pattern on point M''. All changes that occur during the appearance of the yaw angle are marked with a superscript $\cdot''$. When yaw α' and pitch θ'' angles appear in the helicopter position, the transitions from the $0'x'y'z'$ coordinate system to $0x'''y'''z'''$ and to the variables with the superscript $\cdot'''$ occur. In the case of the presence of yaw α', pitch θ'', and roll χ''' angles, the coordinate system changes to $0x^{IV}y^{IV}z^{IV}$. To understand coordinate transforms and an R angular position evaluation more clearly, we introduced a conditional sphere with the center at point $0'$ on the geometry in Figure 2. Straight lines $R^{(\cdots)}$ cross the sphere at the points K', K'', K''', or K^{IV} depending on the current angles of the aircraft's position.

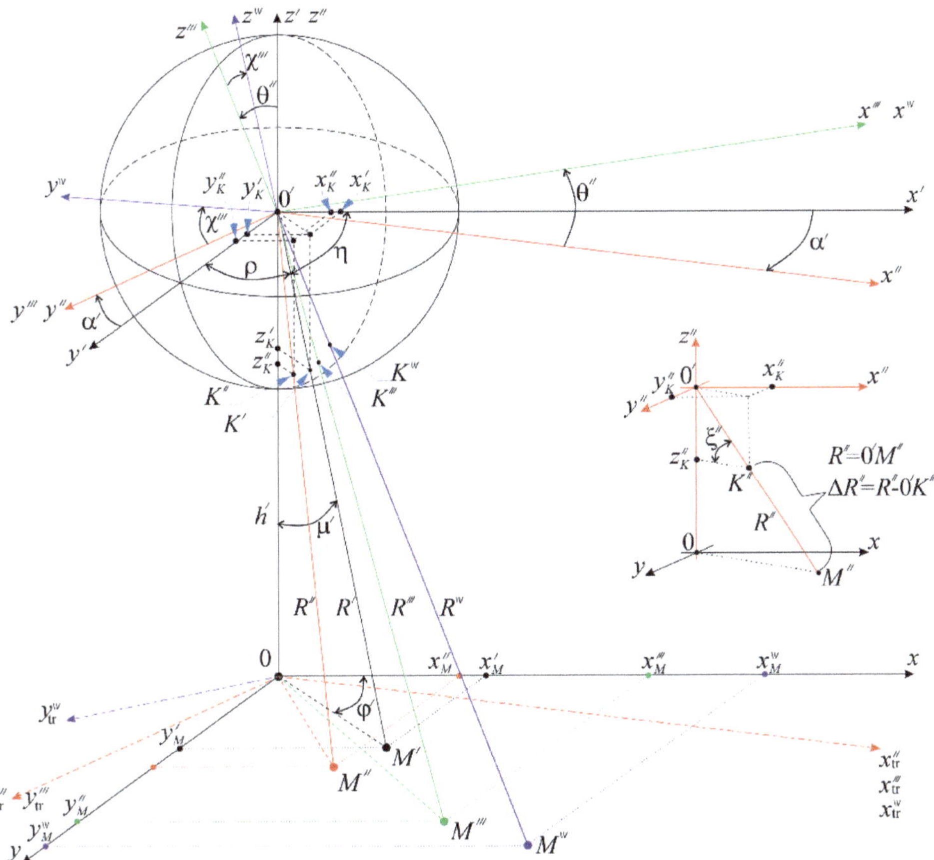

Figure 2. The geometry of the problem, which considers the presence of yaw, pitch, and roll angles in the aircraft movement.

Considering the geometry in Figure 2, possible changes in the yaw α', pitch θ'', and roll χ''' angles in the equation for the range R calculation can be written in the following form [10]:

$$R(\alpha', \theta'', \chi''', h, \mu', \varphi', t) = R\left(t, \vec{r}\right)$$

$$= \frac{h(t)\left[\begin{array}{c}\left(\begin{array}{c}h(t)tg(\mu')\cos(\varphi')\cos(\alpha'(t))\cos(\theta''(t)) - h(t)tg(\mu')\sin(\varphi')\sin(\alpha'(t))\cos(\theta''(t)) \\ -h'(t)\sin(\theta''(t))\end{array}\right)^2 \\ + \left(\begin{array}{c}h(t)tg(\mu')\cos(\varphi')\sin(\alpha'(t))\cos(\chi'''(t)) + h(t)tg(\mu')\sin(\varphi')\cos(\alpha'(t))\cos(\chi'''(t)) \\ +h(t)tg(\mu')\cos(\varphi')\cos(\alpha'(t))\sin(\theta''(t))\sin(\chi'''(t)) \\ -h(t)tg(\mu')\sin(\varphi')\sin(\alpha'(t))\sin(\theta''(t))\sin(\chi'''(t)) + h(t)\cos(\theta''(t))\sin(\chi'''(t))\end{array}\right)^2 \\ + \left(\begin{array}{c}h(t)tg(\mu')\cos(\varphi')\sin(\alpha'(t))\sin(\chi'''(t)) + h(t)tg(\mu')\sin(\varphi')\cos(\alpha'(t))\sin(\chi'''(t)) \\ -h(t)tg(\mu')\cos(\varphi')\cos(\alpha'(t))\sin(\theta''(t))\cos(\chi'''(t)) \\ +h(t)tg(\mu')\sin(\varphi')\sin(\alpha'(t))\sin(\theta''(t))\cos(\chi'''(t)) - h(t)\cos(\theta''(t))\cos(\chi'''(t))\end{array}\right)^2\end{array}\right]^{\frac{1}{2}}}{\left(\begin{array}{c}h(t)tg(\mu')\cos(\varphi')\sin(\alpha'(t))\sin(\chi'''(t)) + h(t)tg(\mu')\sin(\varphi')\cos(\alpha'(t))\sin(\chi'''(t)) \\ -h(t)tg(\mu')\cos(\varphi')\cos(\alpha'(t))\sin(\theta''(t))\cos(\chi'''(t)) \\ +h(t)tg(\mu')\sin(\varphi')\sin(\alpha'(t))\sin(\theta''(t))\cos(\chi'''(t)) - h(t)\cos(\theta''(t))\cos(\chi'''(t))\end{array}\right)}. \tag{1}$$

In Equation (1), we assume that, during the flight, the height and angles of the yaw, pitch, and roll can change (that is, they are functions of time).

2.2. Models of Transmitted and Received Signals. Observation Equation

In this section, we formulate the requirements of the transmitted signal. To measure the flight velocity and altitude, the signal must have an ambiguity function that provides a high resolution in distance and speed. Complex signals usually have such characteristics. For our task, choosing a signal with a linear frequency modulation is reasonable [17,18]:

$$s(t) = A(t)\mathrm{Re}\,\exp\left(j2\pi\left(f_0 t + \frac{\alpha t^2}{2}\right)\right), \tag{2}$$

where $A(t)$ is the signal envelop; f_0 is the emitted signal frequency; $\alpha = (F_{max} - F_{min})T^{-1}$; T is the pulse duration; and F_{max} and F_{min} are the maximum and minimum frequencies in the spectrum of operating frequencies, respectively.

After radiation, the reflection from the underlying surface and registration by the receiving antenna signal (2) has the following form:

$$s_i(t) = \mathrm{Re}\int\limits_{D_i} \left|\dot{G}\left(\vec{r}\right)\right|^2 \dot{F}\left(\vec{r}\right) A\left(t - t_d\left(t, \vec{r}\right)\right) \exp\left(j2\pi\left[f_0\left(t - t_d\left(t, \vec{r}\right)\right) + 0.5\alpha\left(t - t_d\left(t, \vec{r}\right)\right)^2\right]\right) d\vec{r} \tag{3}$$

where the integration occurs over the irradiated antenna radiation pattern $\dot{G}\left(\vec{r}\right)$ area D_i of the underlay surface (we assume that the radiation patterns of the transmission and receiving antennas are the same); $t_d\left(t, \vec{r}\right)$ is the signal delay time, which occurs as a result of its propagation from the antenna to the underlying surface elements and in the reverse direction; $\dot{F}\left(\vec{r}\right) = \left|\dot{F}\left(\vec{r}\right)\right| \exp\left(j\xi\left(\vec{r}\right)\right)$ is the complex reflection coefficient of the underlying surface; $\xi\left(\vec{r}\right)$ is the random phase offset that occurs when the signal is reflected from the underlying surface; and t is the current time.

The delay time in Equation (2) is determined according to the following equation:

$$t_d\left(t, \vec{r}\right) = 2R_{fl}\left(t, \vec{r}\right)c^{-1} \tag{4}$$

where $R_{fl}\left(t, \vec{r}\right)$ is the current range related to the velocity of the aircraft. $R_{fl}\left(t, \vec{r}\right)$ is analytically presented in the following form:

$$R_{fl}\left(t, \vec{r}\right) = \left(\begin{array}{l} R^2\left(t, \vec{r}\right) + \left(R\left(t, \vec{r}\right) \dfrac{\sin\left(\frac{\Delta\theta_{pa}}{2}\right)}{\sin\left(\frac{\pi}{2} - \mathrm{acos}\left(\dfrac{h}{R\left(t, \vec{r}\right)\cos\left(\mathrm{atan}\left(\frac{a(t)}{b(t)}\right)\right)}\right) - \frac{\Delta\theta_{pa}}{2}\right)} - Vt \right)^2 \\ -2R\left(t, \vec{r}\right)\left(R\left(t, \vec{r}\right) \dfrac{\sin\left(\frac{\Delta\theta_{pa}}{2}\right)}{\sin\left(\frac{\pi}{2} - \mathrm{acos}\left(\dfrac{h}{R\left(t, \vec{r}\right)\cos\left(\mathrm{atan}\left(\frac{a(t)}{b(t)}\right)\right)}\right) - \frac{\Delta\theta_{pa}}{2}\right)} - Vt \right)\cos(\phi), \end{array} \right)^{\frac{1}{2}}$$

where $\Delta\theta_{pa}$ is the beam width of the radiation pattern, and

$$\begin{array}{c} a(t) = h'(t)\mathrm{tg}(\mu')\sin(\varphi' + \alpha'(t))\cos(\chi'''(t)) + h'(t)\mathrm{tg}(\mu')\cos(\varphi' + \alpha'(t))\sin(\theta''(t))\sin(\chi'''(t)) \\ +h'(t)\cos(\theta''(t))\sin(\chi'''(t)), \\ b(t) = h'(t)\mathrm{tg}(\mu')\sin(\varphi' + \alpha'(t))\sin(\chi'''(t)) - h'(t)\mathrm{tg}(\mu')\cos(\varphi' + \alpha'(t))\sin(\theta''(t))\cos(\chi'''(t)) \\ -h'(t)\cos(\theta''(t))\cos(\chi'''(t)). \end{array}$$

The observation equation (at the receiver input) is written as an additive mixture of the received signal and the internal noise of the receiver:

$$u_i(t) = s_i(t) + n_i(t),\tag{5}$$

where $n_i(t)$ is white Gaussian noise with a power spectral density of $0.5N_0$. The index i specifies multichannel reception with several receivers. A simulation will justify the number of channels.

3. Results

3.1. Signal Processing Algorithm Synthesis. Simulation Results

We will solve the stated problem using the maximum likelihood method. To do this, we wrote the likelihood functional in the following form [19,20]:

$$p\left(u_i(t)\big|R_{fl,i}\left(t,\vec{r}\right)\right) = k\exp\left\{-\frac{1}{N_0}\int_0^T\left(u_i(t) - s_i\left(t,R_{fl,i}\left(t,\vec{r}\right)\right)\right)^2 dt\right\},\tag{6}$$

where k is some variable that does not depend on the parameter being estimated.

To define the new signal processing algorithm, we wrote the likelihood equation in the following form:

$$\frac{\delta\ln p\left(u_i(t)\big|R_{fl,i}\left(t,\vec{r}\right)\right)}{\delta R_{fl,i}\left(\vec{r}\right)} = \frac{\delta k}{\delta R_{fl,i}\left(\vec{r}\right)} - \frac{1}{N_0}\frac{\delta}{\delta R_{fl,i}\left(\vec{r}\right)}\int_0^T\left(u_i(t) - s_i\left(t,R_{fl,i}\left(t,\vec{r}\right)\right)\right)^2 dt = 0.\tag{7}$$

The estimated parameter $R_{fl,i}\left(\vec{r}\right)$ is related to the flight velocity. This relationship can be established by estimating the Doppler frequency from the exponent argument in Equation (3) considering Equation (4):

$$f_D = -\frac{2}{c}\frac{dR_{fl}\left(t,\vec{r}\right)}{dt}(f_0 + \alpha t) - \alpha\frac{2R_{fl}\left(t,\vec{r}\right)}{c}\left(1 - \frac{2}{c}\frac{dR_{fl}\left(t,\vec{r}\right)}{dt}\right).$$

and using a formal description of the Doppler frequency of a signal with linear frequency modulation:

$$f_D = \frac{2V_r\left(t,\vec{r}\right)}{c}(f_0 + \alpha t).$$

By equating the right-hand sides of the equations for f_D and neglecting unimpactful terms, we obtain the following relationship between velocity and the current range:

$$V_r\left(t,\vec{r}\right) = -\frac{dR_{fl}\left(t,\vec{r}\right)}{dt},\tag{8}$$

which has a clear physical meaning.

By solving the likelihood Equation (7), we obtain the following equation:

$$\int_0^T u_i(t)\frac{\delta}{\delta R_{fl,i}\left(\vec{r}\right)}s_i\left(t,R_{fl,i}\left(t,\vec{r}\right)\right)dt = \int_0^T s_i\left(t,R_{fl,i}\left(t,\vec{r}\right)\right)\frac{\delta}{\delta R_{fl,i}\left(\vec{r}\right)}s_i\left(t,R_{fl,i}\left(t,\vec{r}\right)\right)dt,\tag{9}$$

where the left part is the signal processing algorithm and the right part is the result of averaging the radar effect, which implements the signal processing algorithm. Here, the variational derivative of the signal can be represented as

$$\frac{\delta}{\delta R_{fl,i}\left(\vec{r}\right)} s_i\left(t, R_{fl,i}\left(t, \vec{r}\right)\right)$$

$$= \mathrm{Re}\int_{D_i} \dot{F}\left(\vec{r}\right)\left|\dot{G}\left(\vec{r}\right)\right|^2 \left[\frac{\delta}{\delta R_{fl,i}\left(\vec{r}\right)} A\left(t - \frac{2R_{fl,i}\left(t,\vec{r}\right)}{c}\right)\right]$$

$$\times \exp\left(j2\pi\left[f_0\left(t - \frac{2R_{fl,i}\left(t,\vec{r}\right)}{c}\right) + \frac{\alpha\left(t - 2R_{fl,i}\left(t,\vec{r}\right)c^{-1}\right)^2}{2}\right]\right) d\vec{r}$$

$$+ \mathrm{Re}(-j)2\pi\dot{F}\left(\vec{r}\right)\left|\dot{G}\left(\vec{r}\right)\right|^2 A\left(t - \frac{2R_{fl,i}\left(t,\vec{r}\right)}{c}\right)\left\{\frac{2f_0}{c} + \frac{2\alpha t}{c} - \frac{4\alpha R_{fl,i}\left(t,\vec{r}\right)}{c^2}\right\}$$

$$\times \exp\left(j2\pi\left[f_0\left(t - \frac{2R_{fl,i}\left(t,\vec{r}\right)}{c}\right) + \frac{\alpha\left(t - 2R_{fl,i}\left(t,\vec{r}\right)c^{-1}\right)^2}{2}\right]\right).$$

Having solved Equation (9), we can estimate the range $\hat{R}_{fl}\left(t, \vec{r}\right)$. Further, according to Equation (7) and the geometry shown in Figure 1, we obtained the radial velocity $V_r\left(t, \vec{r}\right)$ and the absolute velocity of the helicopter. The absolute velocity has the following form:

$$V_{hel}\left(t, \vec{r}\right) = \frac{\frac{d}{dt}R_{fl}\left(t, \vec{r}\right)}{\cos\left(\frac{\pi}{2} - \mathrm{acos}\left(\frac{h(t)}{R\left(t,\vec{r}\right)\cos\left(\mathrm{atan}\left(\frac{a(t)}{b(t)}\right)\right)}\right)\right)}. \tag{10}$$

The velocity Equation (9) does not show the direction, only its absolute value. Estimating all of the helicopter's velocity (V_x, V_y, V_z) components with one antenna radiation pattern is difficult, as shown in Figure 1. Therefore, let us consider three rays, as shown in Figure 3, and perform a corresponding simulation.

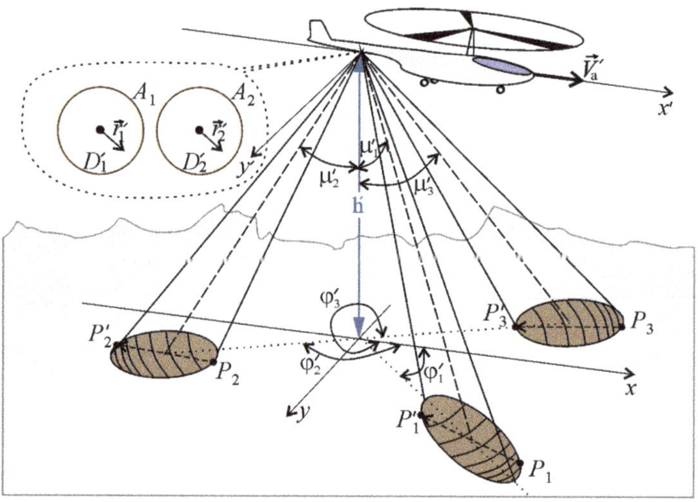

Figure 3. Geometry of the problem with three beams for the three components of the helicopter's velocity vector and altitude.

Figures 1 and 3 show a different number of radiation patterns. In Figure 3, we assumed that the apertures of the transmitting A_1 and receiving A_2 antennas formed three radiation pattern beams in different directions. All of the beams are characterized by the parameters introduced in Figure 1. The subscript under each variable indicates which radiation pattern is under consideration. The angle between the beams is always fixed to $2\mu'$ and, to simplify the calculations, satisfying the condition $\mu'_1 = \mu'_2 = \mu'_3 = \mu'$ is advisable.

In the case of three beams, the helicopter velocity is measured according to Equation (10) in every beam. At the same time, the two components of the velocity vector can be determined as follows:

$$\begin{aligned} V_x(t) &= \tfrac{V_1(t)-V_2(t)}{2}, \\ V_y(t) &= \tfrac{V_1(t)-V_3(t)}{2}, \end{aligned} \tag{11}$$

where the velocity subscript in the right part indicates the number of beams according to Figure 3.

Now, let us determine the equation for determining the flight height and the third velocity vector component. Based on the geometry (Figure 3), the following formula can be written to determine a helicopter's flight height:

$$h(t) = \frac{R_2(\cdot,t)R_3(\cdot,t)}{\sqrt{R_2^2(\cdot,t) + R_3^2(\cdot,t) - 2R_2(\cdot,t)R_3(\cdot,t)\cos(2\mu')}}. \tag{12}$$

The third component of the velocity vector can be found in the following form:

$$V_z(t) = \frac{dh(t)}{dt}. \tag{13}$$

The beams can be placed in any way, but to measure low velocities, placing them mirror-like relative to the longitudinal and transverse axis is advisable, as shown in Figure 3. With this arrangement, doubling (see the numerators in Equation (11)) each component of the velocity vector while taking the measurements is possible and allows one to determine low velocities. We can prove this with simulations, the results of which are shown in Figures 4–7. For the first simulation, we utilized the following initial data: the law of height change was $h(t) = 1000 + 2t$; the glide angle was $\alpha'(t) = 0$; the angle of attack was $\theta''(t) = 0$; the pitch angle was $\chi'''(t) = 0$; and the velocity components were $V_x = 10\,\text{m/s}$ and $V_y = 0\,\text{m/s}$. Figure 4 shows the changes in the current range for each of the beams. Figure 5 describes the measured velocities along each of the beams. Figure 6 depicts three velocity components and Figure 7 shows an analysis of the height estimation.

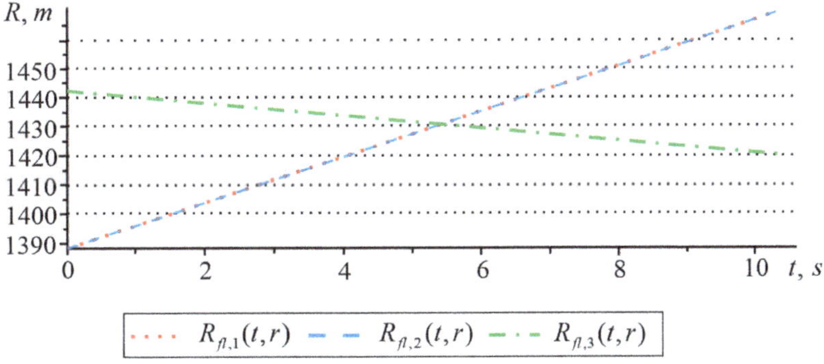

Figure 4. Variation in current distances along each of the beams (first simulation).

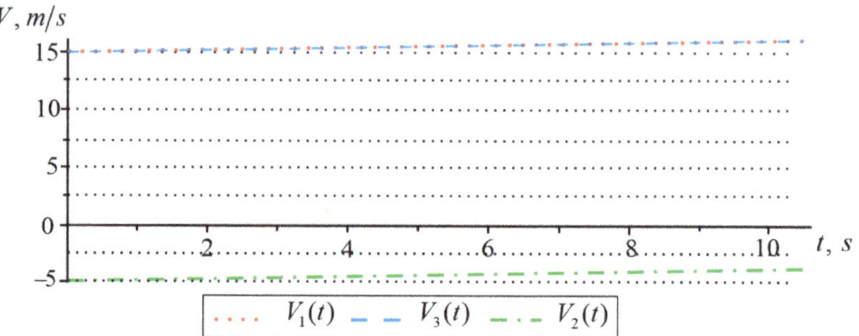

Figure 5. Measured absolute velocities along each of the beams (first simulation).

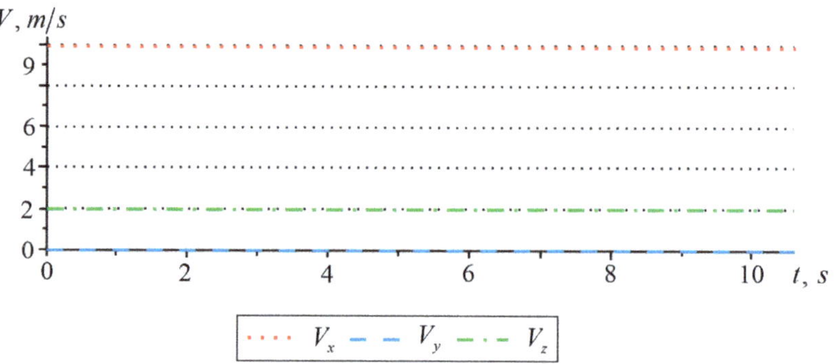

Figure 6. Estimates of the velocity vector components (first simulation).

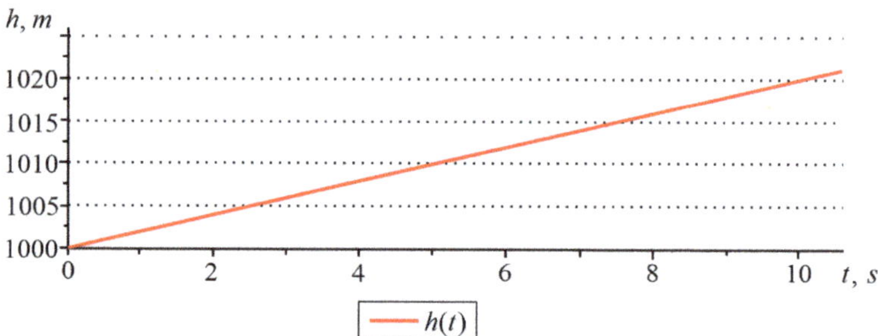

Figure 7. Height estimation (first simulation).

In the case of the flight height uniformly increasing, the absolute values of the velocity along each of the beams (Figure 5) did not correspond to the absolute velocity value. However, by analyzing Figure 6, one can see that the estimation of the velocity vector components during uniform movement was close to the true values, and the absolute error was less than 1.5%. We performed a range estimation according to Equation (12). The obtained results also corresponded to the stated initial data in the simulation.

Considering more complex helicopter flight trajectories when the yaw, attack, and pitch angles differ from zero is of interest. For this simulation, we used the following initial data: the height change law was $h(t) = 1000 + 2t$; the functional dependencies of the glide angle were $\alpha'(t) = (2t)°$; the attack angle was $\theta''(t) = (0.2t)°$; the pitch angle was $\beta'''(t) = 5°$; and the velocity vector modulus was $V = 10.6\,m/s$. Figures 8–11 depict the obtained graphs, which are similar to the results shown in Figures 4–7.

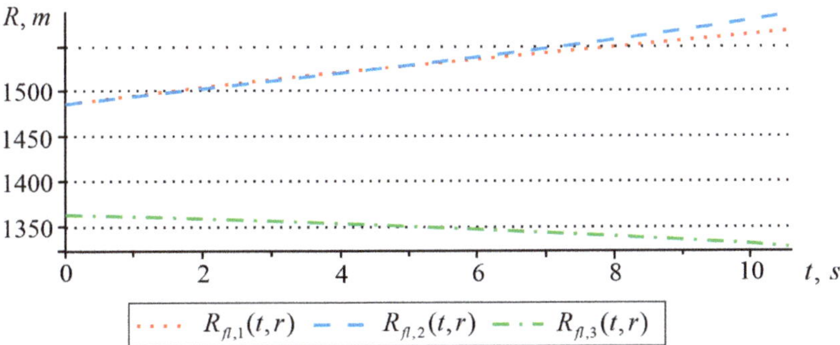

Figure 8. Measured change in current distances along each of the beams (second simulation).

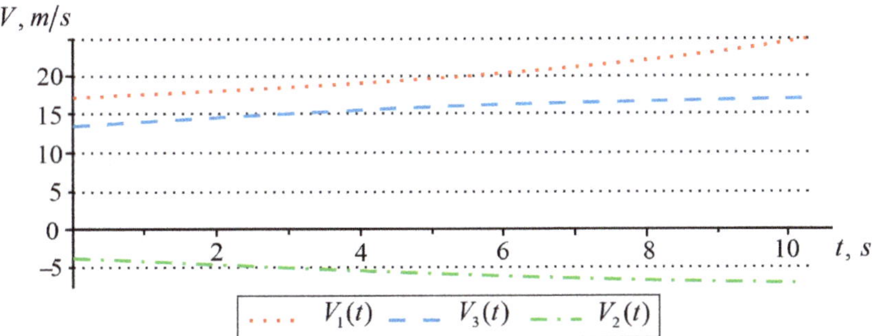

Figure 9. Measured absolute velocities along each of the beams (second simulation).

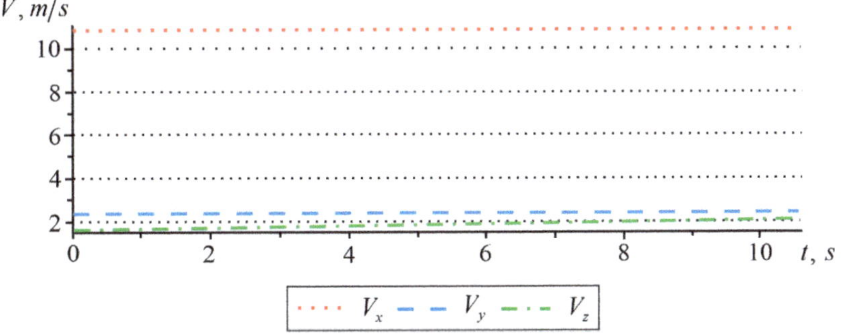

Figure 10. Estimates of the velocity vector components (second simulation).

In the presence of changes in the yaw, attack, and pitch angles during movement, the components of the velocity vector had more complex dependencies than was observed during uniform movement. This was because the ranges along each beam nonlinearly changed.

Considering the case of a vertical takeoff of a helicopter ($h(t) = V_z t$, $V_z = 3\,\mathrm{m/s}$, $t \geq 0$, $V_x = V_y = 0$). Figure 12 shows the estimated height and Figure 13 shows the estimates of the three velocity components.

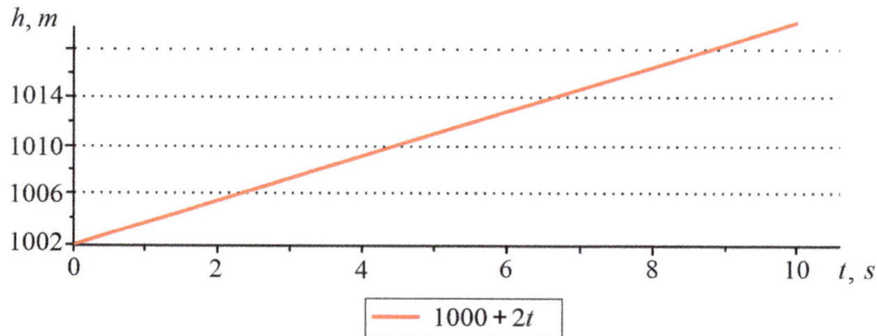

Figure 11. Height estimation (second simulation).

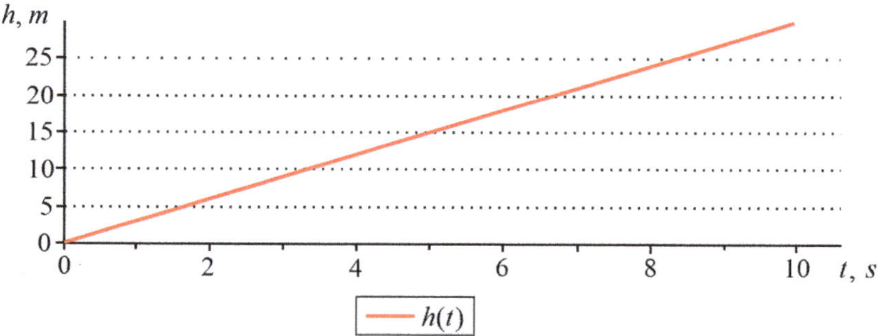

Figure 12. Estimation of height during vertical takeoff.

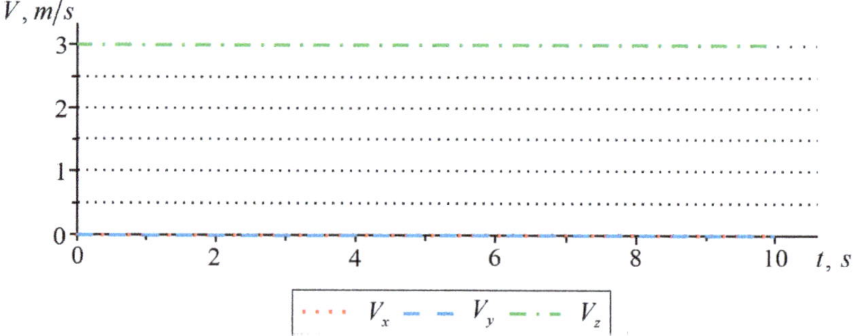

Figure 13. Estimates of the components of the velocity vector during vertical takeoff.

Now, we consider the landing option [21]. Figure 14 shows the simulated changes in the yaw, roll, and attack angles while landing. The height of the aircraft linearly decreased according to the law $h(t) = 50 - 5t$, as shown in Figure 15. The modulus of the velocity vector at the beginning of the simulation was 110 m/s. Figures 16 and 17 show the variation in distances along each of the rays and the changing components of the velocity vector.

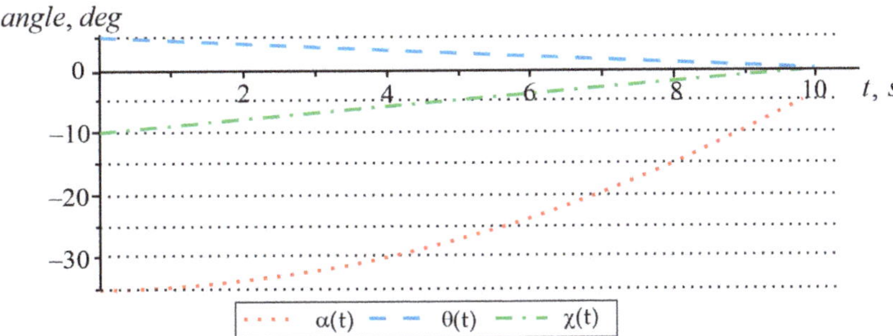

Figure 14. Changes in the angles of sliding, roll, and attack while a helicopter is landing.

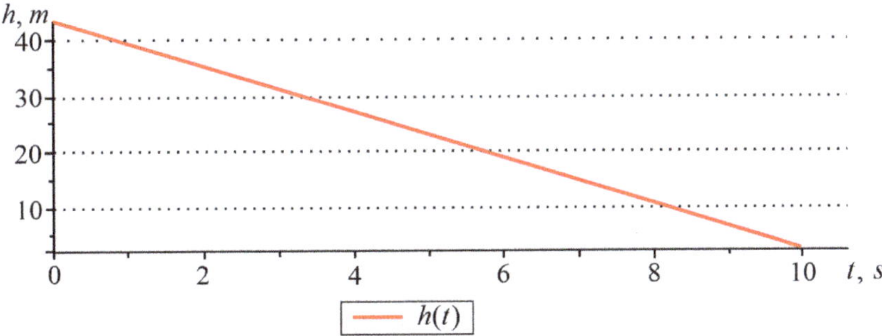

Figure 15. The height while a helicopter is landing.

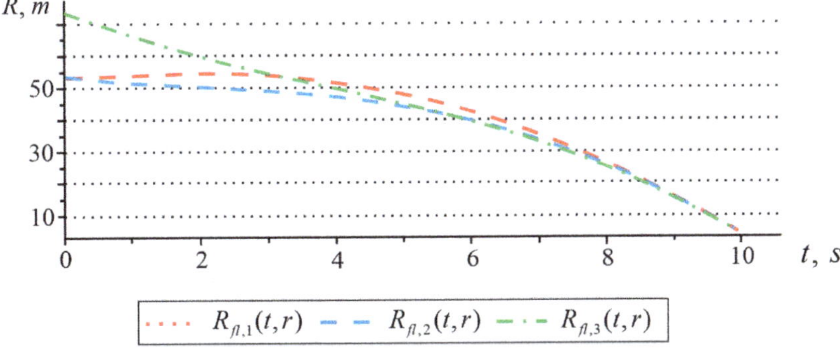

Figure 16. Calculated changes in distances along each of the beams while a helicopter is landing.

The obtained simulation results confirmed the adequacy of the performance of the derived equations and algorithms.

3.2. Structural Diagram of the Radar

We developed a structural diagram (Figure 18) of the radar by measuring a full vector of the velocity and flight height components according to the calculations.

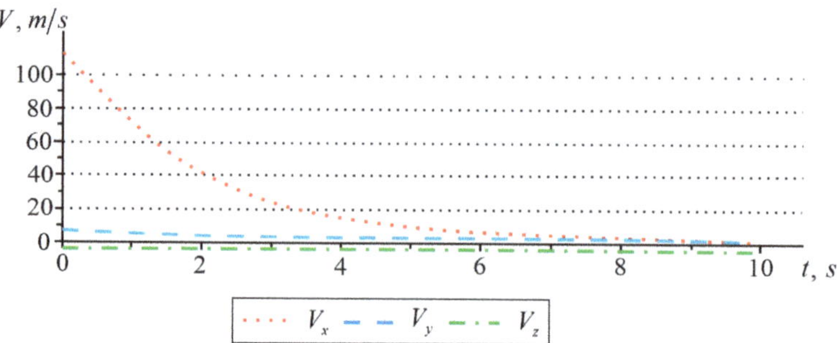

Figure 17. Estimates of the velocity vector components while a helicopter is landing.

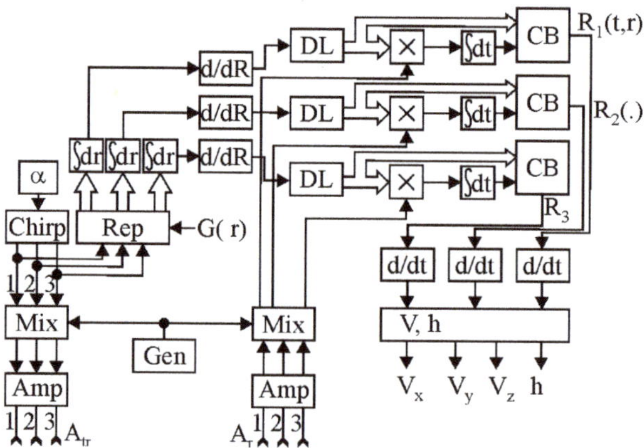

Figure 18. Structural diagram of the radar for the velocity vector components and altitude measurement.

The scheme has three transmitting and three receiving channels. It works as follows. The transmitter is represented by a signal generator with linear frequency modulation (chirp block with a setting block), mixers (mix block), and amplifiers (amp block). We assumed that three different signals were formed (three beams are enough to solve the problem), and we separated them in terms of frequency to exclude the possibility of channel mixing. The signals were transmitted through antennas A_{tr}. Antennas A_r received the signals reflected by the underlying surface. After amplification in the amp block, the signals were transferred to an intermediate frequency in the mix block. Further, the signals were multiplied with the derivatives of the reference signals in the «×» blocks and then sent to the integrators. Derivatives of the reference signals entered the second input of the multipliers. Blocks of repeaters (Rep) were involved during the formation of the reference signals, which resulted in the reflection of the signal from a spatially extended area, integrators on spatial coordinates, blocks of variational derivatives, d/dR calculations, and delay lines (DLs). Then, from the outputs of the integrators, the processes were sent to the comparison block (CB), and distance estimates were formed at the outputs as a function of spatial and temporal coordinates. These range estimates passed through the differentiation blocks were converted into radial velocities (see Equation (8)) along each of the radiation patter beams and were sent to the block to calculate the components of the velocity vector and flight height according to Equations (11)–(13).

4. Discussion

One of the promising directions for modern aircraft development is increasing their autonomy, which requires the simultaneous real-time monitoring of many aircraft parameters and the use of a considerable number of different sensors and systems. At the same time, aircrafts have a remarkable limitation regarding the payload, which can be installed on board without deteriorating the tactical and technical characteristics of the machine. Therefore, developing multifunctional on-board systems that can simultaneously monitor several parameters and characteristics of the aircraft using a minimum amount of equipment is an urgent task. A possible direction for the creation of such systems is the development of new and advanced signal-processing algorithms. We obtained algorithms to calculate the full vector of velocities $\{V_x, V_y, V_z\}$ and flight altitude of an aircraft. These algorithms can be implemented in one multifunctional system.

We obtained the algorithm in Equation (9) as a result of the solution to likelihood Equation (7), considering the relationship between the velocity and the current range Equation (8). This algorithm allows one to calculate the velocity of an aircraft based on the measured range $\hat{R}_{fl}\left(t, \vec{r}\right)$. At the same time, in the presence of only one beam of the antenna radiation pattern, the obtained algorithm does not allow one to separately determine the aircraft velocity components $\{V_x, V_y, V_z\}$. Therefore, we created a transition to the geometry of the problem depicted in Figure 3, which involves the simultaneous use of three radiation patterns. The simultaneous use of three beams allowed us to directly calculate the components of the velocity along the y and x axis according to Equation (10), as well as the current flight height according to Equation (11). We calculated the third component of the velocity based on the height according to Equation (12).

We confirmed the general efficiency of the proposed algorithms with the simulation. Scholars should pay special attention to the calculated distances and velocities along each of the radiation pattern beams. Thus, during the second simulation, with a linear change in the glide $\alpha'(t)$ and attack $\theta''(t)$ angles, as well as height $h(t)$, we observed nonlinear changes in distances and velocities, as depicted in Figures 8 and 9. This fully corresponded to the real nonlinear range change dependence with a linear change in the vertical angle at which the radar probed the surface. Figures 5, 9 and 17 highlight the velocity $V_2(t)$, which had a negative value. This can be explained by the fact that the second radiation pattern beam on the geometry in Figure 3 was directed to the opposite side in relation to the aircraft movement direction. That is, the Doppler frequency behind this beam had a negative value, and it led to a negative measured velocity. The same effect could occur in real velocity measurement radars, and this also confirmed the correctness of the obtained results.

Based on the calculations and geometry of the problem shown in Figure 3, we developed a structural diagram of the on-board meter of the velocity and height vector components. Considering modern achievements in the field of radio element bases, the proposed scheme can be fully implemented. At the same time, when implementing such a system, paying attention to prospective frequency ranges, which are currently in the millimeter wave range, is advisable [22,23]. The implementation of an on-board system in this range in the future will allow scholars to remarkably reduce the weight and dimensions of the result system, which is fundamental for use on an aircraft, and the ability to measure the current height of the carrier will allow researchers to replace classic on-board radio altimeters.

We plan to continue this research in several directions in the future. First, we will obtain marginal errors when using the obtained algorithms for the velocity vector and height estimation. Additionally, we are working to create a radar that implements the proposed algorithms. As a result, we plan to determine the technical requirements for the hardware and signal processing speed and to obtain the first practical results of the velocity vector and flight altitude estimation on a real helicopter.

5. Conclusions

We explored the problem of synthesizing a signal-processing algorithm to estimate a current beam range, three components $\{V_x, V_y, V_z\}$ of a helicopter's velocity vector, and flight height. As a result, we solved several issues. We derived the equation of the distance to the underlying surface along the radiation pattern beam. Its general form allowed us to analyze the distance for any carrier position above the underlying surface. This is an important result because modern helicopters have many diverse flight modes from hovering in place to flying in the reverse direction. For the first time, we propose the idea of calculating aircraft velocity vector components by measuring the distances along the beams. The results of numerical modeling showed that three transmitting and receiving channels were enough to solve the velocity vector component and altitude problem. We formulated the requirements for the signal type selection. We propose the use of a waveform with linear frequency modulations that allows one to obtain high-resolution distance and velocity measurements. We obtained equations and a structural diagram of a multifunctional radar for the flight height and three velocity vector components.

Author Contributions: Conceptualization, V.P., I.P. and O.O.; methodology, V.P. and O.O.; software, E.T. and M.P.; validation, A.H., K.B. and O.K.; formal analysis, V.P. and A.H.; investigation, O.O. and I.P.; resources, E.T.; data curation, N.S.; writing—original draft preparation, V.P. and N.S.; writing—review and editing, O.K. and K.B.; visualization, M.P.; supervision, A.H.; project administration, V.P.; funding acquisition, K.B. All authors have read and agreed to the published version of the manuscript.

Funding: This work was funded by the Ministry of Education and Science of Ukraine, and the state registration numbers of the projects are 0122U200469 and 0121U109598.

Institutional Review Board Statement: Not applicable.

Informed Consent Statement: Not applicable.

Data Availability Statement: Not applicable.

Conflicts of Interest: The authors declare no conflict of interest.

References

1. Nikolić, D.; Drajić, D.; Čiča, Z. Multifunctional Radars as Primary Sensors in IoT Based Safe City Solutions. In Proceedings of the 2021 15th International Conference on Advanced Technologies, Systems and Services in Telecommunications (TELSIKS), Nis, Serbia, 20–22 October 2021; pp. 273–278. [CrossRef]
2. Gurbuz, A.C.; Mdrafi, R.; Cetiner, B.A. Cognitive Radar Target Detection and Tracking With Multifunctional Reconfigurable Antennas. *IEEE Aerosp. Electron. Syst. Mag.* **2020**, *35*, 64–76. [CrossRef]
3. Fan, Z.; Shi, H.; Dong, S.; Zhang, L.; Wang, C.; Wu, Y. Design of a Space-borne Multifunctional Reconfigurable Terminal. In Proceedings of the 2020 IEEE 6th International Conference on Computer and Communications (ICCC), Chengdu, China, 11–14 December 2020; pp. 960–964.
4. Müller, T.; Marquardt, P.; Brüggenwirth, S. A Load Balancing Surveillance Algorithm For Multifunctional Radar Resource Management. In Proceedings of the 2019 20th International Radar Symposium (IRS), Ulm, Germany, 26–28 June 2019; pp. 1–9. [CrossRef]
5. Fränken, D.; Liegl, A. Integration of Multi-Band Passive and Multi-Functional Active Radar Data. In Proceedings of the 2022 IEEE Radar Conference (RadarConf22), New York, NY, USA, 21–25 March 2022; pp. 1–6. [CrossRef]
6. Kiat, W.P.; Mok, K.M.; Lee, W.K.; Goh, H.G.; Achar, R. An energy efficient FPGA partial reconfiguration based micro-architectural technique for IoT applications. *Microproc. Microsys.* **2020**, *73*, 102966. [CrossRef]
7. Rohr, D.; Studiger, M.; Stastny, T.; Lawrance, N.R.J.; Siegwart, R. Nonlinear Model Predictive Velocity Control of a VTOL Tiltwing UAV. *IEEE Robot. Autom. Lett.* **2021**, *6*, 5776–5783. [CrossRef]
8. Lins, R.G.; Givigi, S.N.; Kurka, P.R.G. Velocity Estimation for Autonomous Vehicles Based on Image Analysis. *IEEE Trans. Instrum. Meas.* **2016**, *65*, 96–103. [CrossRef]
9. Mohamed, S.A.S.; Haghbayan, M.-H.; Westerlund, T.; Heikkonen, J.; Tenhunen, H.; Plosila, J. A Survey on Odometry for Autonomous Navigation Systems. *IEEE Acc.* **2019**, *7*, 97466–97486. [CrossRef]
10. Pavlikov, V.; Volosyuk, V.; Tserne, E.; Sydorenko, N.; Prokofiev, I.; Peretiatko, M. Radar for Aircraft Motion Vector Components Measurement. In Proceedings of the 2022 IEEE 2nd Ukrainian Microwave Week, Kharkiv, Ukraine, 14–18 November 2022.
11. Volosyuk, V.; Pavlikov, V.; Zhyla, S.; Tserne, E.; Odokiienko, O.; Humennyi, A.; Popov, A.; Uruskiy, O. Signal Processing Algorithm for Monopulse Noise Noncoherent Wideband Helicopter Altitude Radar. *Computation* **2022**, *10*, 150. [CrossRef]
12. Zhyla, S.; Volosyuk, V.; Pavlikov, V.; Ruzhentsev, N.; Tserne, E.; Popov, A.; Shmatko, O.; Havrylenko, O.; Kuzmenko, N.; Dergachov, N.; et al. Statistical synthesis of aerospace radars structure with optimal spatio-temporal signal processing, extended observation area and high spatial resolution. *Radioel. Comp. Syst.* **2022**, *1*, 178–194. [CrossRef]

3. Volosyuk, V.; Pavlikov, V.; Nechyporuk, M.; Zhyla, S.; Kosharskyi, V.; Tserne, E. Structure Optimization of the Multi-Channel On-Board Radar with Antenna Aperture Synthesis and Algorithm for Power Line Selection on the Background of the Earth Surface. In Proceedings of the 2020 IEEE International Conference on Problems of Infocommunications, Science and Technology (PIC S&T), Kyiv, Ukraine, 6–9 October 2020; pp. 775–778. [CrossRef]

4. Volosyuk, V.; Zhyla, S.; Pavlikov, V.; Tserne, E.; Sobkolov, A.; Shmatko, O.; Belousov, K. Mathematical description of imaging processes in ultra-wideband active aperture synthesis systems using stochastic sounding signals. *Radioel. Comp. Syst.* **2021**, *4*, 166–182. [CrossRef]

5. Fiedler, H.; Boerner, E.; Mittermayer, J.; Krieger, G. Total Zero Doppler Steering—A New Method for Minimizing the Doppler Centroid. *IEEE Geosc. Rem. Sens. Lett.* **2005**, *2*, 141–145. [CrossRef]

6. *Helicopter Flying Handbook (FAA-H-8083-21B)*; United States Department of Transportation, Federal Aviation Administration: Oklahoma City, OK, USA, 2019.

7. Chang, C.-J.; Bell, M.R. Hybrid Filters for Delay-Doppler Resolution Enhancement in Chirp Radar Systems. In Proceedings of the 2021 55th Asilomar Conference on Signals, Systems, and Computers, Pacific Grove, CA, USA, 31 October–3 November 2021; pp. 1053–1060. [CrossRef]

8. Vannicola, V.C.; Hale, T.B.; Wicks, M.C.; Antonik, P. Ambiguity function analysis for the chirp diverse waveform (CDW). In Proceedings of the IEEE 2000 International Radar Conference [Cat. No. 00CH37037], Alexandria, VA, USA, 12 May 2000; pp. 666–671. [CrossRef]

9. Rossi, R.J. *Mathematical Statistics: An Introduction to Likelihood Based Inference*; John Wiley & Sons: New York, NY, USA, 2018.

10. Volosyuk, V.; Kravchenko, V. *Theory of Radio-Engineering Systems of Remote Sensing and Radar*; Fizmatlit: Moscow, Russia, 2008.

11. Zhang, G.; Bian, D.; Zhang, W.; Wu, X.; Zhang, Z.; Yu, H. The research and independent on autonomous safe landing for unmanned helicopter. In Proceedings of the 2017 IEEE International Conference on Unmanned Systems (ICUS), Beijing, China, 7–29 October 2017; pp. 434–437. [CrossRef]

12. Ikram, M.Z.; Ahmad, A.; Wang, D. High-accuracy distance measurement using millimeter-wave radar. In Proceedings of the 2018 IEEE Radar Conference (RadarConf18), Oklahoma City, OK, USA, 23–27 April 2018; pp. 1296–1300. [CrossRef]

13. Futatsumori, S.; Amielh, C.; Miyazaki, N.; Kobayashi, K.; Katsura, N. Helicopter Flight Evaluations of High-Voltage Power Lines Detection Based on 76 GHz Circular Polarized Millimeter-Wave Radar System. In Proceedings of the 2018 15th European Radar Conference (EuRAD), Madrid, Spain, 26–28 September 2018; pp. 218–221. [CrossRef]

 *computation*

Article

Numerical Assessment of Terrain Relief Influence on Consequences for Humans Exposed to Gas Explosion Overpressure

Yurii Skob [1,*], **Sergiy Yakovlev** [2], **Kyryl Korobchynskyi** [1] **and Mykola Kalinichenko** [3]

[1] Mathematical Modelling and Artificial Intelligence Department, National Aerospace University "Kharkiv Aviation Institute", 61070 Kharkiv, Ukraine
[2] Institute of Information Technology, Lodz University of Technology, 90-924 Lodz, Poland
[3] Aircraft Engine Manufacturing Department, National Aerospace University "Kharkiv Aviation Institute", 61070 Kharkiv, Ukraine
* Correspondence: y.skob@khai.edu

Abstract: This study aims to reconstruct hazardous zones after the hydrogen explosion at a fueling station and to assess an influence of terrain landscape on harmful consequences for personnel with the use of numerical methods. These consequences are measured by fields of conditional probability of lethal and ear-drum injuries for people exposed to explosion waves. An "Explosion Safety®" numerical tool is applied for non-stationary and three-dimensional reconstructions of the hazardous zone around the epicenter of the explosion of a premixed stoichiometric hemispheric hydrogen cloud. In order to define values of the explosion wave's damaging factors (maximum overpressure and impulse of pressure phase), a three-dimensional mathematical model of chemically active gas mixture dynamics is used. This allows for controlling the current pressure in every local point of actual space taking into account the complex terrain. This information is used locally in every computational cell to evaluate the conditional probability of such consequences for human beings, such as ear-drum rupture and lethal outcome, on the basis of probit analysis. To evaluate the influence of the landscape profile on the non-stationary three-dimensional overpressure distribution above the Earth's surface near the epicenter of an accidental hydrogen explosion, a series of computational experiments with different variants of the terrain is carried out. Each variant differs in the level of mutual arrangement of the explosion epicenter and the places of possible location of personnel. The obtained results indicate that any change in working-place level of terrain related to the explosion's epicenter can better protect personnel from the explosion wave than evenly leveled terrain, and deepening of the explosion epicenter level related to working place level leads to better personnel protection than vice versa. Moreover, the presented coupled computational fluid dynamics and probit analysis model can be recommended to risk-managing experts as a cost-effective and time-saving instrument to assess the efficiency of protection structures during safety procedures.

Keywords: gas mixtures; explosion pressure wave; overpressure; impulse; probit analysis; probit function; negative impact conditional probability

Citation: Skob, Y.; Yakovlev, S.; Korobchynskyi, K.; Kalinichenko, M. Numerical Assessment of Terrain Relief Influence on Consequences for Humans Exposed to Gas Explosion Overpressure. *Computation* **2023**, *11*, 19. https://doi.org/10.3390/computation11020019

Academic Editor: Ali Cemal Benim

Received: 28 December 2022
Revised: 24 January 2023
Accepted: 27 January 2023
Published: 30 January 2023

1. Introduction

It is well known that hydrogen is one of the most explosive gases [1]. Therefore, increasing the use of hydrogen in the industry creates high risks of accidents, which lead to severe social and economic consequences [2]. Even nonsignificant violations of safety precautions or accidental equipment failures can cause hydrogen release into the atmosphere, mixing with air, causing the formation of a flammable gas mixture and an explosion that generates pressure waves propagated away from the accident epicenter (Figure 1) [3,4]. Sometimes, a wave propagation regime may change to detonation [5].

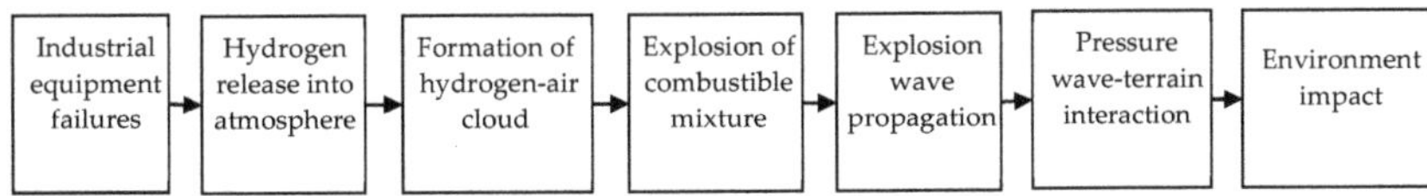

Figure 1. The development of a technogenic accident.

One of the most important and difficult-to-model phases of this complex physical process is the release of a dangerous admixture into the atmosphere. A comprehensive review of mathematical models of gas release during different industrial processes is represented in [3]. The hazardous zones are reconstructed from the fields of explosive concentrations for hydrogen and propane. The high-resolution computational fluid dynamic (CFD) models for flammable gas emissions provide noninvasive and direct quantitative evidence that may influence the safety procedures prepared by regulatory agencies in refining the safety limits in a cost-effective and time-saving manner [3]. A flammable gas mixture is considered as a potential hazard that can explode, with a potential shock-impulse impact to the environment. In our study, the release process is actually omitted, and a premixed hemispheric stoichiometric hydrogen cloud is considered in order to concentrate only on explosion pressure wave generation, its propagation through space with different reliefs, and probable interaction with humans at specific locations that can cause severe injuries for them.

Explosion waves make a shock-impulse impact on the environment, threatening the life and health of industrial workers, destroying infrastructure, and damaging equipment placed at industrial sites. Because of such accidents, social, material, and financial losses can be of catastrophic proportions.

In order to ensure the safety of working conditions on industrial sites, it is necessary to develop and apply protective equipment that can prevent or reduce, to an acceptable level, the possible harmful consequences caused by hydrogen–air explosions [4]. The effectiveness of these protection methods can be tested experimentally [6]. However, a full-scale physical experiment with a hydrogen explosion is difficult to implement, cumbersome, and too expensive. That is why a computational experiment based on computer information systems [7] implementing the considered accident scenarios (Figure 2) is widely used in practice. Thus, an engineering problem of mathematical modeling of physical processes of the considered emergency scenario is relevant.

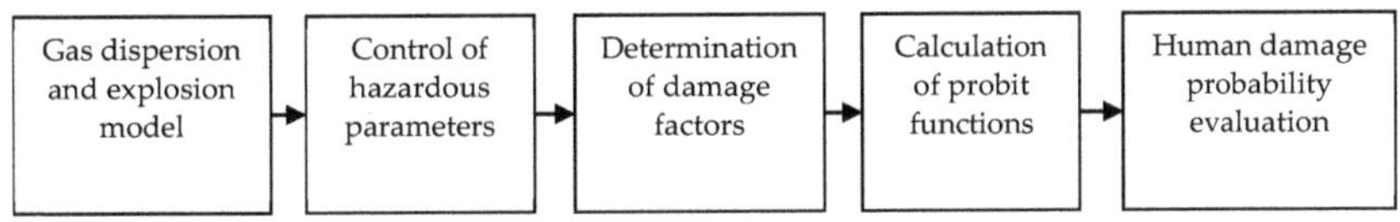

Figure 2. Accident consequences probabilistic evaluation scheme.

The scope of this research covers the problems of development of an environmental science technology based on computational fluid dynamics and probit analysis theory that can be used by safety experts to assess risk distribution around industrial objects where potentially dangerous flammable gases are used in technological processes. The *main aim* of this study is to numerically evaluate an influence of terrain landscape on the distribution of probable harmful consequences for personnel of a hydrogen fueling station caused by accidentally released and exploded hydrogen.

2. Review of the Literature

A mathematical model of the explosion of a hydrogen–air mixture cloud at a hydrogen fueling station site is considered in this paper. An influence of the terrain shape near the accidental hydrogen explosion in open space on the formation of a shock-impulse load and the resulting fields of the conditional probability of damage to working personnel are analyzed. A state of the gas-dynamic environment at the site before an accident can be described as a set of normal values of overpressure, temperature, velocity vector, the chemical composition of the atmosphere. During an accidental explosion, these parameters become locally temporarily disturbed, and excess values of hazardous parameters form damaging factors that have harmful effects on the human body. Sometime after the accident, the environment returns to an unperturbed steady state again.

The purpose of this work is to use an effective mathematical model of the considered hydrogen explosion processes, for three-dimensional prediction and analysis of non-stationary distribution of damaging factors in order to determine the fields of the conditional probability of human damage based on probit analysis methodology.

An adequate description of the physical processes of dispersion of chemically reacting gases, mixing them with air, and further spreading the mixture into an open space [7], tunnel [8], or closed ventilated space [9] is possible only using the Navier–Stokes system of non-stationary equations for compressible gas [10]. Currently, numerical simulation of turbulent flows is carried out by solving the Reynolds–Favre-averaged Navier–Stokes equations, supplemented by a model of turbulence [11]. However, most turbulence models do not describe with an equal degree of adequacy the various types of flows that can appear [12]. This is especially true for currents with intense flow breaks and/or large pressure and temperature gradients.

In work [13], it is indicated that modern engineering methods for predicting the consequences of accidents on chemically hazardous objects (such as [14,15]) implement the Gauss model or the analytical solution of the mass transfer equation and do not take into account the blockage of the calculated space by impenetrable objects. The use of numerical kinematic models [16] to assess territorial risk is also limited to cases of impurity dispersion over a flat surface. Some papers take into account the complex terrain in the process of solving the mass transfer equation by the finite-difference method [13,15], but either there is no consideration for the three-dimensional nature of the flow around obstacles [13] or the effect of compressibility of the flow is not taken into account, which does not allow for using these mathematical models to calculate effects of all damaging factors (explosion shock wave load, thermal radiation, toxic dose), which may be present simultaneously during accidents.

In addition, modern techniques for assessing the technogenic impact on the environment are mainly based on a deterministic approach [17], and during probabilistic consequences assessment based on probit analysis, the table-view dependence of probability on the probit function is used for expert analysis [18]. It is not possible to apply this approach automatically in a computer system to obtain non-stationary fields of damaging factors and probability of damage, and it requires an improvement in computational methods and techniques.

Therefore, there is a need to build effective mathematical models and computational schemes for numerical modeling of three-dimensional flows of multicomponent gas mixtures, taking into account the complicated terrain shape in actual calculation space, compressibility, and chemical interaction effects, which allow for determining the full set of flow hazardous parameters for various scenarios of man-made accidents, calculating the damaging factors (including the shock-impulse load) and building space–time fields of human damage conditional probability needed to assess individual risk.

3. Problem Statement

Summarizing the literature review, we propose to solve the problem of assessment of how the terrain configuration around the epicenter of a gas explosion influences the safety situation at the technogenic object using a solution of the joint problem of gas dynamics of a chemically reacting gas mixture and the safety of a person who is under the influence of an explosion shock wave. For this purpose, firstly, the direct problem of flammable gas release and explosion is considered in an open space under normal environmental conditions using a three-dimensional system of equations that describes the motion of the multicomponent chemically interactive gas mixture in the near-Earth atmosphere layer. It allows for obtaining time-dependent spatial information about the harmful-to-the-environment factor and shock-impulse distribution described by overpressure and impulse at the front of the explosion wave. Secondly, using the means of probit analysis, we can determine the conditional probability of the negative impact that causes the explosion to the environment in every control point of space at any moment of time.

Repeating this calculation process for different configurations of terrain relief, which kind of terrain is more harmful or safe for the personnel can be easily found, comparing the value of impact consequences in a specific working place. This information can be used during the process of determining where the best location of the technogenic object would be considered.

4. Materials and Methods

4.1. Method of Assessing the Impact Caused by an Explosion Wave

It is necessary to determine the peculiarities of the influence of terrain shape near an explosion accident on the spatial and temporal distribution of the shock-impulse load and the probability of personnel harm damage during an explosion of a hydrogen–air cloud at a fueling station site with a two-level landscape based on a mathematical model of the considered physical processes [19].

An accidental release of hydrogen at an industrial site is usually accompanied by the formation of a hydrogen–air mixture, which can explode under the influence of external factors. The resulting explosion wave spreads through the site, causing a shock-impulse load on humans and leading to harmful consequences for their health (Figure 1).

The harmful damaging impact of the shock wave according to a probabilistic assessment approach is determined by the maximum overpressure ΔP_+ (relative to atmospheric pressure P_0) of the wave front and compression phase impulse I_+ (Figure 3).

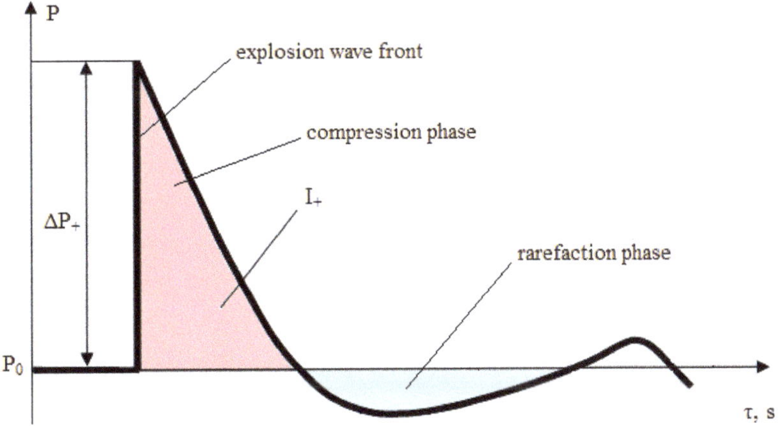

Figure 3. The typical profile of an explosion wave.

The values of these indicators in each control point can be used to determine the individual risk of negative impact on personnel. The risk assessment of the harmful effects of damaging factors on the human body at the accident site is one of the main stages of the safety analysis process of an industrial object. It allows for drawing conclusions about the acceptability of risk and for evaluating the effectiveness of protective facilities. The probability of a specific scenario for the development of an accident P_s depends on the statistical probability of the occurrence of such an accident P_a and the conditional injury probability of an affected person P_c, which can be obtained using mathematical modeling.

The conditional probability P of harmful impact on a person that is under the influence of an explosion shock wave depends on the probit function Pr, which is the upper limit of a definite integral of the normal distribution law with mathematical expectation 5 and variance 1:

$$P = \frac{1}{\sqrt{2\pi}} \int_{-\infty}^{Pr} e^{-\frac{1}{2}(t-5)^2} dt, \tag{1}$$

where t is an integral degree of impact.

For instance, the probability of human lethal damage caused by overpressure can be estimated by the following ratio [20]:

$$Pr_1 = 5 - 0.26 \ln\left[(17,500/\Delta P_+)^{8.4} + (290/I_+)^{9.3}\right] \tag{2}$$

The probit function for rupturing human eardrums depends on the level of overpressure only and can be found by the formula [21]:

$$Pr_2 = -15.6 + 1.93 \ln \Delta P_+ \tag{3}$$

In order to automate the computational process of analysis and prediction—the table of discrete values of the "probit function probability" that is usually used in engineering practice—this dependence is replaced by a generalized piecewise cubic Hermitian spline [22]. The characteristics of such a spline allow one to avoid possible oscillations of the approximated function in the intervals.

4.2. Explosion Mathematical Model and Calculation Algorithm

For a series of comparative computational experiments, in order to evaluate the influence of the two-level terrain shape on the distribution of the wave overpressure at the possible location of the working place, we use a mathematical model of an instantaneous explosion of hydrogen–air mixture [12–14].

It is assumed that the main factor influencing the physical processes under consideration is the convective transfer of mass, momentum, and energy. Therefore, it is sufficient to use the simplified Navier–Stokes equations, which are obtained by dropping the viscous terms in the mixture motion equations (Euler approach with source terms) [13].

The computational domain is a parallelepiped located in the right Cartesian coordinate system (Figure 4). It is divided into spatial cells whose dimensions are determined by the scale of the characteristic features of the area (roughness of streamlined surface, dimensions of objects).

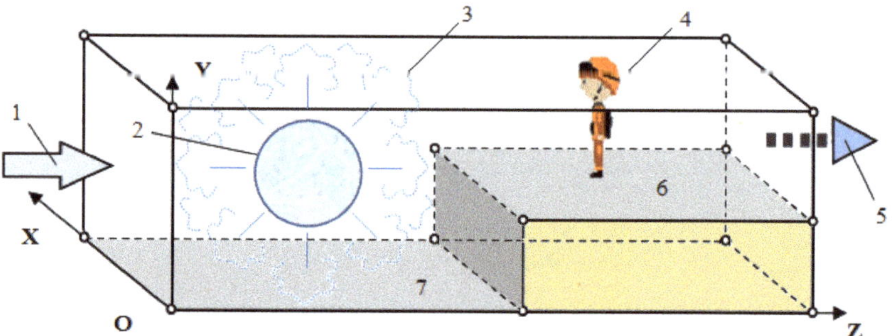

Figure 4. A computer model of the hydrogen–air cloud explosion: 1: inlet air; 2: hydrogen cloud; 3: combustion products; 4: personnel; 5: output mixture; 6: working place terrain level; 7: explosion terrain level.

According to the mathematical explosion model, the following boundary conditions are used: (1) at the entrance to the calculation area, total enthalpy, entropy function, and wind speed vector direction angles are set. Flow parameters here are determined with the involvement of the ratio for the "left" Riemann invariant; (2) at the exit from the calculation area, atmospheric pressure is set with the involvement of the ratio for the "right" Riemann invariant; (3) on the surfaces of solid bodies, "no flow" condition is set.

The following initial conditions are used: (1) in all gaseous "air" computational cells of the calculation area, the parameters of the atmosphere environment are set; (2) in all the cells occupied by the flammable cloud, the gas mixture flow parameters are set with relative mass concentration of the admixture $Q \leq 1$.

It is assumed that the instantaneous chemical reaction takes place in all elementary volumes of the computational grid, where the hydrogen concentration is within the limits of ignition ($Q_{min} \leq Q \leq Q_{max}$). This means that the parameters of the two-component mixture (air and fuel) in the control volume immediately obtain new values of the parameters of the three-component mixture (air, combustion products, and residues of fuel). In other words, it is assumed that the flame front propagates with infinite velocity [17].

A computer solution of the fundamental equations of gas dynamics for a mixture supplemented by the mass conservation laws of admixtures in the integral form is obtained using the explicit Godunov's method [23]. To approximate the Euler equations, the first-order finite-difference scheme is used. Central differences of second order are used for the diffusion source terms in the conservation equations of admixtures. Simple interpolation of the pressure is applied in the vertical direction. Godunov's method is characterized by a robust algorithm that is resistant to large disturbances of the flow parameters (e.g., pressure), which allows for obtaining a solution for modeling large-scale explosions of gas mixtures in calculation spaces of various types of configuration [24].

The mathematical model was validated with respect to Fraunhofer ICT experimental data for hydrogen and propane explosions [25].

The software "Explosion Safety$^{®}$" (ES) [26] was used to analyze the explosion of a hydrogen cloud and dispersion of the combustion products processes, to forecast the pressure history at control points of human location and to evaluate safety differences between the various terrain options of the calculated space. The software can also be used to forecast the environmental impact of toxic spills [27]. It allows for calculating the density, velocity, pressure, temperature of the mixture, concentration of the mixture components (combustible gas, air, and combustion products), and the heat release rate within each control volume of the mixture at each discrete time step. The computer had the following characteristics: Intel$^{®}$ Core™ i7-360QM CPU @ 2.40 GHz 2.40 GHz, 16.0 Gb RAM, Windows 7. CPU time for each experiment was about 15 min.

5. Experiments

A computer simulation of the explosion of a cloud of the hydrogen–air mixture resulting from an accidental release from a destructed dispensing cylinder at a hydrogen fueling station is carried out. The calculated area is shown in Figure 5. The computational experiment is carried out at air velocity q = 0.0 m/s, ambient temperature 293 K, and pressure 101,325 Pa at the entrance to the considered area. The dimensions of the computational domain and other specific sizes are the following: the length L_z = 31.2 m, the height L_y = 14.0 m, the width L_x = 20.2 m, and the height of the first ground level Y_1 = 4.0 m. The second-level part of the site begins from Z_2 = 13.2 m and has a changeable height H.

The cloud of the hydrogen is located at a distance of Z_1 = 10.1 m from the origin of the computational domain; the radius of the cloud is R = 2.88 m. Two control points P1 and P2 at the distances Z_{p1} = 3.2 m and Z_{p2} = 7.1 m from an explosion epicenter C are established. They are located in characteristic places of the second-level part of the industrial site, where overpressure history is monitored.

Five options of the design scheme V1–V5 are considered (Table 1) in order to assess the influence of the site terrain shape on overpressure and damage probability fields. The options differ only by the height H of the second terrain level.

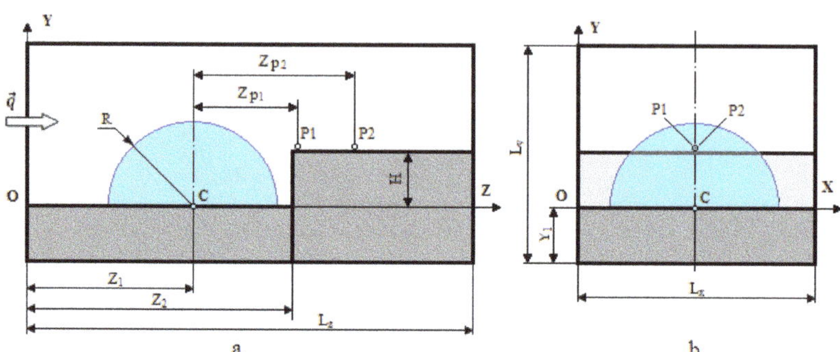

Figure 5. A scheme of the calculated area in planes YOZ (**a**) and YOX (**b**).

Table 1. Types of calculation scheme.

Variants	V1	V2	V3	V4	V5
Scheme					
H, m	4.0	2.0	0.0	−2.0	−4.0

As a result of the hydrogen–air mixture explosion, a cloud of combustion products with high pressure and temperature is formed. The process of combustion products dispersion takes place. It is accompanied by shock wave propagation from an explosion epicenter. During the calculation process, it is possible to monitor the 3D pressure distribution (Figure 6) in order to collect all the needed information to calculate the damage probability fields.

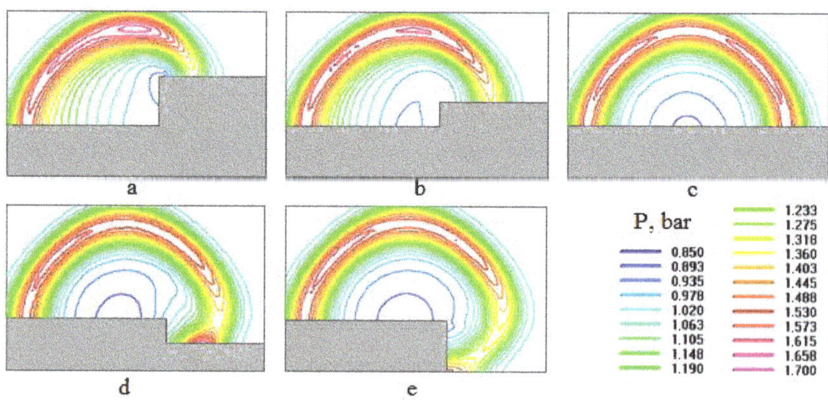

Figure 6. Pressure distribution in the plane YOZ at t = 0.01 s: (**a–e**) options V1–V6.

The overpressure history at the control points P1 and P2 for different design scheme options V1–V5 are presented in Figure 7. It is obvious that the most dangerous variant of the landscape terrain corresponds to variant V3, where both terrain parts of the industrial site are at the same level (height H = 0 m). Any other option of the calculation scheme leads to a decrease in both maximum overpressure and compression phase area that means less shock-impulse loads on people standing in control points P1 and P2. This trend can be noticed also from the comparison of pressure distribution in plane XOY at some moment of time (0.01 s) after the explosion (Figure 6).

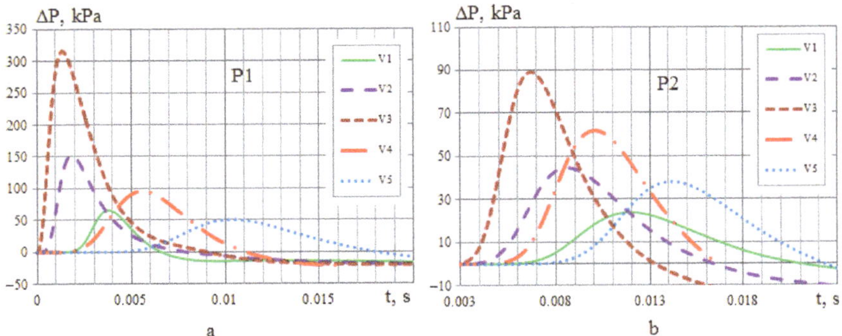

Figure 7. Overpressure history at the control points P1 (**a**) and P2 (**b**).

The collected data allow us to extract all the information needed to evaluate the damaging factors of the explosion shock wave (maximum overpressure (Figure 8) and compression phase impulse (Figure 9)) and to calculate the values of the conditional probability of lethal consequences (Figure 10) according to formula (2) as well as eardrum rupture (Figure 11) according to formula (3) at control points P1 and P2 for different terrain options V1–V5.

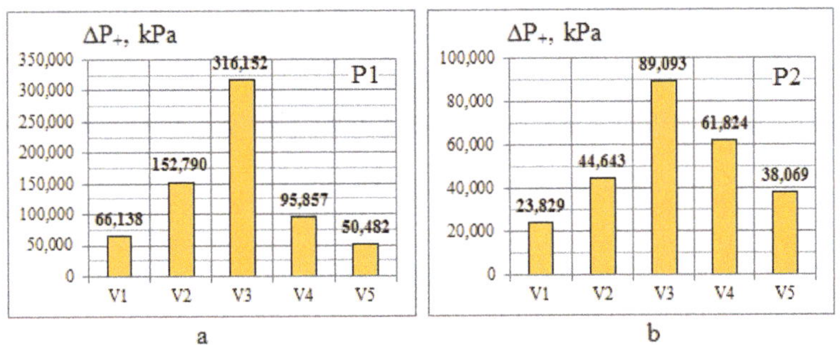

Figure 8. Maximum overpressure at control points: (**a**) point P1; (**b**) point P2.

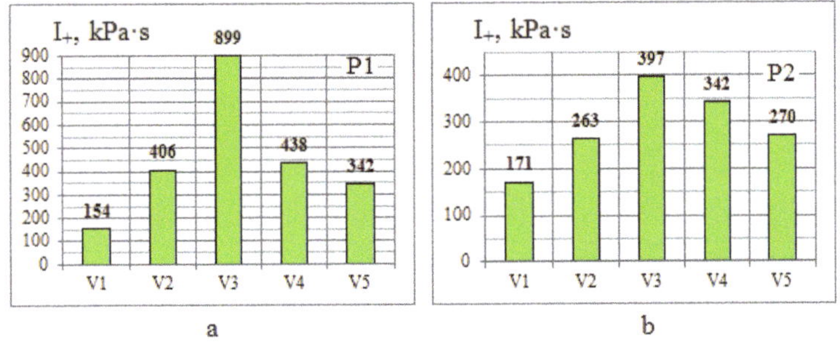

Figure 9. Compression phase impulse I_+ at control points: (**a**) point P1; (**b**) point P2.

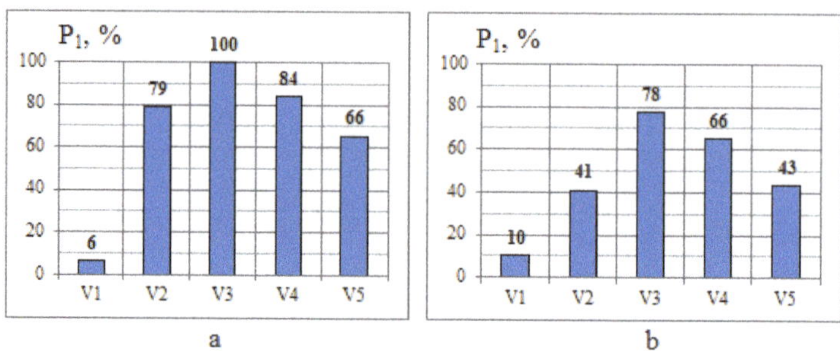

Figure 10. Lethal probability at control points: (**a**) point P1; (**b**) point P2.

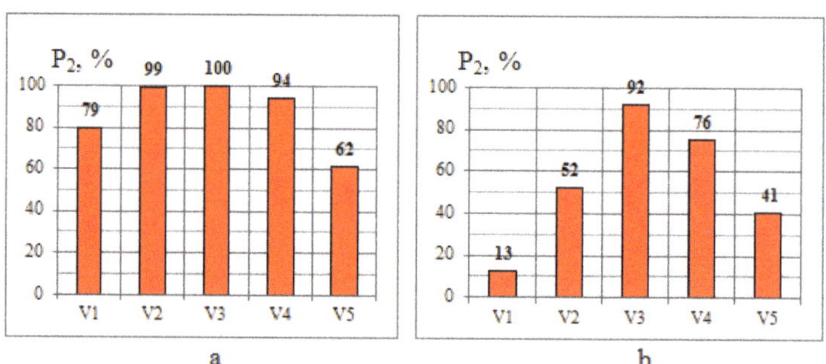

Figure 11. Eardrum rupture probability at control points: (**a**) point P1; (**b**) point P2.

The lethal consequence conditional probability in the most exposed to overpressure vertically centered plane YOZ, and on the ground of the second part of the industrial site (possible working places location) in plane XOZ are displayed in Figures 12–16.

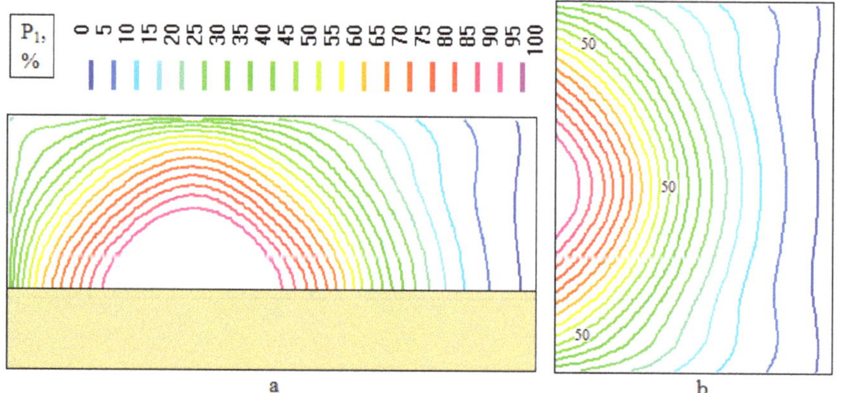

Figure 12. Lethal probability fields for option V3: (**a**) plane YOZ; (**b**) plane XOZ (a working place).

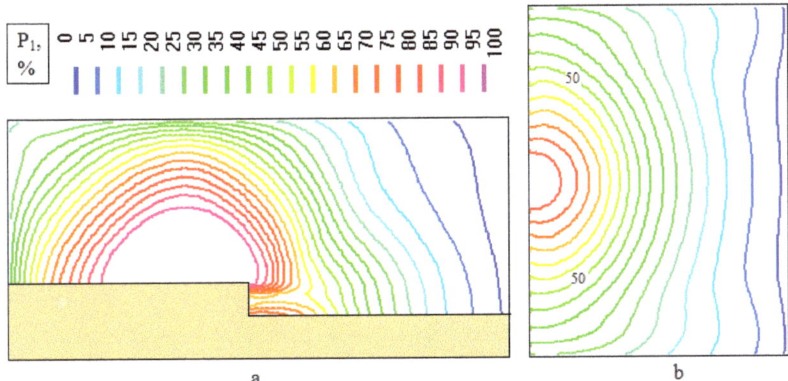

Figure 13. Lethal probability fields for option V4: (**a**) plane YOZ; (**b**) plane XOZ (a working place).

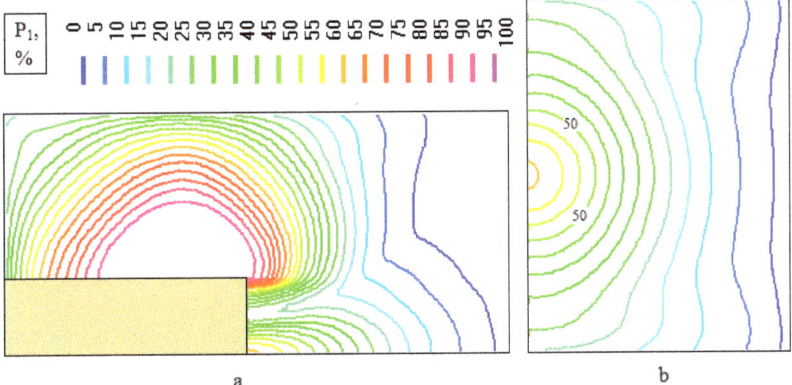

Figure 14. Lethal probability fields (option V5): (**a**) plane YOZ; (**b**) plane XOZ (level two).

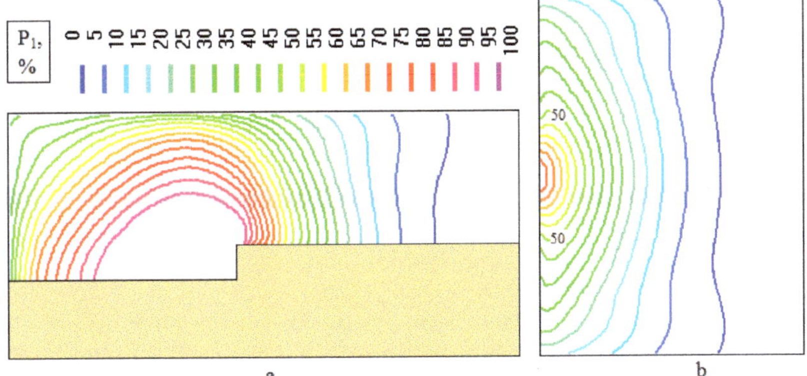

Figure 15. Lethal probability fields (option V2): (**a**) plane YOZ; (**b**) plane XOZ (level two).

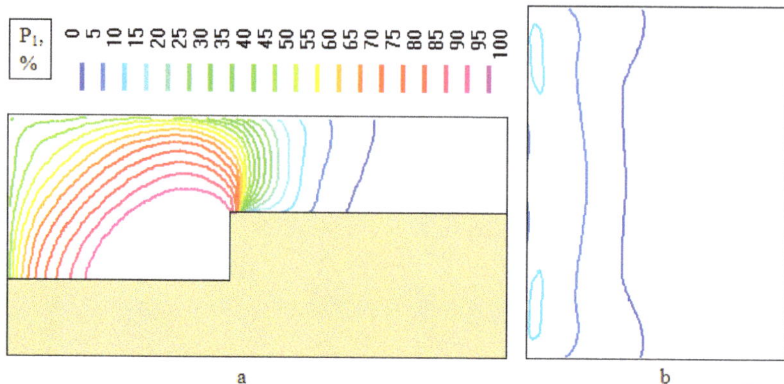

Figure 16. Lethal probability fields for option V1: (**a**) plane YOZ; (**b**) plane XOZ (a working place).

In order to compare different schemes of the terrain landscape, the area of zone S_{50} on the surface of the second part of the industrial site where the lethal consequences conditional probability is greater than 50% (which is considered dangerous for humans) is calculated (Figure 17).

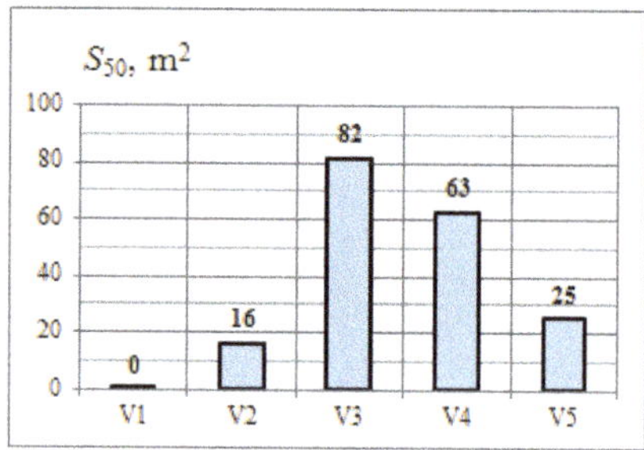

Figure 17. A dangerous area S_{50} at the working place level.

6. Results

A numerical simulation of the pressure wave propagation from the epicenter of the stoichiometric hydrogen cloud explosion along space, with different Earth reliefs, enabled us to reconstruct the hazard zones for humans potentially located at two distant control points (Figure 5). Five options of an explosion epicenter and control point Earth levels were considered (Table 1): two options with deeper located explosion level, one evenly leveled option, and two options with deeper located control points. For all variants of relief, the following parameters were obtained:

- Pressure 3D field during explosion wave propagation (Figure 6);
- Overpressure history in control points P1 and P2 (Figure 7);
- Maximum overpressure of an explosion wave in control points (Figure 8);
- Impulse of explosion wave pressure phase in control points (Figure 9);
- Lethal (Figure 10) and ear-drum rupture (Figure 11) probability in control points;
- Lethal conditional probability 3D fields (Figures 12–16);
- Hazardous zone area at control points level where lethal probability is greater than 50% (Figure 17).

Non-stationary pressure distributions in vertical cross-section YOZ (Figure 6) can give us a clue as to why probable consequences for humans in control points behave in such a manner. Deepening the relief level of the accident epicenter related to control points level (Figure 6a,b) leads to higher obstacles in front of the pressure wave and a bigger reflecting effect. Deepening the relief level of control points related to the accident epicenter level (Figure 6d,e) leads to break-away and reattachment to the Earth of the pressure wave and a bigger expansion effect. The most dangerous option is an evenly leveled relief (Figure 6c) because the control points are openly exposed to the shock-impulse load.

From previous pressure fields, overpressure history in control points P1 and P2 can be collected (Figure 7) and processed in order to calculate maximum overpressure of an explosion wave (Figure 8), which characterizes the shock effect for exposed humans and the impulse of the explosion wave pressure phase (Figure 9), which reflects the timespan of the load and assesses the probable consequences for humans as lethal (Figure 10) and ear-drum rupture (Figure 11) injuries. Lethal conditional probability 3D fields (Figures 12–16) help us to evaluate the dangerous zone area (starting from the edge of the control points level) (Figure 17), which can be used as an additional measure of relief variant safety characteristics.

It can be seen from the presented overpressure, impulse and consequences diagrams that the biggest shock loads correspond to option V3 with equal terrain levels of two parts of the industrial site. This leads to the highest values of the ear-drum rupture conditional probabilities (Figure 11) for both control points. Some decrease in shock loads in control point P2 (compared to point P1 at the edge of the level-two part) can be explained by the more distant location of this point from the explosion epicenter.

Any other design scheme of the level-two part gives a decrease in shock loads that is especially noticeable for control point P2 (Figures 7b, 8b and 11b). Options V4 and V5, which correspond to deeper levels of the part two terrain, give less protective effect for point P2 than for the corresponding options V2 and V1 with higher levels of the part two terrain. It can be explained that in options V2 and V1, part of the explosion wave meets an obstacle and is reflected backward. For control point P1 (Figures 7a, 8a and 11a), the deepening makes a bigger effect in options V4 and V5 comparable to options V2 and V1, maybe because of the less intensive expansion process around the convex corners of the terrain.

Very similar behavior can be seen in compression phase impulse distribution (Figures 7 and 9). Higher levels of the terrain (options V2 and V1) create bigger protection in control points from impulse loads than deeper levels (options V4 and V5).

The total effect from maximum overpressure and impulse loads can be clearly seen in Figure 10 for lethal probabilities in control points, and in Figures 12–16 for the fields of this consequence parameter. Higher levels of options V2 and V1 better protect humans than deeper variants of terrain, especially in point P2 (Figures 10b, 12b, 13b, 14b, 15b and 16b). This conclusion can be confirmed by such safety characteristics as an area S_{50} of dangerously high values (> 50%) of the conditional lethal probability (Figure 17) on the surface of the level-two part of the industrial site (working place). It is clearly seen that higher-level variants V2 and V1 protect the working place much more effectively than deeper level variants V4 and V5 of landscape in relation to variant V3 with evenly leveled terrain.

7. Discussion

A large-scale field experiment [6] is the most adequate way to reconstruct hazardous zones of an accidental release and explosion of flammable gases at industrial objects of high risk. Experimentally measured flow parameters such as admixture concentration, temperature and pressure of explosion products can be used by safety experts to analyze and assess harmful consequences for humans and industrial constructions around an explosion epicenter. However, field experiments are very time-consuming and cost-ineffective, depend on environment conditions, and cannot really be used to carry out such series of experiments to compare different options of relief shape, as this study does. That is why mathematical modeling of release, dispersion, and explosion processes is an effective alternative way to obtain all the needed information about flow parameters with much wider opportunities to experiment with different environment conditions, landscape shape, flammable and/or toxic gases, and various scenarios of accidents. A validation of capabilities of the presented computational fluid dynamics model to reproduce a large-scale hydrogen explosion in open atmosphere [28,29] against the results of intercomparison exercises on capabilities of other CFD models [7], evidence that the presented model adequately describes an explosion process and propagation of an explosion wave in open space. The CFD models analyzed in work [3] (LES, RANS, and FDS models) are based on Navier–Stokes equations. These models can be very useful while thoroughly predicting hydrogen distribution during release and dispersion processes (in open space [7], in tunnels [8], and with pressure relief vents [9]), but they consume huge computer resources and require careful selection of a turbulence model, which is different for different types of flow conditions. These advanced models give very similar results to our CFD model's results in predicting maximum overpressure and impulse at the shock wave front during a large-scale hydrogen explosion [28,29], which is crucially important in probit analysis to evaluate harmful consequences to the environment caused by the explosion. Our CFD model represents the Euler approach with source terms (simplified Navier–Stokes equations obtained by dropping the viscous terms in the mixture motion equations) [13], does not require turbulence model selection, consumes much less computer resources, and is very useful in comparing experimental series.

The presented methodology provides a mathematical tool to evaluate whether the differently leveled terrain at an industrial open space, where an accidental explosion of hydrogen takes place, can change the safety level for humans located near the epicenter of an explosion. The numerical analysis of a three-dimensional pressure field's history during an explosion's wave propagation enabled us to quantify the effect of different options of terrain shape on probable consequences for people exposed to explosion shock-impulse impact. With the use of probit analysis, incorporated into the CFD model, it was possible to present diagrams of conditional probability of harmful injuries to humans at work places during the explosion for different options of terrain. This was in line with work [11], where lethal probability was obtained on the base of results of fluent CFD modeling of hydrogen non-premixed combustion in an enclosure with one vent and sustained release. However, consequences for humans were assessed using "overpressure on impulse" diagrams that provide probability isolines, and this technique did not allow for building three-dimensional fields of impact probability and for making a transition to individual risk assessment in future safety evaluation processes.

Limitations to the Study

The presented methodology provides satisfactory results for open space explosion wave propagation but may encounter some problems in evaluating shock-impulse consequences in narrow tunnels and small premises where multiple explosion wave reflections take place, and it would be difficult to extract the impulse of the explosion wave from the model. The CFD model was used only for the assessment of harmful consequences zones induced by an explosion of a premixed hemispheric cloud, and only one flammable gas (hydrogen) was analyzed. In the future, we would like to consider other flammable gases and include into consideration an admixture releasing process before invoking an explosion. There could be different release scenarios involved, such as evaporation from a spilled liquid spot or jet emissions from a destroyed high-pressure storage vessel. The released admixture or explosion products were assumed to be nontoxic. In the future, we would like to consider a coupled scenario of dangerous zones formation (accidental explosion of flammable gas mixture and dispersion of toxic admixture) to predict the combined consequences for humans.

8. Conclusions

The purpose of this study was to designate risky zones for humans after an accidental explosion of a premixed stoichiometric hemispherical hydrogen cloud placed at differently leveled terrains. Currently, hydrogen is widely used in transport that requires refueling at filling stations where hydrogen is kept in high-pressure vessels. Therefore, it is important to be aware of risks of accidental hydrogen release, formation of a flammable mixture with air, and its explosion with the generation of a shock wave that propagates along the ground surface, which can cause injuries to humans at working places. In order to determine these hazardous zones and probable environmental consequences in control points and to evaluate the influence of terrain shape on the scale of consequences, the ES software using computational fluid dynamics and probit analysis was applied. For the purposes of this study, a series of five simulations were made. They differed by two-level terrain options where the hydrogen cloud was placed at the first-level plane, and two control points (human locations) were placed at the second-level plane. The options of terrain configuration were compared on pressure three-dimensional field evolution during the explosion wave propagation, the values of maximum overpressure and impulse of the first pressure phase of the explosion wave, conditional probability for eardrum rupture and lethal outcome in control points, and the dangerous area value at the level-two plane where the lethal probability was greater than 50%. During every simulation, the same environment and hydrogen cloud parameters were applied.

It was obtained that higher-leveled working places in relation to the possible explosion epicenter terrain level could give better protection than deeper-leveled places.

In conclusion, incorporating the probit analysis procedure into the CFD model provides a powerful instrument to intercompare computer experiments, and it can be used by safety experts to develop measures to reduce the risk of considered accidents at industrial sites and to analyze the efficiency of protection structures. Further improvement of this methodology is possible in the direction of enhancing the accuracy of the gas-dynamics mathematical model and in considering a combination of accidental scenarios, taking into account various influencing factors.

Author Contributions: Conceptualization, Y.S. and S.Y.; methodology, Y.S. and S.Y.; software, K.K. and M.K.; validation, K.K. and M.K.; formal analysis, K.K. and M.K.; investigation, Y.S., K.K. and M.K.; resources, K.K. and M.K.; writing—original draft preparation, K.K. and M.K.; writing—review and editing, Y.S. and S.Y.; visualization, K.K. and M.K.; supervision, Y.S. and S.Y.; project administration, K.K. and M.K. All authors have read and agreed to the published version of the manuscript.

Funding: This research received no external funding.

Institutional Review Board Statement: Not applicable.

Informed Consent Statement: Not applicable.

Data Availability Statement: Generated data and test tasks are used.

Conflicts of Interest: The authors declare no conflict of interest.

References

1. Pramod, T. *Advanced Mine Ventilation: Respirable Coal Dust, Combustible Gas and Mine Fire Control*; Woodhead Publishing: Sawston, UK, 2018; p. 528.
2. Nolan, D.P. *Handbook of Fire and Explosion Protection Engineering Principles: For Oil, Gas, Chemical and Related Facilities*. Burlington; Gulf Professional Publishing, Elsevier: Amsterdam, The Netherlands, 2011; p. 340.
3. Salamonowicz, Z.; Krauze, A.; Majder-Lopatka, M.; Dmochowska, A.; Piechota-Polanczyk, A.; Polanczyk, A. Numerical Reconstruction of Hazardous Zones after the Release of Flammable Gases during Industrial Processes. *Processes* **2021**, *9*, 307. [CrossRef]
4. Jianwei, C. *Explosions in Underground Coal Mines. Risk Assessment and Control*; Springer: Cham, Switzerland, 2018; p. 208.
5. Fedorov, A.V.; Khmel', T.A.; Lavruk, S.A. Exit of a heterogeneous detonation wave into a channel with linear expansion. I. Propagation regimes. *Combust. Explos. Shock. Waves* **2017**, *53*, 585–595. [CrossRef]
6. Schneider, H. Large Scale Experiments: Deflagration and Deflagration to Detonation within a partial Confinement similar to a lane. In Proceedings of the 1st International Conference on Hydrogen Safety, Pisa, Italy, 8–10 September 2005; p. 12.
7. García, J.; Baraldi, D.; Gallego, E.; Beccantini, A.; Crespo, A.; Hansen, O.; Høiset, S.; Kotchourko, A.; Makarov, D.; Migoya, E. An intercomparison exercise on the capabilities of CFD models to reproduce a large-scale hydrogen deflagration in open atmosphere. *Int. J. Hydrogen Energy* **2010**, *35*, 4435–4444. [CrossRef]
8. Baraldi, D.; Kotchourko, A.; Lelyakin, A.; Yanez, J.; Middha, P.; Hansen, O.; Gavrikov, A.; Efimenko, A.; Verbecke, F.; Makarov, D.; et al. An inter-comparison exercise on CFD model capabilities to simulate hydrogen deflagrations in a tunnel. *Int. J. Hydrogen Energy* **2009**, *34*, 7862–7872. [CrossRef]

9. Baraldi, D.; Kotchourko, A.; Lelyakin, A.; Yanez, J.; Gavrikov, A.; Efimenko, A.; Verbecke, F.; Makarov, D.; Molkov, V.; Teodorczyk, A. An inter-comparison exercise on CFD model capabilities to simulate hydrogen deflagrations with pressure relief vents. *Int. J. Hydrogen Energy* **2010**, *35*, 12381–12390. [CrossRef]
10. Andersson, B.; Andersson, R. *Computational Fluid Dynamics for Engineers*, 1st ed.; Cambridge University Press: Cambridge, UK, 2012; p. 97.
11. Molkov, V.; Shentsov, V.; Brennan, S.; Makarov, D. Hydrogen non-premixed combustion in enclosure with one vent and sustained release: Numerical experiments. *Int. J. Hydrogen Energy* **2014**, *39*, 10788–10801. [CrossRef]
12. Sreenivas, J. *Computational Fluid Dynamics for Engineers and Scientists*; Springer: New York, NY, USA, 2018; p. 402.
13. Biliaiev, M.M.; Kalashnikov, I.V.; Kozachyna, V.A. Raschet territorialnogo riska pri terakte: Ekspress model. Territorial risk assessment after terrorist act: Express model. *Nauka Prohres Transp. Sci. Transp. Prog.* **2018**, *1*, 7–14.
14. Methods for assessing the consequences of accidental explosions of fuel-air mixtures. In *Safety Manual*; Scientific and Technical Center for Research of Industrial Safety Problems: Moscow, Russia, 2015; Volume 27, p. 44.
15. Biliaiev, M.; Muntian, L.Y. Numerical simulation of toxic chemical dispersion after accident at railway. *Sci. Transp. Progress. Bull. Dnipropetrovsk Natl. Univ. Railw. Transp.* **2016**, *2*, 7–15. [CrossRef] [PubMed]
16. Tashvigh, A.A.; Nasernejad, B. Soft computing method for modeling and optimization of air and water gap membrane distillation—A genetic programming approach. *Desalination Water Treat.* **2017**, *76*, 30–39. [CrossRef]
17. Skob, Y.; Ugryumov, M.; Granovskiy, E. Numerical assessment of hydrogen explosion consequences in a mine tunnel. *Int. J. Hydrogen Energy* **2020**, *46*, 12361–12371. [CrossRef]
18. Dadashzadeh, M.; Kashkarov, S.; Makarov, D.; Molkov, V. Risk assessment methodology for onboard hydrogen storage. *Int. J. Hydrogen Energy* **2018**, *43*, 6462–6475. [CrossRef]
19. Skob, Y.; Ugryumov, M.; Dreval, Y. Numerical Modelling of Gas Explosion Overpressure Mitigation Effects. *Mater. Sci. Forum* **2020**, *1006*, 117–122. [CrossRef]
20. Instituut voor Milieu- en Energietechnologie TNO Prins Maurits Laboratorium TNO. Methods for the determination of possible damage to people and objects resulting from releases of hazardous materials (CPR-16E). In *Green Book/Director-General of Labour*, 1st ed.; Committee for the Prevention of Disasters: Den Haag, The Netherlands, 1992; p. 337.
21. Pietersen, C. Consequences of accidental releases of hazardous material. *J. Loss Prev. Process. Ind.* **1990**, *3*, 136–141. [CrossRef]
22. Knott, G.D. *Interpolating Cubic Splines*; Springer: New York, NY, USA, 2012; p. 254.
23. Toro, E.F. *Riemann Solvers and Numerical Methods for Fluid Dynamics*, 3rd ed.; Springer: Berlin, Germany, 2009; p. 724.
24. Pichugina, O.; Yakovlev, S. Euclidean Combinatorial Configurations: Continuous Representations and Convex Extensions. In *Lecture Notes in Computational Intelligence and Decision Making*; Lytvynenko, V., Ed.; Springer: Cham, Switzerland, 2019; pp. 65–80. [CrossRef]
25. Skob, Y.; Ugryumov, M.; Dreval, Y.; Artemiev, S. Numerical Evaluation of Safety Wall Bending Strength during Hydrogen Explosion. *Mater. Sci. Forum* **2021**, *1038*, 430–436. [CrossRef]
26. Skob, Y.; Ugryumov, M. Computer system of engineering analysis and forecast "Explosion Safety" for safety assessment during an explosion of gas-air mixture. *Off. Bull. Copyr.* **2017**, *45*, 236. (In Ukrainian)
27. Skob, Y.; Ugryumov, M.; Granovskiy, E. Numerical Evaluation of Probability of Harmful Impact Caused by Toxic Spill Emergencies. *Environ. Clim. Technol.* **2019**, *23*, 1–14. [CrossRef]
28. Skob, Y. Numerical modeling of gas mixture explosions in the atmosphere. *Aerosp. Tech. Technol.* **2007**, *3*, 72–78. (In Russian)
29. Skob, Y. Mathematical Modelling of Gas Mixture Deflagration Fire in Enclosed Space. *Bull. V. Karazin Kharkiv Natl. Univ.* **2009**, *863*, 218–236. (In Russian)

article

Statistical Theory of Optimal Stochastic Signals Processing in Multichannel Aerospace Imaging Radar Systems

Valeriy Volosyuk and Semen Zhyla *

Aerospace Radio-Electronic Systems Department, National Aerospace University "Kharkiv Aviation Institute", 61070 Kharkiv, Ukraine
* Correspondence: s.zhyla@khai.edu

Abstract: The work is devoted to solving current scientific and applied problems of the development of radar imaging methods. These developments are based on statistical theory of optimal signal processing. These developments allow researchers to create coherent high-resolution information-enriched images as well as incoherent images. These methods can be practically applied in multichannel aerospace radars through the proposed programs and algorithms. Firstly, the following models of stochastic signals at the output of multichannel registration regions of scattered electro-magnetic fields, internal noise, and observation equations are developed and their statistical characteristics investigated. For the considered models of observation equations, the likelihood functional is defined. This definition is an important stage in optimizing spatial and temporal signal processing. These signals are distorted by internal receiver noises in radar systems. Secondly, by synthesising and analysing methods of measuring a radar cross section, the problem of incoherent imaging by aerospace radars with planar antenna array is solved. Thirdly, the obtained optimal mathematical operations are physically interpreted. The proposed interpretation helps to implement a quasi-optimal algorithm of radar cross section estimation in aerospace radar systems. Finally, to verify the proposed theory, a semi-natural experiment of real radio holograms processing was performed. These radio holograms are digital recordings of spatial and temporal signals by an airborne synthetic aperture radar (SAR) system. The results of the semi-natural experiment are presented and analysed in the paper. All the calculations, developments and results in this paper can be applied to new developments in areas such as remote sensing or non-destructive testing.

Keywords: multi-channel and multi-view radio engineering systems; radar cross section; optimal methods; spatio-temporal signal processing

Citation: Volosyuk, V.; Zhyla, S. Statistical Theory of Optimal Stochastic Signals Processing in Multichannel Aerospace Imaging Radar Systems. *Computation* **2022**, *10*, 224. https://doi.org/10.3390/computation10120224

Academic Editor: Demos T. Tsahalis

Received: 10 October 2022
Accepted: 13 December 2022
Published: 18 December 2022

Publisher's Note: MDPI stays neutral with regard to jurisdictional claims in published maps and institutional affiliations.

1. Introduction

The use of spatio-temporal signal processing methods in radar systems located on mobile platforms makes it possible to form high-precision and highly informative radar images in any weather conditions or time of day. Due to its advantages radar images are used for nondestructive testing of various objects, aerospace research of the Earth's surface, ship tracking and illegal vessel detection, surface digital elevation model reconstruction and geohazards monitoring.

Radar imaging methods have been used for more than 70 years. Radar imaging developments, in both theory and practice, can be divided into three periods. The first period was associated with the search for heuristic solutions for the construction of airborne radars. It also was devoted to the accumulation of basic knowledge in the field of theory and practice of high-precision radar imaging. The second is a period of rapid implementation of methods for optimizing signal processing algorithms and of synthesis of optimal and quasi-optimal radio engineering systems. Most of the methods and algorithms of this second period were based on mathematical statistics, statistical decision theory and optimal estimates of the parameters of probability distributions. The third period is associated with the rapid development of a high-precision and small-sized element base of radio measuring devices and with improvements in such elements as phased antenna arrays, low-noise amplifiers, power amplifiers, DSPs and FPGAs, all of which are necessary to implement the optimal algorithms synthesized in the second period. These developments in the manufacturing of radio-electronic assemblies and blocks have rapidly increased in the last decade so there is now a lag in the theoretical development of methods and algorithms. These modern methods do not use the full potential of the existing radio engineering equipment, therefore, they need further optimization and development which takes into account the technical capabilities of advanced radio electronics.

The development of the theory must be based on the works of V. A. Kotel'nikov [1], D. Midlton [2], F. Woodward [3], P. A. Bakut [4], I. A. Bol'shakov [4], S. Ye. Fal'kovich [5–7], V. K Volosyuk [8], L. Gutkin [9], B. R. Levin [10], Ya. D. Shirman [11], A. A. Kuriksha [4], G. P. Tartakovskiy [4], V. Repin [4], I. Ya. Kremer [12], R.L. Stratonovich [13–15], V. Tikhonov [16,17]. In the most of these works, the research was performed in the context of fixed ground spatially distributed systems and antenna arrays, however, the methodology of their optimization can also be applied to mobile airborne radars with aerospace-based antenna array. It should also be noted that the methods for synthesizing radar apertures in mobile airborne systems obtained in mentioned works are considered to a greater extent in relation to signals that can be described by a functionally deterministic model. At the same time, it was proved [18–21] that the roughness of the investigated surfaces, the random placement of scattering elements in space and the inhomogeneity of the electro-physical composition of objects lead to stochastic scattering of signals with a random distribution of phases and amplitudes. The result of the coherent processing of such electro-magnetic fields leads to the appearance of speckle noise and requires secondary processing and filtration. The secondary processing result is proportional to the radar cross section of surfaces and objects. Such approaches to processing stochastic signals and radar cross section estimation in aerospace radars are not optimal.

From the analysis of onboard radar systems and the achievements of the statistical theory of optimization of signal processing the following contradiction follows: on the one hand, today there are all the necessary technical solutions and element basis for creating combined high-precision and information-enriched radar images in multichannel multi-glide radio systems. On the other hand, there is no statistical theory of optimal stochastic signals processing in multichannel aerospace imaging radar systems which would make it possible to overcome the existing limits of accuracy and achieve potential qualitative characteristics of the functioning of multichannel onboard radar. This work reveals results from research that allow this contradiction to be overcome.

2. Materials and Methods

2.1. Models of Signals, Noises and Observation Equation

The determination of signal and noise models is one of the key issues in the development of the theory. For further conclusions, it is advisable to consider the generalized geometry of surface and object imaging in multichannel aerospace radars in Figure 1.

In Figure 1D is the surface on which the boundary conditions are phenomenologically stated in the form of the scattering coefficient $\dot{F}(\vec{r},\lambda)$. λ is the possible parameter that is interesting to estimate. There are a lot of electrodynamic models of the surface that describe the relations between surface parameters λ and scattering coefficient $\dot{F}(\vec{r},\lambda)$. The most interesting parameters, λ, in remote sensing are real and imaginary parts of the complex permittivity, slopes and heights of the relief, soil density, and soil moisture.

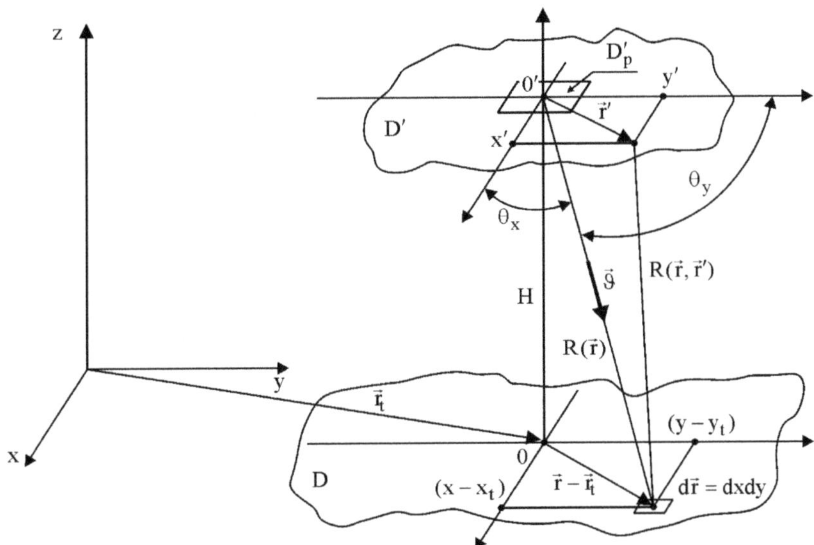

Figure 1. Generalized geometry of surfaces and objects imaging in multichannel aerospace radars.

D' is the area of observation, and the primary (before processing) description of the field in a coherent form taking into account its amplitude and phase dependences. Within region D' there is a physical rectangular region, D'_p, of registration of scattered signals, which in general may have an arbitrary shape. The coordinates of D include a coherent image that takes the influence of various inhomogeneities, such as inhomogeneities of chemical composition, inhomogeneities of the refractive index gradient and irregularities of the real surface, into account in its amplitude-phase structure. The region emitting the probing signal may be outside of D' or may be aligned with the origin of region D'. The only limitation is the wide emission angle of the probing signal, particularly omnidirectional radiation.

The model of the probing signal has the following form:

$$\dot{s}_p(t) = \dot{A}(t)e^{j\omega_0 t},\tag{1}$$

where

$$\dot{A}(t) = \sum_i \dot{S}_i(t - iT_n),\tag{2}$$

is the complex envelope in the form of periodic serial of pulses $\dot{S}_i(t)$ (simple rectangular, chirp, phase–code manipulation, etc.) with pulse repetition period T_n; $\dot{S}_i(t)$ is the complex envelope of one pulse; $\omega_0 = 2\pi f_0$, f_0 is the carrier frequency.

The probing signal propagates and reaches the observation area with coordinates $\vec{r} = (x, y) \in D$, is reflected therefrom (scattered on its inhomogeneities), and is then received by registration area D'. The received space–time signal should be described using a phenomenological approach [22]. On the one hand, this approach is based on the diffraction theory, Kirchhoff's and Rayleigh-Sommerfeld theorems, and combines in its structure already known, well-defined electrodynamic models of test surfaces [23–25]; on the other hand, it is based on the Huygens-Fresnel principle [26,27] which allows it to describe underlying surfaces and objects of complex shape with an arbitrary distribution of electrophysical parameters and statistical characteristics. The model of the received signal is as follows:

$$\dot{s}(t, \vec{r}_t, \vec{r}', \vec{\lambda}(\vec{r})) = \int_D \dot{F}(\vec{r}, \vec{\lambda}(\vec{r}))\, \dot{s}_0(t, \vec{r}, \vec{r}_t, \vec{r}')\, d\vec{r}, \tag{3}$$

where $\dot{F}(\vec{r}, \vec{\lambda}(\vec{r}))$ is the specific complex scattering coefficient of some small element dr of surface D; $\vec{\lambda}(\vec{r})$ is the estimated parameter;

$$\dot{s}_0(t, \vec{r}, \vec{r}_t, \vec{r}') = \varepsilon \dot{I}(\vec{r}')\dot{A}(t - t_{del}(\vec{r}, \vec{r}_t, \vec{r}'))\exp\left\{j2\pi f_0(t - t_{del}(\vec{r}, \vec{r}_t, \vec{r}'))\right\} \tag{4}$$

is the unit signal scattered from element dr of surface D with specific complex scattering $\dot{F}(\vec{r}, \vec{\lambda}(\vec{r})) = 1$; ε is the attenuation of the field in the path of propagation; $\dot{I}(\vec{r}')$ is the amplitude-phase distribution in the region of signal registration; $t_{del}(\vec{r}, \vec{r}_t, \vec{r}')$ is the delay time of the field when it is distributed from point $\vec{r}$ to point $\vec{r}'$ taking into account the shift of the registration area to the point $\vec{r}_t$.

The received signals are always observed against the background of the internal noise $n(t, \vec{r}')$ of the receiver, which is approximated by white Gaussian processes with a correlation function:

$$R_n(t_1, t_2, \vec{r}'_1, \vec{r}'_2) = \left\langle n(t_1, \vec{r}'_1)n(t_2, \vec{r}'_2)\right\rangle = \frac{1}{2}N_{0n}\delta(t_1 - t_2)\delta(\vec{r}'_1 - \vec{r}'_2). \tag{5}$$

We assume that the spectral noise density in each element of the registration area is the same.

For coherent radio vision systems, the observation equation can be presented as an additive mixture of useful signals and noise:

$$u(t, \vec{r}_t, \vec{r}') = \mathrm{Re}\,\dot{s}(t, \vec{r}_t, \vec{r}', \vec{\lambda}(\vec{r})) + n(t, \vec{r}'). \tag{6}$$

When solving the optimization problem of processing the received oscillations according to the model (6), several cases of determining the structure of the complex scattering coefficient and parameters $\vec{\lambda}(\vec{r})$ are considered. In one case, $\dot{F}(\vec{r}, \vec{\lambda}(\vec{r}))$ is considered to be a deterministic function, which is a complex coherent image of the surface and objects of observation. Otherwise, it is a stochastic process, the correlation function of which is related to the radar cross section (RCS) as follows:

$$\sigma^0(\vec{r}, \vec{\lambda}(\vec{r})) = \int_D \left\langle \dot{F}(\vec{r}, \vec{\lambda}(\vec{r}))\dot{F}^*(\vec{r} + \Delta\vec{r}, \vec{\lambda}(\vec{r} + \Delta\vec{r}))\right\rangle e^{-jq_\perp \Delta\vec{r}}d\Delta\vec{r} = \int_D \dot{R}_F(\vec{r}, \Delta\vec{r}, \vec{\lambda}(\vec{r}), \vec{\lambda}(\vec{r} + \Delta\vec{r}))e^{-jq_\perp \Delta\vec{r}}d\Delta\vec{r}, \tag{7}$$

where $\dot{R}_F(\vec{r}, \Delta\vec{r}, \vec{\lambda}(\vec{r}), \vec{\lambda}(\vec{r} + \Delta\vec{r}))$ is the spatial correlation function; $q_\perp$ is the scattering vector. For such problems, all of the necessary information about the observation area is concentrated not in the coefficient $\dot{F}(\vec{r}, \vec{\lambda}(\vec{r}))$ itself, but in the spectral statistical characteristic $\sigma^0(\vec{r}, \vec{\lambda}(\vec{r}))$ (spectral power density) of a statistically inhomogeneous random process. It is a parameter of the observation area to be evaluated. This parameter is not in the observation equation, but in its statistical characteristics—the correlation function $R_u(t_1, t_2, \vec{r}_{t1}, \vec{r}_{t2}, \vec{r}'_1, \vec{r}'_2)$ and the function $W_u(t_1, t_2, \vec{r}_{t1}, \vec{r}_{t2}, \vec{r}'_1, \vec{r}'_2)$ inverse to the correlation function.

The correlation function of Equation (6) is written as follows:

$$\begin{aligned} R_u(t_1, t_2, \vec{r}_{t1}, \vec{r}_{t2}, \vec{r}'_1, \vec{r}'_2) &= \left\langle u(t_1, \vec{r}_{t1}, \vec{r}'_1)u(t_2, \vec{r}_{t2}, \vec{r}'_2)\right\rangle = R_s(t_1, t_2, \vec{r}_{t1}, \vec{r}_{t2}, \vec{r}'_1, \vec{r}'_2) + R_n(t_1, t_2, \vec{r}'_1, \vec{r}'_2) \\ &= \tfrac{1}{2}\mathrm{Re}\int_D \sigma^0(\vec{r}, \vec{\lambda}(\vec{r}))\dot{s}_0(t_1, \vec{r}, \vec{r}_{t1}, \vec{r}'_1)\dot{s}_0^*(t_2, \vec{r}, \vec{r}_{t2}, \vec{r}'_2)d\vec{r} + \tfrac{1}{2}N_{0n}\delta(t_1 - t_2)\delta(\vec{r}'_1 - \vec{r}'_2). \end{aligned} \tag{8}$$

2.2. Bases of Statistical Theory

In this paper it is noted that both the received signals and noises are of a random nature and are statistically similar. We will perform the optimal processing of such signals using the modern achievements of the theory of statistical decisions and estimates of the parameters of probability distributions, which will allow us not only to synthesize optimal algorithms, but also to design the corresponding structural schemes. In this subsection we will consider the bases of the theory of optimization of the stochastic space–time signals processing in onboard radars, the features of constructing probability density functionals, and methods for calculating the potential characteristics of algorithms and systems. In the formulation of the optimization problem, the target direction of the work, the initial data and a priori information are noted. An optimization criterion is selected from the analysis of the problem statement. It is known [5,8,10] that in conditions of parametric a priori uncertainty, a minimum of a priori information requires the maximum likelihood method.

The essence of the maximum likelihood functional method is to find an estimated parameter λ_i that maximizes the likelihood function $P[\vec{u}(t)|\vec{\lambda}]$. Function $P[\vec{u}(t)|\vec{\lambda}]$ is a conditional probability density function of a vector random process $\vec{u}(t) = \|u_1(t), u_2(t), u_3(t), \ldots u_M(t)\|$ at a fixed value of the vector of parameters, $\vec{\lambda}$, which can have both constant values and be a function of time $\vec{\lambda}(t)$ or spatial $\vec{\lambda}(\vec{r})$ (angular $\vec{\lambda}(\vartheta)$) coordinates. Instead of function $P[\vec{u}(t)|\vec{\lambda}]$, its logarithm is more often maximized.

To find optimal estimates of the parameters $\vec{\lambda}$, the following system of equations is solved:

$$\left. \frac{\partial \ln P[\vec{u}(t)|\vec{\lambda}]}{\partial \lambda_i} \right|_{\vec{\lambda}=\vec{\lambda}_{\text{true}}} = 0 \tag{9}$$

where $\partial/\partial\lambda_i$ is the partial derivative operator, and $\vec{\lambda}_{\text{true}}$ is the true value of the parameter $\vec{\lambda}$.

When solving problems of estimating parameters of spatially extended objects, the parameters are functions of spatial coordinates $\vec{\lambda}(\vec{r})$. Taking into account geometry in Figure 1, the system of Equation (9) will have the following form:

$$\left. \frac{\delta \ln P[\vec{u}(t, \vec{r}_t, \vec{r}')|\vec{\lambda}(\vec{r})]}{\delta \lambda_i(\vec{r})} \right|_{\vec{\lambda}(\vec{r})=\vec{\lambda}_{\text{true}}(\vec{r})} = 0, \tag{10}$$

where $\delta/\delta\lambda_i(\vec{r})$ is the operator of the variational (functional) derivative. The method for calculating the variational derivatives is considered in [28].

One of the most important stages in solving optimization problems is the construction of a likelihood function. In [8] a technique for constructing a wide class of likelihood functions is presented. The likelihood function for observation Equation (6) with statistical characteristics (8) is written in the form:

$$P[u(t, \vec{r}_t, \vec{r}')|\sigma^0(\vec{r})] = k[\sigma^0(\vec{r})] \exp \left\{ -\frac{1}{2} \int_T \int_T \int_{D_t} \int_{D_t} \int_{D'} \int_{D'} u(t_1, \vec{r}_{t1}, \vec{r}'_1) \right.$$
$$\left. \times W(t_1, t_2, \vec{r}_{t1}, \vec{r}_{t2}, \vec{r}'_1, \vec{r}'_2, \sigma^0(\vec{r})) u(t_2, \vec{r}_{t2}, \vec{r}'_2) dt_1 dt_2 d\vec{r}_{t1} d\vec{r}_{t2} d\vec{r}'_1 d\vec{r}'_2 \right\}, \tag{11}$$

where $k[\sigma^0(\vec{r})]$ is the coefficient that depends on the energy parameter $\sigma^0(\vec{r})$, and $W(t_1, t_2, \vec{r}_{t1}, \vec{r}_{t2}, \vec{r}\,'_1, \vec{r}\,'_2, \sigma^0(\vec{r}))$ is a function inverse to the correlation function $R_u(t_1, t_2, \vec{r}_{t1}, \vec{r}_{t2}, \vec{r}\,'_1, \vec{r}\,'_2)$ that can be found from the integral equation:

$$\underset{T\ D_t D'}{\int \int \int} R_u(t_1, t_2, \vec{r}_{t1}, \vec{r}_{t2}, \vec{r}\,'_1, \vec{r}\,'_2, \sigma^0(\vec{r}))W_u(t_2, t_3, \vec{r}_{t2}, \vec{r}_{t3}, \vec{r}\,'_2, \vec{r}\,'_3, \sigma^0(\vec{r}))d\vec{r}\,'_2 d\vec{r}_{t2}dt_2$$
$$= \delta(t_1 - t_3)\delta(\vec{r}_{t1} - \vec{r}_{t3})\delta(\vec{r}\,'_1 - \vec{r}\,'_3). \tag{12}$$

In order to estimate the limiting errors of the underlying surface parameters estimation, it is necessary to calculate the trace of the operator which is inverse to the Fisher operator:

$$\rho = \int_D \mathrm{tr}\ \Phi_{\mu\nu}^{-1}(\vec{r}, \vec{r}_1)\, d\vec{r} \ \Bigg|_{\vec{r}_1 = \vec{r}}, \tag{13}$$

where $\mathrm{tr}(\cdot)$ is the symbol of the trace,

$$\underline{\Phi}(\vec{r}, \vec{r}_1) = -\left\langle \frac{\delta^2 \ln P[u(t, \vec{r}_t, \vec{r}\,')|\vec{\lambda}(\vec{r})]}{\delta\lambda_\mu(\vec{r})\delta\lambda_\nu(\vec{r}_1)} \right\rangle$$

is the Fisher operator.

In the case of estimating one parameter, Expression (13) has the following form:

$$\sigma_\lambda^2 = \int_D -\frac{1}{\left\langle \frac{\delta^2}{\delta\lambda^2(\vec{r})} \ln P[u(t, \vec{r}_t, \vec{r}\,')|\lambda(\vec{r})] \right\rangle}\, d\vec{r}\ \Bigg|_{\vec{r}_1 = \vec{r}}, \tag{14}$$

where $\langle\cdot\rangle$ is the sign of statistical averaging.

3. Results

The proposed statistical theory can be applied to the problem of optimal radar cross section estimation in onboard imaging radars with planar antenna arrays.

3.1. Geometry of the Surface Sensing

In Figure 2 the geometry of the surface sensing is shown from the aerospace carrier of the radar system with a planar antenna array.

Using a small area around the phase center of the registration area (in the case of discrete idealization, one element of the antenna array) the signal (1) is emitted in the direction of the investigated surface in a wide sector of angles.

The received signals are stochastic and have the following form:

$$\dot{s}(t, \vec{r}\,') = \int_D \dot{F}(\vec{r})\dot{s}_0(t, \vec{r}, \vec{r}\,')d\vec{r}, \tag{15}$$

where

$$\dot{s}_0(t, \vec{r}, \vec{r}\,') = \dot{S}_0(t, \vec{r}, \vec{r}\,') \exp(j2\pi f_0 t), \tag{16}$$

$$\dot{S}_0(t, \vec{r}, \vec{r}\,') = \varepsilon \dot{I}(\vec{r}\,') \exp\left(j2k\vec{\vartheta}(\vec{r}, t)\vec{r}\,' \right)\dot{A}\left(t - \frac{2R_0(\vec{r}, t)}{c} \right)$$
$$\times \exp\left(-j2k\left(\frac{V^2(t - t_0)^2}{2R_0(\vec{r}, t_0)} \sin^2\theta_x(\vec{r}, t_0) - V(t - t_0)\cos\theta_x(\vec{r}, t_0) \right) \right). \tag{17}$$

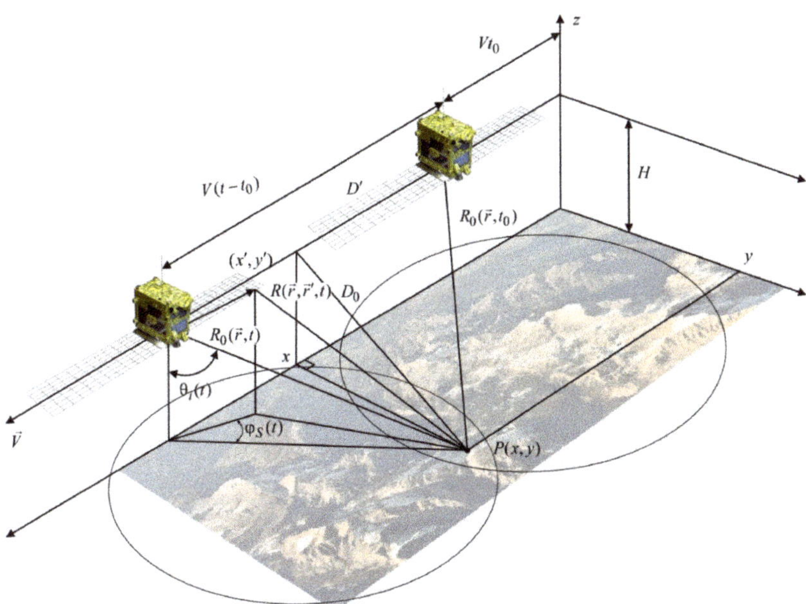

Figure 2. Geometry of surface sensing.

Expression (15), taking into account (17), describes the reflected signal from each point $P(x, y)$ at each point of the registration area with coordinates (x', y') in the process of rectilinear motion of the aircraft at a constant speed V.

For a statistical description of the received signals and noises, we will write down the observation equation:

$$u(t, \vec{r}') = \mathrm{Re}\,\dot{s}(t, \vec{r}') + n(t, \vec{r}'),\tag{18}$$

where $n(t, \vec{r}')$ is a white Gaussian noise.

Correlation function of observation Equation (18):

$$R_u(t_1, t_2, \vec{r}'_1, \vec{r}'_2) = \left\langle u(t_1, \vec{r}'_1)u(t_2, \vec{r}'_2) \right\rangle = R_s(t_1, t_2, \vec{r}'_1, \vec{r}'_2) + R_n(t_1, t_2, \vec{r}'_1, \vec{r}'_2)$$
$$= \tfrac{1}{2}\mathrm{Re}\int_D \sigma^0(\vec{r}, \vec{\lambda}(\vec{r}))\dot{s}_0(t_1, \vec{r}, \vec{r}'_1)\dot{s}_0^*(t_2, \vec{r}, \vec{r}'_2)d\vec{r} + \tfrac{1}{2}N_{0n}\delta(t_1 - t_2)\delta(\vec{r}'_1 - \vec{r}'_2).\tag{19}$$

3.2. Problem Statement

According to the reception of stochastic reflected oscillations $\dot{s}(t, \vec{r}')$ of each element of the antenna array D' which are observed against the background of additive Gaussian noise $n(t, \vec{r}')$, it is necessary to optimally estimate the radar cross section $\sigma^0(\vec{r})$ of the underlying surface as a statistical characteristic of the complex scattering coefficient $\dot{F}(\vec{r})$.

3.3. Solution of the Optimization Problem

We obtain the optimal estimation algorithm by the method of maximum likelihood with the likelihood function

$$P[u(t, \vec{r}')\,|\,\sigma^0(\vec{r})] = k[\sigma^0(\vec{r})]\exp\left\{ -\frac{1}{2}\int_T\int_T\int_{D'}\int_{D'} u(t_1, \vec{r}'_1)\,W(t_1, t_2, \vec{r}'_1, \vec{r}'_2, \sigma^0(\vec{r}))\,u(t_2, \vec{r}'_2)dt_1 dt_2\,d\vec{r}'_1 d\vec{r}'_2 \right\}.\tag{20}$$

Since $\sigma^0(\vec{r})$ depends on the coordinates $\vec{r}$, the problem of finding the maximum of Function (20) is solved by variational methods.

Since the exponent and its power are monotonically related to each other, instead of the likelihood Function (20) derivation, it is possible to take the derivative of its logarithm and equate the obtained result to zero:

$$\left. \frac{\delta \ln P[\vec{u}(t, \vec{r}')\,|\,\sigma^0(\vec{r})]}{\delta \sigma^0(\vec{r})} \right|_{\sigma^0(\vec{r})=\sigma^0_{true}(\vec{r})} = 0. \tag{21}$$

To obtain a variational derivative $\frac{\delta}{\delta \sigma^0(\vec{r})}$, the estimate $\sigma^0(\vec{r})$ can be represented as a sum

$$\hat{\sigma}^0(\vec{r}) = \sigma^0_{opt}(\vec{r}) + \delta\sigma^0(\vec{r}),$$

where $\sigma^0_{opt}(\vec{r})$ is the optimal value, and $\delta\sigma^0(\vec{r})$ is the variation of the estimate, which is some of its arbitrary small deviation from the optimal value

$$\delta\sigma^0(\vec{r}) = \alpha\gamma(\vec{r}),$$

$\gamma(\vec{r})$ is an arbitrary function describing the deviation of the desired estimate $\hat{\sigma}^0(\vec{r})$ in a short distance from a small parameter α from its optimal value $\sigma^0_{opt}(\vec{r})$. Instead of the variational derivative with respect to the function $\sigma^0(\vec{r})$, we will find a partial derivative with respect to the degree of deviation α.

As a result of taking the derivative, we obtain the following equation:

$$-\int_T \int_T \int_{D'} \int_{D'} \frac{dR_u(t_1, t_2, \vec{r}'_1, \vec{r}'_2, \sigma^0_{opt}(\vec{r})+\alpha\gamma(\vec{r}))}{d\alpha} W(t_1, t_2, \vec{r}'_1, \vec{r}'_2, \sigma^0_{opt}(\vec{r}) + \alpha\gamma(\vec{r}))d\vec{r}'_1 d\vec{r}'_2 dt_1 dt_2$$
$$= \int_T \int_T \int_{D'} \int_{D'} u(t_1, \vec{r}'_1) \frac{dW(t_1, t_2, \vec{r}'_1, \vec{r}'_2, \sigma^0_{opt}(\vec{r})+\alpha\gamma(\vec{r}))}{d\alpha} u(t_2, \vec{r}'_2)d\vec{r}'_1 d\vec{r}'_2 dt_1 dt_2. \tag{22}$$

This equation is called the likelihood equation. Its exact definition was obtained in [8]. It should be noted that in this equation does not use an equal sign, but rather a sign equating the left part to the right part. The right part is the main algorithm that corresponds to the result of processing the received signal $u(t, \vec{r}')$. Observation equations have a random nature because of internal noise. The result of $u(t, \vec{r}')$ will be random too. The left part is a deterministic process. It can be shown that the left-hand side of this likelihood equation is the average result (mathematical expectation) of the right-hand side.

We write the inverse correlation function as follows:

$$W(t_1, t_2, \vec{r}'_1, \vec{r}'_2, \sigma^0(\vec{r})) - \int_T \int_{D'} \int_T \int_{D'} W(t_1, t_0, \vec{r}'_1, \vec{r}'_0, \sigma^0(\vec{r}))$$
$$\times \left(\frac{1}{2}\mathrm{Re}\int_D \sigma^0(\vec{r})\dot{s}_0(t_3, \vec{r}, \vec{r}_3')\dot{s}_0^*(t_4, \vec{r}, \vec{r}_4')d\vec{r} + \frac{N_{0n}}{2}\delta(t_3 - t_4) \right) W(t_4, t_2, \vec{r}'_4, \vec{r}'_2, \sigma^0(\vec{r}))d\vec{r}'_3 dt_3 d\vec{r}'_4 dt_4$$
$$= \frac{1}{2}\mathrm{Re}\int_D \sigma^0(\vec{r})\int_T \int_{D'} \int_T \int_{D'} W(t_1, t_3, \vec{r}'_1, \vec{r}'_3, \sigma^0(\vec{r}))\dot{s}_0(t_3, \vec{r}, \vec{r}_3')\dot{s}_0^*(t_4, \vec{r}, \vec{r}_4')W(t_4, t_2, \vec{r}'_4, \vec{r}'_2, \sigma^0(\vec{r}))d\vec{r}'_3 dt_3 d\vec{r}'_4 dt_4\, d\vec{r}$$
$$+ \int_T \int_{D'} \int_T \int_{D'} W(t_1, t_3, \vec{r}'_1, \vec{r}'_3, \sigma^0(\vec{r}))\frac{N_{0n}}{2}\delta(t_3 - t_4)\delta(\vec{r}_3' - \vec{r}_4')W(t_4, t_2, \vec{r}'_4, \vec{r}'_2, \sigma^0(\vec{r}))d\vec{r}'_3 dt_3 d\vec{r}'_4 dt_4. \tag{23}$$

The derivative of the inverse correlation function has the following form:

$$\frac{dW(t_1, t_2, \vec{r}'_1, \vec{r}'_2, \sigma^0_{opt}(\vec{r})+\alpha\gamma(\vec{r}))}{d\alpha} = -\int_T \int_{D'} \int_T \int_{D'} W(t_1, t_3, \vec{r}'_1, \vec{r}'_3, \sigma^0(\vec{r}))$$
$$\times \frac{dR_u(t_3, t_4, \vec{r}'_3, \vec{r}'_4, \sigma^0_{opt}(\vec{r})+\alpha\gamma(\vec{r}))}{d\alpha} W(t_4, t_2, \vec{r}'_4, \vec{r}'_2, \sigma^0(\vec{r}))d\vec{r}'_3 dt_3 d\vec{r}'_4 dt_4. \tag{24}$$

Derived from the direct correlation function:

$$\frac{dR_u(t_3,t_4,\vec{r}\,'_3,\vec{r}\,'_4,\sigma^0_{opt}(\vec{r})+\alpha\gamma(\vec{r}))}{d\alpha} = \frac{d}{d\alpha}\left\{\frac{1}{2}\mathrm{Re}\int_D \left[\sigma^0_{opt}(\vec{r}) + \alpha\gamma(\vec{r})\right]\dot{s}_0(t_3,\vec{r},\vec{r}_3')\dot{s}_0^*(t_4,\vec{r},\vec{r}_4')d\vec{r}\right\}$$
$$= \frac{1}{2}\mathrm{Re}\int_D \gamma(\vec{r})\dot{s}_0(t_3,\vec{r},\vec{r}_3')\dot{s}_0^*(t_4,\vec{r},\vec{r}_4')d\vec{r}. \tag{25}$$

Substituting (25) into (24), we obtain:

$$\frac{dW(t_1,t_2,\vec{r}\,'_1,\vec{r}\,'_2,\sigma^0_{opt}(\vec{r}) + \alpha\gamma(\vec{r}))}{d\alpha} = -\frac{1}{2}\mathrm{Re}\int_D \gamma(\vec{r})\dot{s}_{0W}(t_1,\vec{r}_1',\sigma^0(\vec{r}))\dot{s}_{0W}^*(t_2,\vec{r}_2',\sigma^0(\vec{r}))d\vec{r}, \tag{26}$$

where $\dot{s}_{0W}(t_1,\vec{r}_1',\sigma^0(\vec{r})) = \int_T\int_{D'} W(t_1,t_3,\vec{r}\,'_1,\vec{r}\,'_3,\sigma^0(\vec{r}))\dot{s}_0(t_3,\vec{r},\vec{r}_3')d\vec{r}\,'_3 dt_3$ is the reference signal in the synthesized optimal method of recovery of radar cross section. The
presence of an inverse correlation function under the sign of the integral indicates that the
introduction of this signal into the processing algorithm involves inverse filtering of the
received oscillations, i.e., their decorrelation or bleaching (approaching white noise).

Substituting the obtained expressions (26) and (25) into the likelihood Equation (20),
we obtain:

$$\mathrm{Re}\int_D \gamma(\vec{r})\left(\frac{1}{4}\int_D \sigma^0(\vec{r}_1)\left|\dot{\Psi}_W(\vec{r},\vec{r}_1)\right|^2 d\vec{r}_1 + \frac{N_{0n}}{2}E_W(\vec{r}) - \frac{1}{2}\left|\dot{Y}(\vec{r})\right|^2\right)d\vec{r} = 0, \tag{27}$$

where

$$\dot{\Psi}_W(\vec{r},\vec{r}_1) = \int_T\int_{D'}\dot{s}_0(t_1,\vec{r},\vec{r}\,'_1)\dot{s}_{0W}^*(t_1,\vec{r}_1,\vec{r}\,'_1)d\vec{r}\,'_1 dt_1 \tag{28}$$

is the uncertainty function (hardware function) of scatterometric SAR and planar antenna
array, which determines its resolution. It is a spatial pulse characteristic of the radar system,
i.e., the reaction (result of processing) of the SAR to a point source $d\vec{r}$, which creates a
reflected single signal $\dot{s}_0(t_1,\vec{r},\vec{r}\,')$,

$$E_W(\vec{r}) = \frac{1}{2}\int_T\int_{D'}\left|\dot{s}_{0W}(t_3,\vec{r},\vec{r}\,'_3)\right|^2 d\vec{r}\,'_3 dt_3 \tag{29}$$

is the reference signal energy $\dot{s}_{0W}(t_1,\vec{r}_1',\sigma^0(\vec{r}))$,

$$\dot{Y}(\vec{r}) = \int_T\int_{D'} u(t_1,\vec{r}\,'_1)\dot{s}_{0W}\left[t_1,\vec{r}\,'_1,\sigma^0(\vec{r})\right]d\vec{r}\,'_1 dt_1 \tag{30}$$

is the optimal output effect of modified scatterometric SAR.

Since the function $\gamma(\vec{r})$ was defined as arbitrary, the zero equality of the integral (27)
is possible only when the expression under the integral is equal to zero, i.e.,

$$\dot{Y}(\vec{r}) = \int_T\int_{D'} u(t_1,\vec{r}\,'_1)\dot{s}_{0W}\left[t_1,\vec{r}\,'_1,\sigma^0(\vec{r})\right]d\vec{r}\,'_1 dt_1. \tag{31}$$

For simplification and more effective implementation of this algorithm, it is expedient
to pass in it to complex envelopes, therefore

$$\dot{Y}(\vec{r}) = \frac{1}{2}\int_T\int_{D'}\dot{U}(t_1,\vec{r}\,'_1)\dot{S}_{0W}^*\left[t_1,\vec{r}\,'_1,\sigma^0(\vec{r})\right]d\vec{r}\,'_1 dt_1. \tag{32}$$

From the obtained expressions it follows that the function $\left|\dot{Y}(\vec{r})\right|^2$ is a consistent and biased estimate (the bias $N_{0n}E_W(\vec{r})$ can be eliminated) of the radar cross section. It does not strive for true value with increasing observation time, even when the process $\dot{F}(\vec{r})$ is statistically homogeneous and the radar cross section is the constant $\sigma^0(\vec{r}) = \sigma^0 = \text{const}$. Note that the estimation of the radar cross section as a component of the energy spectrum $\dot{F}(\vec{r})$ is a spectral estimation. Its consistency can be ensured by further averaging.

Expression (32) is the basis of a modified algorithm for synthesizing apertures in onboard radars with planar antenna array. In contrast to the classical algorithm for synthesizing the aperture, which integrates the product of the received signal and the reference equal to a single signal, the modified algorithm additionally performs decorrelation of signals reflected from Earth's surface, which consists of their integration with $W(t_1, t_3, \vec{r}'_1, \vec{r}'_3, \sigma^0(\vec{r}))$. As a result, the characteristic speckle intervals will be much smaller than in the classical aperture synthesis. The decorrelation procedure provides a certain subdivision and can be performed using an inverse filter with a pulse response $W(t_1, t_3, \vec{r}'_1, \vec{r}'_3, \sigma^0(\vec{r}))$, which is usually used to solve incorrect inverse problems of recovery of various functions and, in particular, images [29].

For further analysis (32) rewrite the unit signal

$$\dot{S}_0(t, \vec{r}, \vec{r}') = \varepsilon \dot{I}(\vec{r}') \exp(j2k\vec{\vartheta}(\vec{r}, t)\vec{r}')\dot{A}\left(t - \frac{2R_0(\vec{r},t)}{c}\right)$$

$$\times \exp\left(-j2k\left(\frac{V^2(t-t_0)^2}{2R_0(\vec{r},t_0)}\sin^2\theta_x(\vec{r}, t_0) - V(t - t_0)\cos\theta_x(\vec{r}, t_0)\right)\right) \tag{33}$$

$$= \dot{S}_0(t, \vec{r})\dot{I}(\vec{r}')\exp\left(j2k\vec{\vartheta}(r, t)\vec{r}'\right),$$

where

$$\dot{S}_0(t, \vec{r}) = \varepsilon\dot{A}\left(t - \frac{2R_0(\vec{r},t)}{c}\right)\exp\left(-j2k\left(\frac{V^2(t - t_0)^2}{2R_0(\vec{r}, t_0)}\sin^2\theta_x(\vec{r}, t_0) - V(t - t_0)\cos\theta_x(\vec{r}, t_0)\right)\right). \tag{34}$$

Taking into account the presented single signal, we will rewrite the optimal output effect as follows:

$$\dot{Y}(\vec{r}) = 0,5\int_T\int_T \dot{S}_0^*(t_3, \vec{r})\int_{D'}\int_{D'} \dot{U}(t_1, \vec{r}'_1)\dot{I}^*(\vec{r}'_3)W(t_1, t_3, \vec{r}'_1, \vec{r}'_3, \sigma^0(\vec{r}))\exp\left(-j2k\vec{\vartheta}(\vec{r}, t_3)\vec{r}'_3\right)d\vec{r}'_3 d\vec{r}'_1 dt_1 dt_3, \tag{35}$$

where

$$\dot{S}_0(t, \vec{r}) = \varepsilon\dot{A}(t - 2R_0(\vec{r}, t)/c)\exp\left\{-j2k\left(\frac{V^2(t - t_0)^2}{2R_0(\vec{r}, t_0)}\right)\sin^2\theta_x(\vec{r}, t_0) - V(t - t_0)\cos\theta_x(\vec{r}, t_0)\right\} \tag{36}$$

is the complex envelope of the unit signal, $\dot{U}(t_1, \vec{r}'_1)$ is the complex envelope of the observation equation, $\vec{\vartheta}(\vec{r}, t)$ is the vector of directing cosines that change in time proportionally to the motion of the aircraft.

The obtained method of processing space–time signals (35) is general and fair for solving many problems of estimating the parameters of useful signals. However, the general expression in solving partial problems acquires a number of new properties that characterize the features of the construction of algorithms for a particular geometry of problems. Assume that the area of the registration of the reflected signals is discrete, consisting of a set of elementary antennas, which together form the antenna array. The model of the observation equation for such a discrete disclosure has the form:

$$u(t, \vec{r}'_m) = \text{Re}\dot{s}(t, \vec{r}'_m) + n(t, \vec{r}'_m), \quad m = \overline{1, M}. \tag{37}$$

In this case the algorithm (35) is written in matrix form

$$\dot{Y}(\vec{r}) = 0,5 \int_T \dot{S}_0^*(t_3, \vec{r}) \int_T W(t_1, t_3, \sigma^0(\vec{r})) \left(\vec{U}^T(t_1) \underline{W}[t_1, \sigma^0(\vec{r})] \dot{E}(\vec{r}, t_1) \right) dt_1 dt_3, \quad (38)$$

where $\vec{U}(t_1)$ is the vector-column of the received signals (dimension m $\times$ 1), $(\cdot)^T$ is the transposition sign, $\underline{W}[t_1, \sigma^0(\vec{r})]$ is the matrix of spatial matching filter with dimension m $\times$ n, $\dot{E}(\vec{r}, t_1) = \dot{I}_n^* \exp\left(-j2k\vartheta(\vec{r}, t_1)\vec{r}'_n\right)$ is the vector-column of amplitude phase distribution with dimension n $\times$ 1.

The essence of processing the received space–time signals $\dot{U}_{mn}(t)$ according to (38) is as follows. Initially, the received oscillations from the output of each element of the antenna array are processed in a spatial filter with a pulse response $\dot{I}_{mn}^*$. Then the space–time signals are bleached in the space filter $W_{mn}(t_1, \sigma^0(\vec{r}))$. The resulting signals are processed in a beam synthesizer, which performs a spatial discrete Fourier transform and forms a beam fan in all directions. As the aircraft moves, each individual beam at each time moves in space so that its maximum is always directed to a selected point on the surface. The next stage of processing is the decorrelation of signals over time in a filter with a pulse response $W(t_1, t_3, \sigma^0(\vec{r}))$. The coordinated filtering unit performs coherent detection of the amplitudes of the received and decorrelated signals from the directions $\vartheta(\vec{r})$ and coherent accumulation of the reflected signals along the flight path of the aircraft. Coherent phase shift in the reference signal leads to the formation of an artificial aperture, the length of which is equal to the product of the speed of the aircraft and the synthesis time. In this case, the synthesis time is determined by the focusing time on the selected point on the surface.

4. Discussion

To discuss effectiveness of the optimal method of incoherent imaging in radars with planar antenna array and processing of stochastic space–time signals a semi-natural experiment was performed on digital records of real radio-holograms of aircraft synthetic aperture radar. Such data (raw radar images) were obtained according to the program of cooperation with 14 research institutes in China, which specializes in the development of radars. In Figure 3 it is shown the whole "raw images" obtained from the side view radar and in the Figure 4 it is shown part of the radio hologram of the mirror point on the surface. In the digital radio hologram, the pixel number is plotted on the abscissa and ordinate axes.

All of the images in Figure 6 were not filtered; only the signal processing algorithm was implemented. In the image processed by the new method, the level of speckle noise is lower, the boundaries of objects are more clearly observed, and objects are detected (two bright points in Figure 6b and two extended objects in Figure 6d, which are the most likely to be cars moving on the road), which are hidden by noise in Figure 6a,b.

The reference quality metrics were used in order to quantify the quality of processing of received space–time signals in radars with planar antenna array on digital records of real radio holograms. The result of averaging 1000 estimates is shown in Table 1.

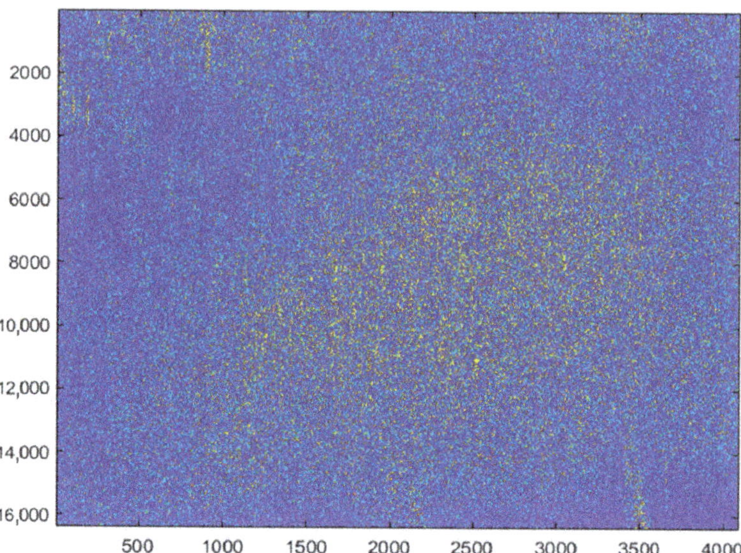

Figure 3. Radar data before processing ("raw" data)—the whole image.

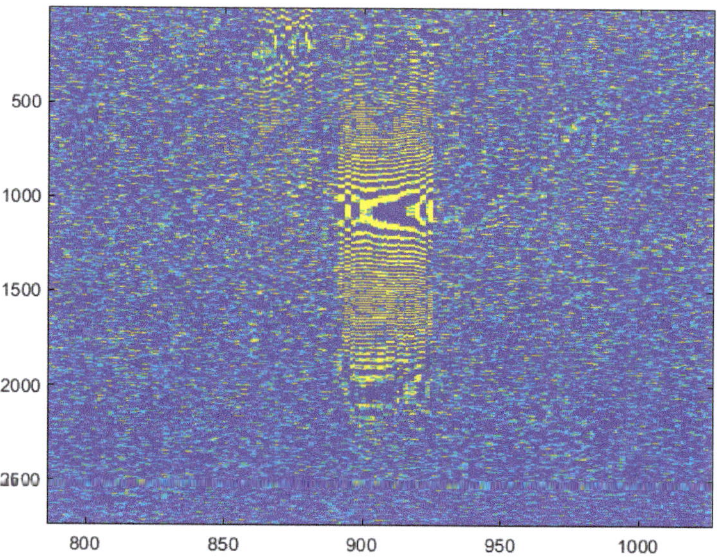

Figure 4. Part of the digital radio hologram of the mirror point on the surface.

The radar image recovered by the proposed method is shown in Figure 5. In Figure 6 the results of comparative analyses are shown—the radar images obtained by the classical method of signal processing (Figure 6a,c) and the new synthesized method (Figure 6b,d).

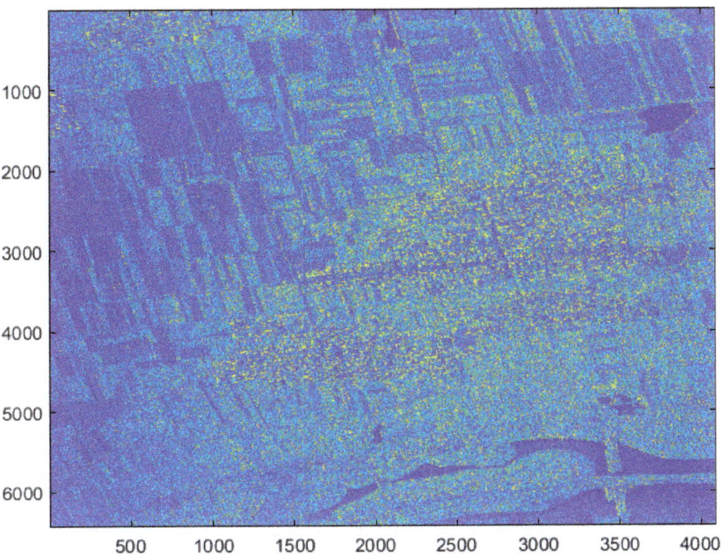

Figure 5. The radar image recovered by the proposed method.

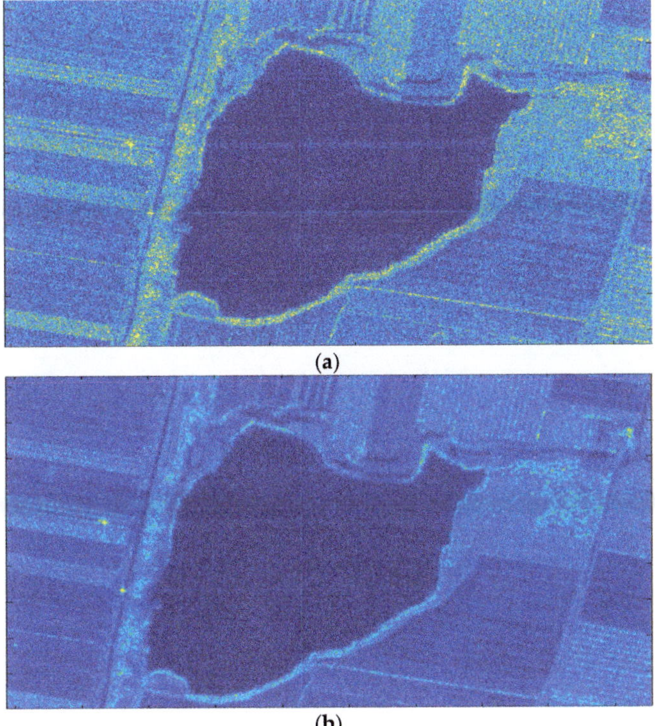

(a)

(b)

Figure 6. *Cont.*

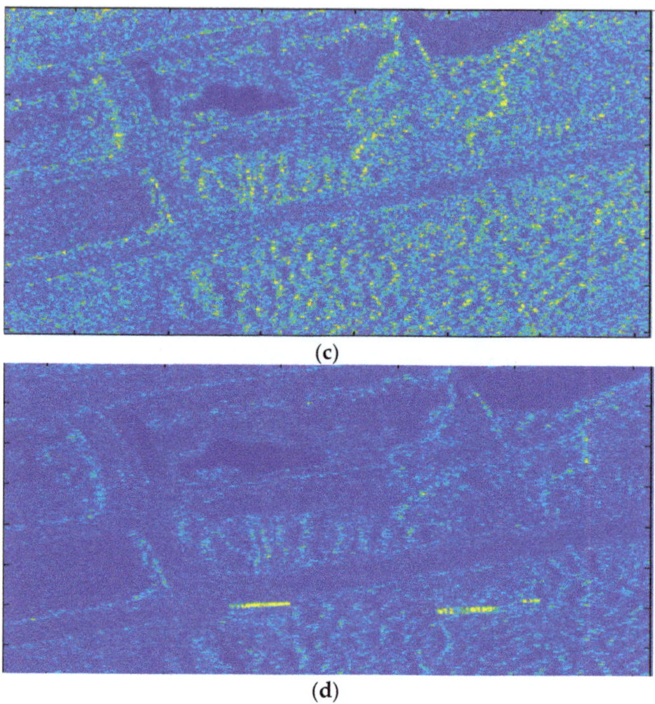

(c)

(d)

Figure 6. The result of the radar imaging: (**a,c**)—parts of the image obtained by the classical method (**b,d**)—more informative image obtained in accordance with the new method of signal processing.

Table 1. Quantitative assessments of the quality of coherent image recovery.

Metrics	Coherent Processing without Decorrelation	The Proposed Optimal Method of Stochastic Signal Processing
MSE	4.1259×10^3	4.0514×10^3
PSNR	11.9756	12.0547
SSIM	0.1245	0.1326

From the analysis of the results given in Table 1 it follows that the proposed method has a higher quality and a smaller amount of speckle noise.

5. Conclusions

The article shows how to overcome the contradiction between existing high-tech achievements of technical solutions of imaging radars and undeveloped statistical theory of optimal stochastic signals processing in such airborne radars. The developed models of signals, noises and observation equations allow researchers to describe geometries of imaging by multichannel and multi-view aerospace radars. It is supposed that receiving signals can be formed in any time, any point of receiving antenna and from any point of the surface. The proposed statistical approach of new methods synthesis enables the potential information to be calculated from observation equation. One of the main advantages of this theory is the analytical expression for marginal errors estimation. The solution of the particular problem of optimal radar cross section estimation in onboard imaging radars gave several new results. In contrast to the known heuristic methods of incoherent imaging it is proposed operation of whitening the received signals in the adaptive decorrelation filter. The proposed decorrelation operation in proportion to the signal-to-noise ratio expands the range of received signals, bringing them closer to white noise, and, thus, provides superresolution while coherent and incoherent imaging. These results were proved by a semi-natural experiment. In sum a statistical theory of radar cross section estimation as an incoherent image of the observation region in aerospace radars with antenna array has been developed.

Author Contributions: Conceptualization, V.V.; methodology, S.Z. All authors have read and agreed to the published version of the manuscript.

Funding: The work was funded by the Ministry of Education and Science of Ukraine. The state registration numbers of the projects are 0120U102082, 0119U100968, 0121U109600, and 0121U109598.

Institutional Review Board Statement: Not applicable.

Informed Consent Statement: Not applicable.

Data Availability Statement: Not applicable.

Conflicts of Interest: The authors declare no conflict of interest.

References

1. Kotel'nikov, V.A. *The Theory of Optimum Noise Immunity*; McGraw-Hill Book, Co.: New York, NY, USA, 1959.
2. Middlton, D. *Introduction to Statistical Communication Theory*; Sovetskoye radio: Moscow, Russia, 1961–1962; Volume 1–2. (In Russian)
3. Woodward, F.M. *Probability Theory and Information Theory with Application to Radar*; Pergamon: New York, NY, USA, 1964.
4. Bakut, P.A.; Tartakovsky, G.P. *Questions of the Statistical Theory of Radar*; Sovetskoye radio: Moscow, Russia, 1963. (In Russian)
5. Falkovich, S.E.; Ponomarev, V.I.; Shkvarko, Y.V. *Optimal Reception of Space-Time Signals in Radio Channels with Scattering*; Radio i svyaz: Moscow, Russia, 1989. (In Russian)
6. Falkovich, S.E.; Kostenko, P.Y. *Fundamentals of the Statistical Theory of Radio Engineering Systems: A Tutorial*; Kharkiv Aviation Institute: Kharkov, Russia, 2005. (In Russian)
7. Falkovich, S.E. *Signal Parameters Estimation*; Sovetskoye radio: Moscow, Russia, 1970. (In Russian)
8. Volosyuk, V.K.; Kravchenko, V.F. *Statistical Theory of Radio Engineering Systems for Remote Sensing and Radar*; Fizmatlit: Moscow, Russia, 1988. (In Russian)
9. Gutkin, L.S. *Theory of Optimal Radio Reception Methods with Fluctuation Interference*; Gosenergoizdat: Leningrad, Russia, 1961. (In Russian)
10. Levin, B.R. *Theoretical Foundations of Statistical Radio Engineering*; Sovetskoye radio: Moscow, Russia, 1969. (In Russian)
11. Shirman Ya, D.; Manzhos, V.N. *Theory and Technique of Processing Radar Information Against the Background of Interference*; Radio i svyaz: Moscow, Russia, 1981. (In Russian)
12. Kremer, I. *Spatial-Temporal Signal Processing*; Moscow Izdatel Radio Sviaz: Moscow, Russia, 2004; 204p.
13. Stratonovich, R. *Selected Questions of Fluctuation Theory in Radio Engineering*; Sov. Radio: Moscow, Russia, 1961. (In Russian)
14. Stratonovich, R.L. On the theory of optimal non-linear filtration of random functions. *Theory Prob. Appl.* **1959**, *4*, 223–225.
15. Stratonovich, R.L. Application of the Markov processes theory to optimum filtration of signals. *Radio Eng. Electron. Phys.* **1960**, *5*, 1–19.
16. Tikhonov, V.I. *Statistical Radio Engineering*; Radio i svyaz: Moscow, Russia, 1982. (In Russian)
17. Tikhonov, V.I. *Optimal Signal Reception*; Radio i svyaz: Moscow, Russia, 1983. (In Russian)
18. Klyatskin, V.I. *Stochastic Equations: Theory and Applications in Acoustics, Hydrodynamics, Magnetohydrodynamics, and Radiophysics, Volume 1: Basic Concepts, Exact Results, and Asymptotic Approximations. Understanding Complex Systems*; Springer International Publishing: Berlin/Heidelberg, Germany, 2015.
19. Tan, S. Multiple Volume Scattering in Random Media and Periodic Structures with Applications in Microwave Remote Sensing and Wave Functional Materials. Ph.D. Thesis, University of Michigan, Ann Arbor, MI, USA, 2016.
20. Tang, K.; Buckius, R.O. A statistical model of wave scattering from random rough surfaces. *Int. J. Heat Mass Transf.* **2001**, *44*, 4059–4073. [CrossRef]
21. Mishchenko, M.I.; Liu, L.; Mackowski, D.W.; Cairns, B.; Videen, G. Multiple scattering by random particulate media: Exact 3D results. *Opt. Express* **2007**, *15*, 2822–2836. [CrossRef] [PubMed]
22. Volosyuk, V.K.; Pavlikov, V.V.; Zhyla, S.S. Phenomenological Description of the Electromagnetic Field and Coherent Images in Radio Engineering and Optical Systems. In Proceedings of the 2018 IEEE 17th International Conference on Mathematical Methods in Electromagnetic Theory (MMET), Kyiv, Ukraine, 2–5 July 2018; pp. 302–305. [CrossRef]
23. Yang, Y.; Chen, K.; Shang, G. Surface Parameters Retrieval from Fully Bistatic Radar Scattering Data. *Remote Sens.* **2019**, *11*, 596. [CrossRef]

24. Zeng, J.Y.; Chen, K.S.; Bi, H.Y.; Zhao, T.J.; Yang, X.F. A comprehensive analysis of rough soil surface scattering and emission predicted by AIEM with comparison to numerical simulations and experimental measurements. *IEEE Trans. Geosci. Remote Sens.* **2017**, *55*, 1696–1708. [CrossRef]

25. Liao, T.H.; Tsang, L.; Huang, S.; Niamsuwan, N.; Jaruwatanadilok, S.; Kim, S.B.; Ren, H.; Chen, K.L. Co-polarized and cross-polarized backscattering from random rough soil surfaces from L-band to Ku-band using numerical solutions of Maxwell's equations with near-field precondition. *IEEE Trans. Geosci. Remote Sens.* **2016**, *54*, 651–662. [CrossRef]

26. TTeperik, V.; Archambault, A.; Marquier, F.; Greffet, J.J. Huygens-Fresnel principle for surface plasmons. *Opt. Express* **2009**, *17*, 17483–17490. [CrossRef] [PubMed]

27. Chu, H.; Qi, J.; Wang, R.; Qiu, J. Generalized Rayleigh-Sommerfeld Diffraction Theory for Metasurface-Modulating Paraxial and Non-Paraxial Near-Field Pattern Estimation. *IEEE Access* **2019**, *7*, 57642–57650. [CrossRef]

28. Klyatskin, V.I. *Stochastic Equations Through the Eye of the Physicist*; Elsevier: Amsterdam, The Netherlands, 2005.

29. Vasilenko, G.I.; Taratorin, A.M. *Image Recovery*; Radio i svyaz: Moscow, Russia, 1986.

Article

Statistical Theory of Optimal Functionally Deterministic Signals Processing in Multichannel Aerospace Imaging Radar Systems

Valeriy Volosyuk and Semen Zhyla *

Aerospace Radio-electronic Systems Department, Kharkiv Aviation Institute, National Aerospace University, 61070 Kharkiv, Ukraine
* Correspondence: s.zhyla@khai.edu

Abstract: The theory of the optimal formation of coherent and incoherent images is developed using the foundations of the statistical theory of optimization of radio engineering information-measuring systems. The main operations necessary for synthesizing optimal methods of spatio-temporal processing of functionally deterministic signals in on-board radio imaging radars with antenna arrays are shown. Models of radio engineering signals and noise have been developed. The statistical and correlation characteristics of spatio-temporal signals and noises in the area of their observation by antenna systems have been investigated. The technique for estimating the limiting errors of the measured characteristics of the studying media is presented. Using the developed theory, a new method for high-resolution radar imaging of the surface from a wide swath was obtained. This method has a new optimal observation mode combining the advantages of several terrain observation modes and fully complies with modern trends in the creation of cognitive radars with the possibility of restructuring the antenna pattern in space and adaptive receiving of reflected signals. The principles of construction and algorithmic support of high-precision airborne radars with an extended observation area are formulated. The effectiveness of the obtained results is investigated by simulation, taking into account the phenomenological approach to the description of electromagnetic fields and coherent images.

Keywords: coherent and incoherent images; multichannel multi-view radio engineering systems; optimal spatio-temporal signal processing algorithms

Citation: Volosyuk, V.; Zhyla, S. Statistical Theory of Optimal Functionally Deterministic Signals Processing in Multichannel Aerospace Imaging Radar Systems. *Computation* **2022**, *10*, 213. https://doi.org/10.3390/computation10120213

Academic Editor: Xiaoqiang Hua

Received: 10 October 2022
Accepted: 1 December 2022
Published: 3 December 2022

Publisher's Note: MDPI stays neutral with regard to jurisdictional claims in published maps and institutional affiliations.

1. Introduction

One of the strategic ways of developing national economic complexes of many countries is space exploration of the Earth and near-Earth space, as well as the planets of the solar system. In particular, the primary method is the creation of aircraft and satellite radio engineering means of remote sensing.

Recently, more attention has been paid to the use of aerospace-based radio equipment for solving problems of monitoring the environment and its ecological protection, namely the development of means for monitoring the state of the Earth's surface and its atmosphere and the degree of pollution, as well as estimates of parameters and statistical characteristics, and the development of means for collecting and transmitting information about the ecological state of the seas, oceans, agricultural lands, and ice cover of the Arctic and Antarctic.

In connection with the extreme importance of solving these problems at the present stage of the development of society, the principles of building such radio-technical means and their functioning are becoming increasingly common in practice and require new methods, devices, and systems for signal processing, especially spatio-temporal ones.

It should be noted that over the past two decades, the element base of radar devices has been significantly improved and new technical solutions have been developed for the construction of antenna arrays, optical sensors, amplifiers, high-speed digital processors and FPGAs. This technical breakthrough has led to an increase in the spatial resolution and information content of coherent images. At the same time, analysis of existing aero space synthetic aperture radars (SARs) shows that the existing systems have reached a certain limit in accuracy, spatial resolution, and the rate of the formation of global coherent images, which cannot be overcome. This is primarily due to the almost absent fundamental theoretical study of the operation of modern multichannel and multi-view airborne radars using the theory of end-to-end optimization of signal processing and structural synthesis of methods, devices, and systems. Such methods should be based on the existing achievements of the statistical theory of optimization of signal processing in measuring systems presented in the works of Falkovich S. E. [1–4], Tikhonov V. I. [5,6], Bakut P.A. [7], Amiantov I.M. [8], Kotelnikov V.A. [9], Gutkin L.S. [10], Levin B.R. [11], Van Tries G. [12], Middleton D. [13], Shirman Ya.D. [14], and others [15–18], as well as in the results of the statistical synthesis of coherent and incoherent images in SAR, specified in the works of Kondratenkov G.S. [19], Reutov A.P. [20], Karavaev V.V. [21], Sazonov V.V. [21], Antipov V.N. [22], Volosyuk V.K. [23], Moreira A. [24], Krieger G. [25], Reigber A. [26], and Charvat G.L. [27] et al.

The works and achievements of scientists in [1–27] cannot be directly applied to the processing of functionally deterministic signals in multichannel aerospace imaging radar systems. This is due to the following features: theoretical and implementation foundations of the spatio-temporal processing of signals by spatially distributed and multi-position systems, in particular, by antenna arrays, which are based on the theory of optimal statistical solutions and estimates of the parameters of probability distributions, are reflected in works [1–14]. To a greater extent, in these works, the results were obtained for stationary ground-based spatially distributed systems and antenna arrays; however, it is advisable to use the methodologies for their development and research in some aspects of their synthesis for the statistical synthesis of mobile airborne radars with aerospace-based antenna arrays in the state of their motion with extremely complex sets of Doppler frequency shifts that are absent in motionless systems when building images of motionless objects.

The works of Kraus J. D. [15], Tseitlin N.M. [16], Thompson A.R. [17], Van Schonveld K. [18] should also be noted because they consider the design of aperture synthesis radio astronomy systems in spatially distributed antenna systems and antenna arrays. However, the systems under consideration are also terrestrial and passive systems for receiving stochastic self-radiation signals of objects of study, the principles of aperture synthesis and imaging, in which they radically differ from the principles of imaging by active radars in aerospace-based antenna arrays.

In the basic works of [19–27], methods for synthesizing apertures in mobile airborne systems are considered to a greater extent in relation to the use of single longitudinal fuselage airborne antenna systems, which do not allow for optimizing processing in airborne phased antenna arrays and realizing multiview surface studies. The possibilities of using antenna arrays are considered mainly for solving specific problems of holographic studies of the atmosphere, subsurface sounding, and mapping of surface relief heights, not for improving the quality and information content of images. At the same time, the obtained results can be used to confirm the reliability of the synthesized methods, algorithms, and structures developed in this dissertation by analyzing special cases while simplifying the problem formulations.

Problem statement. From the analysis of the modern airborne radars and the existing separated theoretical foundations of statistical synthesis comes the following contradiction: on the one hand, technical breakthrough of radar components and units is enough for implementing a new method of rapid high resolution surface imaging. On the other hand, theoretical foundations of statistical synthesis of new methods of signals processing, taking into account multichannel, multi-view, and airborne or spaceborne reception, are absent. It is necessary to develop a new statistical theory of optimal processing of functionally deterministic signals, taking into account their multichannel reception while multi-view observation is made from the moving platforms.

2. Materials and Methods

The development of the statistical theory for the synthesis of new signal processing methods will be based on the maximum likelihood method. The peculiarity of the application of this method in this article is the construction of functionals rather than likelihood functions, which is due to the estimation of spatial functions in the form of coherent radar images of the surface, rather than unit parameters.

2.1. Models of Signals, Noises, and Observation Equation

For the development of the statistical theory of optimal functionally deterministic signals processing, it is necessary, firstly, to determine the models of the probing signal, the received signals by each element of the antenna array, the model of the relationship of the estimated parameters with the received signals, and the models of internal noise and observation equations, which are subject to optimal processing.

The probing signal has the following form:

$$s_t(t) = A(t)\cos(2\pi f_0 t + \varphi) = \mathrm{Re}\left\{\dot{A}(t)e^{j\omega_0 t}\right\}, \tag{1}$$

where $A(t)$ is the envelope of the probing signal, $\dot{A}(t) = A(t)e^{j\varphi}$ is the complex envelope, taking into account some initial phase, φ, f_0 is the central frequency of the spectrum of the probing signal, ω_0 is the angular frequency, and t is the time. $\dot{A}(t)$ characterizes a wide class of radio signals, both simple and complex, with internally pulsed modulation.

The probing signal propagates, reaches the observation area D with coordinates $\vec{r} = (x, y, 0) \in D$, and is scattered on its inhomogeneities. Scattered signals in the registration area can be determined coherently, taking into account the fundamentals of the diffraction theory [28], Kirchhoff's integral theorem, Kirchhoff–Helmholtz integral theorem [29], Rayleigh–Sommerfeld theory [30], and Stratton–Chu formulas [31,32]. From the analysis of these theories, it follows that these fundamentals are sophisticated and gives close results. It is reasonable to use a phenomenological description of the electromagnetic field. On the one hand, it will give a general and understandable description of electromagnetic field calculation, and on the other hand it has a clear mathematical description. Mathematically, the essence of the phenomenological approach can be explained by the form of the received spatio-temporal signal by the multichannel receiving area:

$$\dot{s}(t, \vec{r}') = \int_D \dot{F}(\vec{r})\, \dot{s}_0(t, \vec{r}, \vec{r}')\, d\vec{r}, \tag{2}$$

where $\dot{F}(\vec{r})$ is the general expression for the Kirchhoff–Helmholtz integral theorem:

$$\dot{F}(\vec{r}) = (4\pi)^{-1}(\partial E(\vec{r})/\partial \vec{n}) - (4\pi)^{-1} jk E(\vec{r}) \cos(\vec{n}, \vec{R}), \tag{3}$$

The Rayleigh–Sommerfeld theory:

$$\dot{F}(\vec{r}) = (j\lambda)^{-1} E(\vec{r}) \cos(\vec{n}, \vec{R}), \tag{4}$$

calculations of Rytov S. M., Kravtsov Yu. A. and Tatarskiy V. I. [33]:

$$\dot{F}(\vec{r}) = (2\pi)^{-1} \partial E(\vec{r})/\partial z, \tag{5}$$

$E(\vec{r})$ is the electromagnetic field, $\vec{n}$ is the outer normal to the surface, k is the wavenumber, j is the imaginary unit,

$$\dot{s}_0(t, \vec{r}, \vec{r}') = \varepsilon \dot{I}(\vec{r}') \dot{A}(t - t_d(\vec{r}, \vec{r}')) \exp[j2\pi f_0(t - t_d(\vec{r}, \vec{r}'))] \tag{6}$$

is the unit signal depending on time delay $t_d(\vec{r}, \vec{r}')$, amplitude-phase distribution of receiving area $\dot{I}(\vec{r})$, and attenuation of electromagnetic waves ε, $\vec{r}'$ are the coordinates of the multichannel receiving area.

Function $d\dot{Q}(\vec{r}) = \dot{F}(\vec{r})d\vec{r}$ is the complex scattering coefficient of the surface element $d\vec{r}$ and $\dot{F}(\vec{r}) = d\dot{Q}(\vec{r})/d\vec{r}$ is the specific complex scattering coefficient [23]. This coefficient $\dot{F}(\vec{r})$ takes into account the amplitude and phase structure of the field. $\dot{F}(\vec{r})$ we call a true coherent image of the medium. Such images are widely used in remote sensing problems [23,34,35].

The received signals are always observed against the background of the internal noise of the receiver $n(t, \vec{r}')$, which is approximated by white Gaussian processes with a correlation function:

$$R_n(t_1, t_2, \vec{r}'_1, \vec{r}'_2) = \left\langle n(t_1, \vec{r}'_1)n(t_2, \vec{r}'_2) \right\rangle = 0,5N_{0n}\delta(t_1 - t_2)\delta(\vec{r}'_1 - \vec{r}'_2). \tag{7}$$

The spectral noise density N_{0n} in each element of the AR is usually the same, i.e., $N_{0n} = N_0$.

The observation equation is stated as an additive mixture of reflected useful signals and delta correlated noises:

$$u(t, \vec{r}') = \text{Re}\dot{s}(t, \vec{r}') + n(t, \vec{r}'). \tag{8}$$

Equation (8) will be processed optimally in the information-measuring systems.

2.2. Basic Principles of the Theory

The optimal spatio-temporal signal processing observed in multi-channel radars against the background of internal noise of the receivers will be performed using modern advances in theory, namely on the basis of the criterion of the maximum likelihood function. The essence of this criterion is to find a parameter λ that maximizes the likelihood functional $P[u(t, \vec{r}')|\lambda]$, the conditional probability density functional of a random process $u(t, \vec{r}')$ at a fixed value of the parameter $\lambda(\vec{r})$, which is a function of spatial coordinates $\lambda(\vec{r})$. Instead of a functional $P[u(t, \vec{r}')|\lambda(\vec{r})]$, its logarithm is more often maximized. To find the optimal estimates of the parameter $\lambda(\vec{r})$, it is necessary to solve a system of equations:

$$\left. \frac{\delta P[u(t, \vec{r}')|\lambda(\vec{r})]}{\delta\lambda(\vec{r})} \right|_{\lambda(\vec{r})=\lambda_{true}(\vec{r})} = 0, \tag{9}$$

where $\delta/\delta\lambda(\vec{r})$ is the symbol of the variational (functional) derivative, which is taken at the point of the true value $\lambda_{true}(\vec{r})$ of the parameter $\lambda(\vec{r})$.

For the specified equation of observation (9), the likelihood functional has the following form:

$$P[u(t, \vec{r}')|\lambda(\vec{r})] = \kappa \exp\left\{ -\frac{1}{N_0} \int_T \int_{D'} [u(t, \vec{r}') - \text{Re}\dot{s}(t, \vec{r}')]^2 d\vec{r}' \, dt \right\}, \tag{10}$$

where κ is the coefficient that does not depend on the parameter $\lambda(\vec{r})$ and T is the observation time.

The marginal errors of the estimation of several parameters $\vec{\lambda}(\vec{r})$ are obtained by calculating the trace of the operator inverse to the Fisher operator $\underline{\Phi}(\vec{r}, \vec{r}_1)$:

$$\rho = \int_D \mathrm{tr}\, \underline{\Phi}^{-1}(\vec{r}, \vec{r}_1)\, d\vec{r}\, \bigg|_{\vec{r}_1 = \vec{r}}, \qquad (11)$$

where tr is the matrix trace symbol.

We find the elements of the operator $\underline{\Phi}(\vec{r}, \vec{r}_1)$ by calculating the second mixed variational derivatives of the logarithm of the likelihood functional:

$$\Phi_{\mu\nu}(\vec{r}, \vec{r}_1) = -\left\langle \frac{\delta \ln P[u(t, \vec{r}')|\lambda(\vec{r})]}{\delta\lambda_\mu(\vec{r})\delta\lambda_\nu(\vec{r}_1)} \right\rangle, \qquad (12)$$

where $\langle \cdot \rangle$ is the sign of statistical averaging.

In the case of estimating one parameter, $\lambda(\vec{r})$ expression (11) has the following form:

$$\sigma_\lambda^2 = \int_D -\frac{1}{\left\langle \frac{\delta^2}{\delta\lambda^2(\vec{r})} \ln P[u(t, \vec{r}')|\lambda(\vec{r})] \right\rangle} d\vec{r}\, \bigg|_{\vec{r}_1 = \vec{r}'}, \qquad (13)$$

where $\delta^2/\delta\lambda^2(\vec{r})$ is the symbol of the secondary variational derivative.

Investigation of the marginal errors in the estimation of the parameters of natural environments in remote sensing problems is of great importance for the optimal choice of the conditions for carrying out measurements and the corresponding experiments [36–39]. This concerns the choice of frequency ranges, directions of irradiation and directions of reception of scattered radiation from the investigated medium, the choice of polarization, etc. Analysis of these errors allows us to choose such conditions for measurements, in which the expected real measurement errors will be minimal.

3. Results

Using the developed theory, a new method for high-resolution radar imaging of the surface from an aerial vehicle is obtained.

3.1. Problem Geometry and Received Signal Model

The geometry of the surface sensing from the aircraft is shown in the Figure 1. The aircraft moves with a constant speed V at a height H parallel to the axis x. The parameters V and H are known. The registration area is a rectangular planar antenna array with coordinates $\vec{r} = (x, y, 0) \in D$. It is supposed that the transmitting signal is radiated in a wide band of angles. To implement such a mode, it is possible to use a small area of antenna array (several elements) around the antenna phase center.

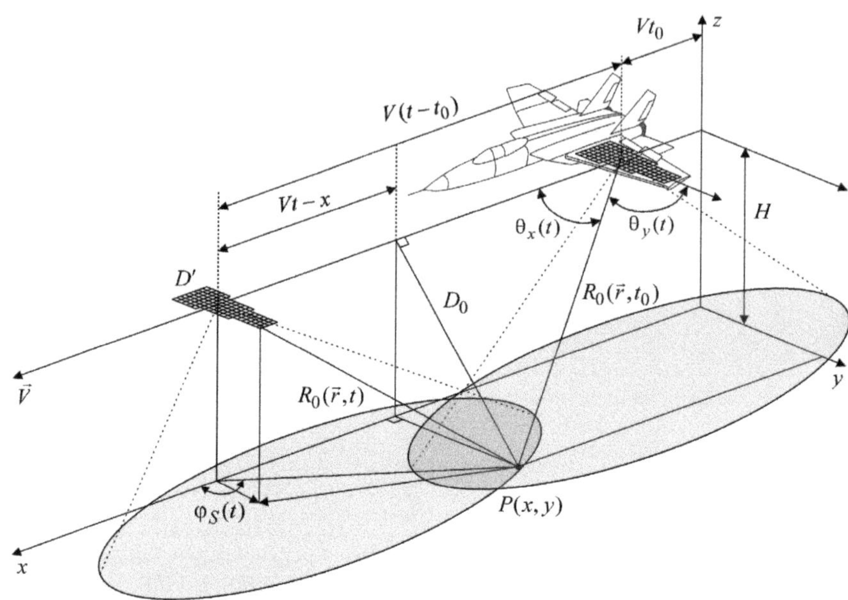

Figure 1. Geometry of surface sensing.

For the stated problem, the unit signal has the following form:

$$\dot{s}_0(t,\vec{r},\vec{r}') = \varepsilon \dot{I}(\vec{r}')\exp\left\{ j2\pi f_0 2\frac{\vec{\vartheta}(\vec{r},t)\vec{r}'}{c}\right\}\dot{A}\left(t - 2\frac{R_0(\vec{r},t)-\vec{\vartheta}(\vec{r},t)\vec{r}'}{c}\right)\times$$
$$\times \exp\left(-j2\pi f_0 2\frac{R_0(\vec{r},t)}{c}\right)\exp(j2\pi f_0 t) = \dot{S}_0(t,\vec{r},\vec{r}')\exp(j2\pi f_0 t), \tag{14}$$

where $\dot{S}_0(t,\vec{r},\vec{r}') = \varepsilon \dot{I}(\vec{r}')e^{j2\pi f_0 2\frac{\vec{\vartheta}(\vec{r},t)\vec{r}'}{c}}\dot{A}\left(t - 2\frac{R_0(\vec{r},t)-\vec{\vartheta}(\vec{r},t)\vec{r}'}{c}\right)e^{-j2\pi f_0 2\frac{R_0(\vec{r},t)}{c}}$, c is the speed of light and $\vec{\vartheta}(\vec{r},t) = \left(\vartheta_x(\vec{r},t) = \cos\theta_x(\vec{r},t),\ \vartheta_y(\vec{r},t) = \cos\theta_y(\vec{r},t)\right)$ is the vector of direction cosines that change in time in proportion to the motion of the aircraft. Angles $\theta_x(\vec{r},t)$ and $\theta_y(\vec{r},t)$ are shown in Figure 1.

Distance $R_0(\vec{r},t)$ can be described mathematically as follows:

$$R_0(\vec{r},t) = R_0(\vec{r},t_0) + \frac{V^2(t-t_0)^2}{2R_0(\vec{r},t_0)}\sin^2\theta_y(\vec{r},t_0) - V(t-t_0)\cos\theta_y(\vec{r},t_0), \tag{15}$$

where t_0 is the initial time of signal processing and $R_0(\vec{r},t_0)$ is the initial distance from phase center of the antenna array to chosen surface point $P(x,y)$.

Revealing $R_0(\vec{r},t_0)$ in the exponents in Expression (14), a number of practically justified approximations can be performed. Firstly, in the practice of radar measurements, it is almost impossible to determine the phase delay with an accuracy of wavelength when the signal passes a double distance $R_0(\vec{r},t_0)$. This is due to the roughness of the surface of the study, the height of which is several (and even hundreds) times greater than the wavelength, multiple reflections of signals in urban areas, random phase rotation of reflected signals from natural vegetation, and other reasons. Therefore, we include the exponent $\exp\left(-j2kR_0(\vec{r},t_0)\right)$ as an unknown quantity in the complex scattering coefficient $\dot{F}(\vec{r})$ in Formula (2). Secondly, it is possible to neglect the delay $2\frac{\vec{\vartheta}(\vec{r},t)\vec{r}'}{c}$ of the envelope signal relative to the phase center of the registration area, as it is much less time $2\frac{R_0(\vec{r},t)}{c}$. At the same time, this delay cannot be neglected in the multiplier $\exp\left\{j2\pi f_0 2\frac{\vec{\vartheta}(\vec{r},t)\vec{r}'}{c}\right\}$, as it is responsible for the alignment of the phases within the aperture of the receiving plane relative to the phase center during the formation of the directing antenna pattern. Expression (15) is eventually converted to the following form:

$$\dot{s}_0(t,\vec{r},\vec{r}') = \varepsilon\,\dot{I}(\vec{r}')\exp\left\{j2k\vec{\vartheta}(\vec{r},t)\vec{r}'\right\}\dot{A}\left(t - \frac{2R_0(\vec{r},t)}{c}\right)\times$$
$$\times\exp\left(-j2k\left(\frac{V^2(t-t_0)^2}{2R_0(\vec{r},t_0)}\sin^2\theta_x(\vec{r},t_0) - V(t-t_0)\cos\theta_x(\vec{r},t_0)\right)\right)\exp(j2\pi f_0 t). \tag{16}$$

Expression (16) describes the reflected signal from each point $P(x,y)$ at each point of the registration area with coordinates (x',y'), while rectilinear motion of the aircraft has a constant speed V.

3.2. Problem Statement

According to the reception of oscillations $\dot{s}(t,\vec{r}')$ by each element of the antenna array, which are observed against the background of additive Gaussian noise $n(t,\vec{r}')$, it is necessary to optimally estimate the specific complex scattering coefficient $\dot{F}(\vec{r})$ of the underlying surface.

3.3. Optimization Problem Solution

The likelihood functional (10) can be written in the following form:

$$P[u(t,\vec{r}')|\lambda(\vec{r})] = \kappa\exp\left\{-\frac{1}{N_0}\int_T\int_{D'}\left[u(t,\vec{r}') - \mathrm{Re}\int_D \dot{F}(\vec{r})\dot{s}_0(t,\vec{r},\vec{r}')d\vec{r}\right]^2 d\vec{r}'dt\right\}. \tag{17}$$

As the desired parameter for estimation is the scattering coefficient $\dot{F}(\vec{r})$, it is necessary to solve the following equation:

$$\frac{\delta P[u(t,\vec{r}')|\dot{F}(\vec{r})]}{\delta\dot{F}(\vec{r})}\Bigg|_{\dot{F}(\vec{r})=\dot{F}_{true}(\vec{r})} = 0. \tag{18}$$

The result of the maximum likelihood functional determination is the following inequality:

$$\int_T\int_{D'} u(t,\vec{r}')\dot{s}_0(t,\vec{r},\vec{r}')d\vec{r}'dt = \frac{1}{2}\int_D \hat{\dot{F}}(\vec{r}_1)\dot{\Psi}^*(\vec{r}_1,\vec{r})d\vec{r}_1, \tag{19}$$

where

$$\dot{\Psi}^*(\vec{r}_1,\vec{r}) = \int_T\int_{D'}\dot{s}_0^*(t,\vec{r}_1,\vec{r}')\dot{s}_0(t,\vec{r},\vec{r}')d\vec{r}'dt \tag{20}$$

is the complex ambiguity function of the measuring system, which characterizes the resolution of radar by spatial coordinates. It takes into account many factors that affect the quality of the formed coherent images: the type of probing signal, the size of the aperture of the receiving area, the complex distribution of the field in the area, and the time of accumulation of reflected signals.

The resulting form of Equation (19) is not simple. The left part of (19) is the optimal signal processing algorithm, and the right part is the optimal estimation of the coherent image smoothed by the ambiguity function.

Using the method of complex envelopes, the inequality (19) is written as follows:

$$\int_T \int_{D'} \dot{U}(t, \vec{r}')\dot{S}_0^*(t, \vec{r}, \vec{r}')d\vec{r}'dt = \frac{1}{2}\int_D \hat{F}(\vec{r}_1)\dot{\Psi}^*(\vec{r}_1, \vec{r})d\vec{r}_1, \tag{21}$$

where $\dot{U}(t, \vec{r}')$ is the complex envelope of the observation equation.

If $\dot{\Psi}^*(\vec{r}_1, \vec{r})$ has a form of spatial delta function, it is possible to restore a true coherent image without distortions.

The left parts of Equations (19) and (21):

$$\dot{Y}(\vec{r}) = \int_T \int_{D'} u(t, \vec{r}')\dot{s}_0(t, \vec{r}, \vec{r}')d\vec{r}'dt \approx \int_T \int_{D'} \dot{U}(t, \vec{r}')\dot{S}_0^*(t, \vec{r}, \vec{r}')d\vec{r}'dt \tag{22}$$

are called correlation integrals, which contain the basic necessary operations on the received signals. If it is assumed that the unit signals under the signs of the integrals are the impulse characteristics of the optimal filters, then operations (22) are called matched filtering operations.

Substituting (16) with (22), we obtain the optimal output effect in a mobile radar system with a planar registration area:

$$\dot{Y}(\vec{r}) = \varepsilon \int_T \int_{D'} \dot{U}(t, \vec{r}')\dot{I}^*(\vec{r}') \exp\left\{-j2k\vec{\vartheta}(\vec{r}, t)\vec{r}'\right\}d\vec{r}' \dot{A}^*\left(t - \frac{2R_0(\vec{r}, t)}{c}\right) \times$$
$$\times \exp\left(j2k\left(0,5V^2(t - t_0)^2R_0^{-1}(\vec{r}, t_0)\sin^2\theta_x(\vec{r}, t_0) - V(t - t_0)\cos\theta_x(\vec{r}, t_0)\right)\right)dt. \tag{23}$$

3.4. Physical Interpretation of the Optimal Method

The essence of processing the received field according to (23) is as follows: firstly, the antenna is focused on each point of the underlying surface $P(x, y)$. To do this, the signals received by each point of the coordinate region D' are delayed for a time $\vec{\vartheta}(\vec{r}, t)\vec{r}'c^{-1}$ and coherently summarize with amplitude-phase distribution $\dot{I}^*(\vec{r}')$. The selection of $\dot{I}^*(\vec{r}')$ can adjust the shape of the antenna pattern. It should be noted that the delay in each channel at the time t_0 is based on the vector of angles $\vec{\vartheta}(\vec{r}, t_0)$. This leads to the formation of a multibeam antenna pattern with the possibility of further processing of the signals in each beam separately. During the movement of the aircraft, at each point in time t, the delay time $\vec{\vartheta}(\vec{r}, t)\vec{r}'c^{-1}$ changes so that the maxima of each beam of antenna pattern is always directed to the selected points of the surface. This type of inspection allows us to increase the observation time and expand the range of viewing angles. The next stage of processing is coherent amplitude detection, which can be implemented in series or parallel circuits. The last multiplier reveals the essence of the classical method of antenna aperture synthesis, which consists of the coherent accumulation of reflected signals along the flight path of the aircraft.

The described processing combines two methods of forming a synthesized antenna aperture using a Spot-Light and Multibeam observation mode of the underlying surface. At the same time, it also realizes the benefits of each of them. The obtained method has the highest spatial resolution in azimuth (along the flight path) due to the constant focusing on the selected area of space and covers a significant area of the surface as a result of the formation of a significant number of partial antenna patterns. The principle of the formation of many rays of the antenna pattern, each of which focuses on the selected area of the underlying surface, is followed by coherent processing of the trajectory signal, as shown in Figure 2.

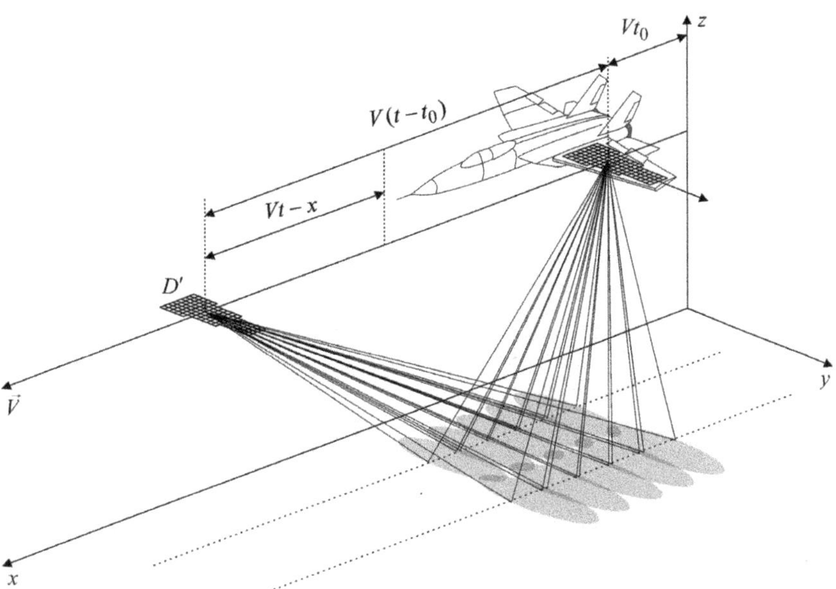

Figure 2. New mode of high resolution and wide-swath surface sensing.

3.5. Heuristical Modification of the Optimal Method

3.5.1. Multibeam Inspection with a Fixed Fan of Rays

The obtained algorithm (23) assumes the rotation of the antenna pattern when focusing on the selected point of the surface. At the same time, it is possible to develop another algorithm, which consists in the formation of a fixed antenna pattern fan, the accumulation of sections of the trajectory signal in each beam, combining the obtained sections or serial connection of the outputs of the diagram-forming circuit to the receiver, and coordinated processing of the trajectory signal in the form of convolution with reference signal.

The effective width of each individual beam in the fan is determined by the linear dimensions of the antenna (D'_x, D'_y) and amplitude-phase distribution $\dot{I}(\vec{r}\,')$. For example, for uniform amplitude distribution and zero phase distribution, the effective width of the antenna pattern at the first zeros along the flight path will be equal to $\Delta\theta_x \approx \lambda/D'_x$. The step between the partial diagrams must also be equal $\Delta\theta_x$ so that the observation angles of the selected point of the surface "flow" are continuous from one beam to another during the movement of the aircraft.

To obtain the analytical form of the proposed modification of the optimal algorithm, the sampling of the entire observation time $(0, T)$ was performed with a step in which the direction of each individual antenna pattern does not change.

The sampling step must be variable because the observation time is determined by the size of the antenna pattern on the surface:

$$\Delta X_i = 2H\frac{\sin(\Delta\theta_x)}{\cos(\Delta\theta_x)}\frac{1}{\left(1 - \frac{\cos(2\theta_{xi})}{\cos(\Delta\theta_x)}\right)} = \frac{2Htg(\Delta\theta_x)}{\left(1 - \frac{\cos(2\theta_{xi})}{\cos(\Delta\theta_x)}\right)}, \tag{24}$$

where θ_{xi} is the fixed direction of the i-th maximum of the antenna pattern and $\Delta\theta_x$ is the width of the antenna pattern in the azimuthal plane.

Taking into account the obtained discrete dimensions of the observation area (24), we write the optimal output effect (23) as follows:

$$\dot{Y}(\vec{r}) = \varepsilon \int_T \sum_{i=0}^{N-1} \Pi(t - iT_i, T_i) \left[\dot{U}_i(t, \vec{r}', i) \dot{A}^* \left(t - \frac{2R_0(\vec{r}, i)}{c} \right) \right] \times$$
$$\times \exp\left(j2k\left(0,5\frac{V^2(t-t_0)^2}{R_0(\vec{r}, t_0)} \sin^2 \theta_x(\vec{r}, t_0) - V(t - t_0) \cos \theta_x(\vec{r}, t_0) \right) \right) dt. \tag{25}$$

In Expression (25), the functions that are under the integral over the variable t are multiplied by a discrete sequence of rectangular pulses $\Pi(t - iT_i, T_i)$ of duration T_i. The total number of pulses N is equal to the number of the generated partial antenna patterns. These pulses are displaced relative to each other unequally with a variable step iT_i. Within the limits T_i, it is assumed that the direction of the maximum of the antenna pattern does not change $\vartheta(\vec{r}, t) \approx \vartheta(\vec{r}, i)$ (i.e., the rotation of the antenna pattern is not performed) and the accumulation of the signal at different angles is due to the expansion of the antenna pattern by the value $\Delta\theta_x$.

We can also assume that within T_i, the distance $R_0(\vec{r}, t) \approx R_0(\vec{r}, i)$. Function

$$\dot{U}_i(t, \vec{r}', i) = \int_{D'} \dot{U}(t, \vec{r}') \dot{I}^*(\vec{r}') \exp\left(-j2k\vartheta(\vec{r}, i)\vec{r}' \right) d\vec{r}' \tag{26}$$

is the voltage at one of the outputs of the diagram-forming circuit antenna array, function

$$\exp\left(j2k\left(0,5\frac{V^2(t - t_0)^2}{R_0(\vec{r}, t_0)} \sin^2 \theta_x(\vec{r}, t_0) - V(t - t_0) \cos \theta_x(\vec{r}, t_0) \right) \right)$$

included in the sign of the time integral are rapidly oscillating and cannot be represented by a constant value within the pulses $\Pi(t - iT_i, T_i)$.

3.5.2. Fixed Beam Fan with Coherent Processing at the Output of Each Channel and Coherent Inter-channel Addition

To analyze this modification of the algorithm, rewrite expression (23) as follows:

$$\dot{Y}(\vec{r}) = \varepsilon \sum_{i=0}^{N-1} \int_T \Pi(t - iT_i, T_i) \left[\dot{U}_i(t, \vec{r}', i) \dot{A}^* \left(t - \frac{2R_0(\vec{r}, i)}{c} \right) \right] \times$$
$$\times \exp\left(j2k\left(0,5\frac{V^2(t-t_0)^2}{R_0(\vec{r}, t_0)} \sin^2 \theta_x(\vec{r}, t_0) - V(t - t_0) \cos \theta_x(\vec{r}, t_0) \right) \right) dt = \sum_{i=0}^{N-1} \dot{Y}(\vec{r}, i), \tag{27}$$

where

$$\dot{Y}(\vec{r}, i) = \varepsilon \int_T \Pi(t - iT_i, T_i) \left[\dot{U}_i(t, \vec{r}', i) \dot{A}^* \left(t - \frac{2R_0(\vec{r}, i)}{c} \right) \right] \times$$
$$\times \exp\left(j2k\left(0,5\frac{V^2(t-t_0)^2}{R_0(\vec{r}, t_0)} \sin^2 \theta_x(\vec{r}, t_0) - V(t - t_0) \cos \theta_x(\vec{r}, t_0) \right) \right) dt \tag{28}$$

is the output effect in each beam of the antenna pattern.

The essence of algorithm (27) is as follows:

(1) The registration of signals at the outputs of the diagram-forming circuit $\dot{U}_i(t, \vec{r}', i)$, coherent detection of amplitudes $\left[\dot{U}_i(t, \vec{r}', i) \dot{A}^* \left(t - \frac{2R_0(\vec{r}, i)}{c} \right) \right]$;

(2) Division of pulses into intervals that correspond to the observation of one area of the surface by different antenna pattern $\Pi(t - iT_i, T_i) \left[\dot{U}_i(t, \vec{r}', i) \dot{A}^* \left(t - \frac{2R_0(\vec{r}, i)}{c} \right) \right]$;

(3) Convolution of individual trajectory signals in different processing channels with the expected reference signal $\exp\left(j2k\left(0,5\frac{V^2(t-t_0)^2}{R_0(\vec{r}, t_0)} \sin^2 \theta_x(\vec{r}, t_0) - V(t - t_0) \cos \theta_x(\vec{r}, t_0) \right) \right)$

with duration T corresponding to the full time of sequential observation of one area of the surface by all antenna pattern;

(4) Adding the results of the agreed processing to form the final output effect $\dot{Y}(\vec{r})$.

The obtained results of spatio-temporal signal processing, according to Algorithms (25) and (27), must be identical.

The proposed modification of the algorithm also requires coherent assembly and accurate phase tracking to obtain the synthesized antenna pattern. At the same time, separate pre-processing of the signals from different outputs of the diagram-forming scheme does not have strict requirements for the overlapping and docking of the antenna pattern, and also allows insignificant shifts in time. That is, there is no need to form one continuous implementation with precise transitions at the joints of the diagrams.

3.5.3. Fixed Beam Fan with Coherent Processing at the Output of Each Channel and Incoherent Inter-channel Addition

If it is not possible to achieve synchronization between signals from individual antenna patterns, i.e., it is not possible from the entire flow of samples to coherently select the desired function $\Pi(t - iT_i, T_i)$, proceed to the algorithm of incoherent averaging of output effects from different channels. For this purpose, it is necessary to limit the intervals of integration in algorithm (27) and to fix a point t_0 in each separate beam

$$\dot{Y}(\vec{r}) = \varepsilon \sum_{i=0}^{N-1} \int_{iT_i}^{(i+1)iT_i} \left[\dot{U}_i(t, \vec{r}', i) \dot{A}^* \left(t - \frac{2R_0(\vec{r}, i)}{c} \right) \right] \times$$
$$\times \exp\left(j2k \left(0,5 \frac{V^2(t-t_{0i})^2}{R_0(\vec{r}, t_{0i})} \sin^2 \theta_x(\vec{r}, t_{0i}) - V(t - t_{0i}) \cos \theta_x(\vec{r}, t_{0i}) \right) \right) dt. \tag{29}$$

According to the obtained analytical expression (29), the convolution will be performed only within the selected area of the antenna pattern and subsequently participate in incoherent averaging with other rays with the index i. This algorithm is already known and practically implemented [40,41] and has the same resolution as in single-beam observation, but is characterized by a reduced level of multiplicative interference (speckle noise) in the image due to incoherent averaging.

3.5.4. Fixed Beam Fan with Coherent Signal Processing at the Output of Each Channel, Doppler Frequency Offset Compensation, and Incoherent Inter-channel Addition

The above signal processing algorithms show the basic optimal operations on the received oscillations and do not specify the type of probing signal $s_t(t) = \text{Re}\left\{ \dot{A}(t)e^{j\omega_0 t} \right\}$, the type of its modulation, coding, etc. At the same time, a significant part of the existing SAR use pulse mode have in their processing algorithms contradictions in the choice of pulse repetition frequency: increasing the frequency allows us to increase the resolution of the SAR azimuth, but leads to ambiguous distance measurement. In this case, the algorithm given in (29) on the extreme antenna patterns at significant angles of deviation from the nadir will be difficult to implement in practice. It is more expedient in each of the channels to perform Doppler frequency compensation due to the deviation of the i-th antenna pattern by an angle $\theta_x(\vec{r}, i)$ and then perform coordinated processing with a reference signal at a lower frequency. As a result (29) should be presented as follows:

$$\dot{Y}(\vec{r}) = \varepsilon \sum_{i=0}^{N-1} \int_{iT_i}^{(i+1)iT_i} \left[\dot{U}_i(t, \vec{r}', i) \exp\left(-j2\pi \frac{2V \cos \theta_x(\vec{r}, t_{0i})}{\lambda} (t - t_{0i}) \right) \right] \times$$
$$\times \dot{A}^* \left(t - \frac{2R_0(\vec{r}, i)}{c} \right) \exp\left(j2k \left(0,5 \frac{V^2(t-t_{0i})^2}{R_0(\vec{r}, t_{0i})} \sin^2 \theta_x(\vec{r}, t_{0i}) \right) \right) dt. \tag{30}$$

Compensation of the Doppler shift of the frequency of the trajectory signal to the coordinated processing allows each channel to use the same pulse repetition frequency and, in the general case, to reduce its value to achieve unambiguous measurements by range.

3.5.5. Single-Beam SAR

Further simplifications of Expression (30) may lead to known modes of forming coherent images in SAR. Assuming one beam and a set of range channels, we will receive the generalized for the front side, strictly lateral, and back side route mode of SAR

$$
\begin{aligned}
\dot{Y}(\vec{r}) = \varepsilon \int_0^T &\left[\dot{U}_i(t, \vec{r}') \exp\left(-j2\pi \frac{2V\cos\theta_x(\vec{r},t_0)}{\lambda}(t - t_0) \right) \right] \times \\
&\times \dot{A}^*\left(t - \frac{2R_0(\vec{r})}{c} \right) \exp\left(j2k\left(0,5\frac{V^2(t-t_0)^2}{R_0(\vec{r},t_0)} \sin^2\theta_x(\vec{r},t_0) \right) \right) dt.
\end{aligned}
\tag{31}
$$

For a strictly lateral review, we obtain the most well-known in practice algorithm

$$
\dot{Y}(\vec{r}) = \varepsilon \int_0^T \dot{U}_i(t, \vec{r}') \dot{A}^*\left(t - \frac{2R_0(\vec{r})}{c} \right) \exp\left(jk \frac{V^2(t-t_0)^2}{R_0(\vec{r},t_0)} \right) dt.
\tag{32}
$$

3.6. Marginal Errors of Estimation

The marginal errors of coherent imaging of the surface can be found from the following expression:

$$
\sigma_{\dot{F}}^2 = \int_D -\left(\left\langle \frac{\delta^2}{\delta \dot{F}^2(\vec{r})} \ln P[u(t, \vec{r}')|\dot{F}(\vec{r})] \right\rangle \right)^{-1} d\vec{r} \ \Bigg|_{\vec{r}_1 = \vec{r}}.
\tag{33}
$$

Calculating the variational derivative of the second order:

$$
\frac{\delta^2}{\delta \dot{F}^2(\vec{r})} \ln P[u(t, \vec{r}')|\dot{F}(\vec{r})] = -2\mu \mathrm{Re}\dot{\Psi}_H^*(\vec{r}_1, \vec{r}),
\tag{34}
$$

the marginal errors have the following form:

$$
\sigma_{\dot{F}}^2 = \int_D -\left(\mu \mathrm{Re}\dot{\Psi}_H^*(\vec{r}_1, \vec{r}) \right)^{-1} d\vec{r} \ \Big|_{\vec{r}_1 = \vec{r}},
\tag{35}
$$

where

$$
\dot{\Psi}_H^*(\vec{r}_1, \vec{r}) = \dot{\Psi}^*(\vec{r}_1, \vec{r})/\dot{\Psi}^*(0,0)
\tag{36}
$$

is the normalized ambiguity function and $\mu = 2\mathrm{Re}\dot{\Psi}^*(0,0)N_0^{-1}$ is a signal-to-noise ratio.

Analysis of expression (35) in the partial case, when $\dot{\Psi}_H^*(\vec{r}_1, \vec{r}) = \dot{\Psi}_H^*(\vec{r}_1 - \vec{r})$, shows that the marginal errors of measurement are inversely proportional to the value μ. For the nonstationary case, the marginal errors will be proportional to the value of the averaged inverse uncertainty function over the aperture synthesis interval in the point $\vec{r}_1 = \vec{r}$.

4. Discussion

To discuss the results and how they can be interpreted, a test model of the surface with anthropogenic objects was developed, taking into account the phenomenological approach to the description of the electromagnetic field and coherent images [42,43]. The model of the height profile of the entire observation area is shown in Figure 3a,b. This is a model of some part of the surface, such as an airfield. In this model, we wanted to combine natural surfaces, such as forest, mountains, and fields, and anthropogenic objects. In particular, it was proposed to place a building, a truck, an aircraft on the runway, a group of tanks, and anti-aircraft missile systems, one of which is located in the forest. Height models for anthropogenic objects are shown in Figure 3c–h.

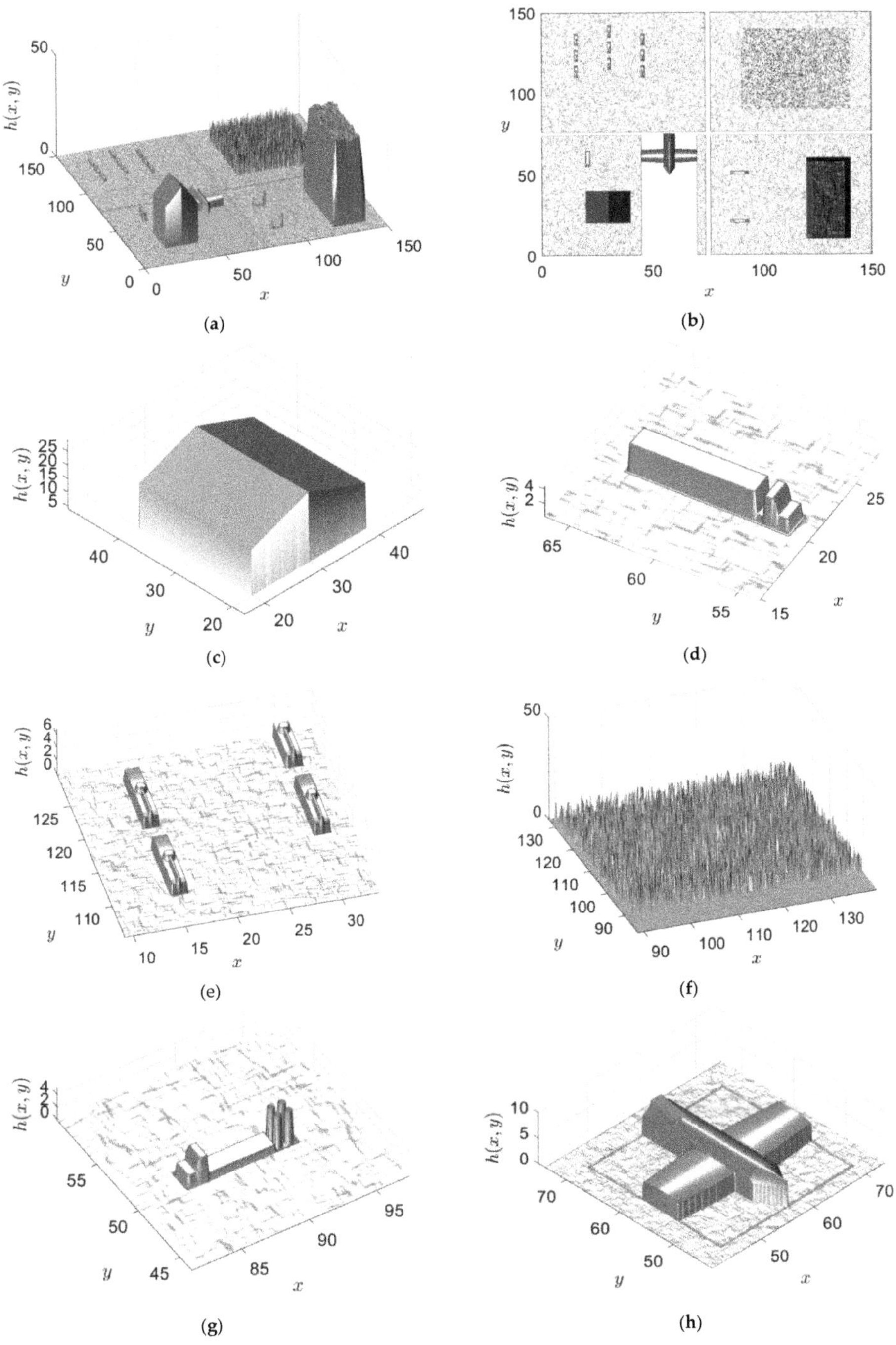

Figure 3. Altitude profile of the simulation model: (**a**) the model of the entire observation area, (**b**) the top projection, (**c**) the model of the building, (**d**) the model of the tractor with trailer, (**e**) the model of tanks, (**f**) the model of the forest, (**g**) the model of anti-aircraft missile system, (**h**) the model aircraft.

The radar cross section of the whole area and the actual part of the complex scattering coefficient for each point of the observation area is shown in Figure 4. The amplitudes of the complex scattering coefficient were chosen from the analysis of existing electrodynamic surface models and the database of radar images of various satellites available on the Internet. The metal roof of the building is the most reflective. Furthermore, the metal fuselage of the aircraft has a large amplitude of reflections. The tractor and other mobile equipment is less noticeable.

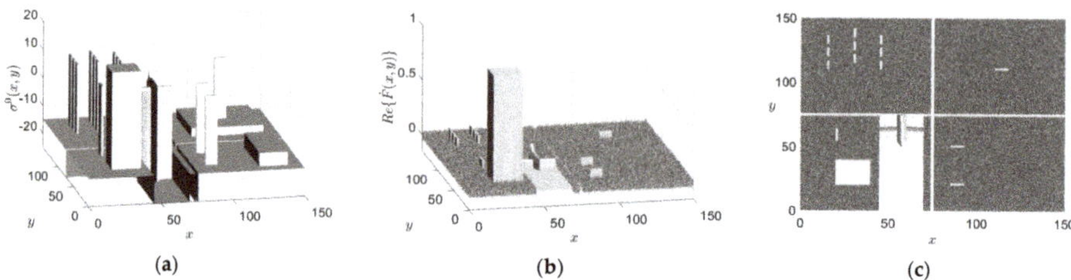

(a)   (b)   (c)

Figure 4. Characteristics of the test surface: (**a**) the radar cross section, (**b**) the real part of the complex scattering coefficient, (**c**) the top projection of $\mathrm{Re}\left\{\dot{F}(x,y)\right\}$.

To test the simulation, the optimal output effect $\left|\dot{Y}(\vec{r})\right|$ was first obtained in an aerospace radar system with a planar antenna array for a single beam with the following parameters: $\theta_x(\vec{r},t) = 0$, $\theta_y(\vec{r},t) = 0$, $f_0 = 10\,\mathrm{GHx}$, $H = 10\,\mathrm{km}$, surface size 150 m × 150 m, the size of the uncertainty function 3 m × 3 m and 1 m × 1 m. Radar images for these parameters are shown in Figure 5a–d.

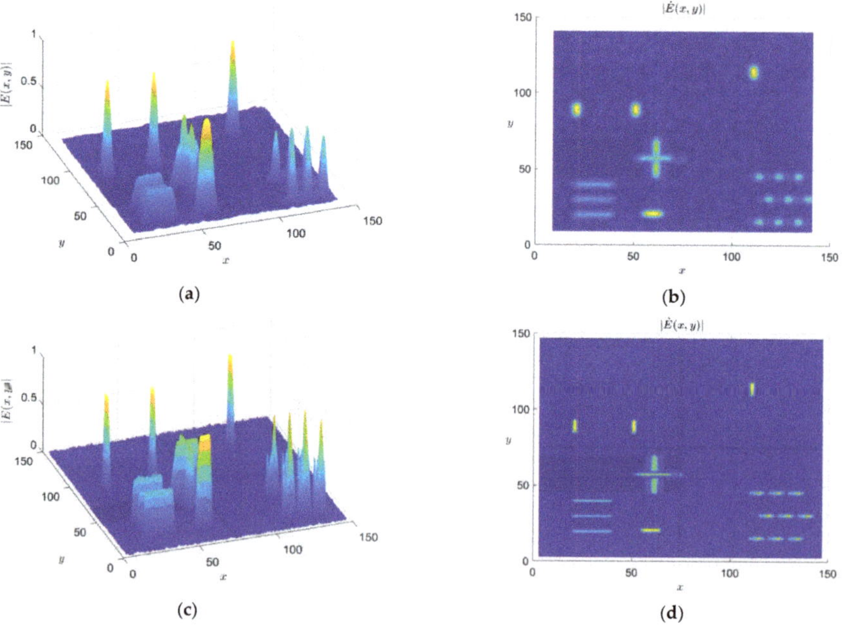

(a)   (b)

(c)   (d)

Figure 5. The optimal output effect of the radar system $\left|\dot{Y}(\vec{r})\right|$: (**a**) for the uncertainty function with dimensions 3 m × 3 m, (**b**) projection of the radar image, when the uncertainty function has dimensions 3 m × 3 m, (**c**) for the uncertainty function with dimensions 1 m × 1 m, (**d**) projection of the radar image, when the uncertainty function has dimensions 1 m × 1 m.

It follows from the analysis of the radar images in Figure 5 that their detail depends on the width of the uncertainty function (36).

For more practical situations, radar images were simulated with the same parameters, but for single-beam observation $(\theta_x(\vec{r},t) = 90°, \theta_y(\vec{r},t) = (20° \div 20,5°))$, incoherent addition of two beams $(\theta_x(\vec{r},t) = (90°,60°), \theta_y(\vec{r},t) = (20° \div 20,5°))$, incoherent addition of three beams $(\theta_x(\vec{r},t) = (90°,60°,120°), \theta_y(\vec{r},t) = (20° \div 20,5°))$, and the optimal coherent sum of three beams scanning is in the Spot-light mode. All of the results of the radar images simulation are shown in Figure 6.

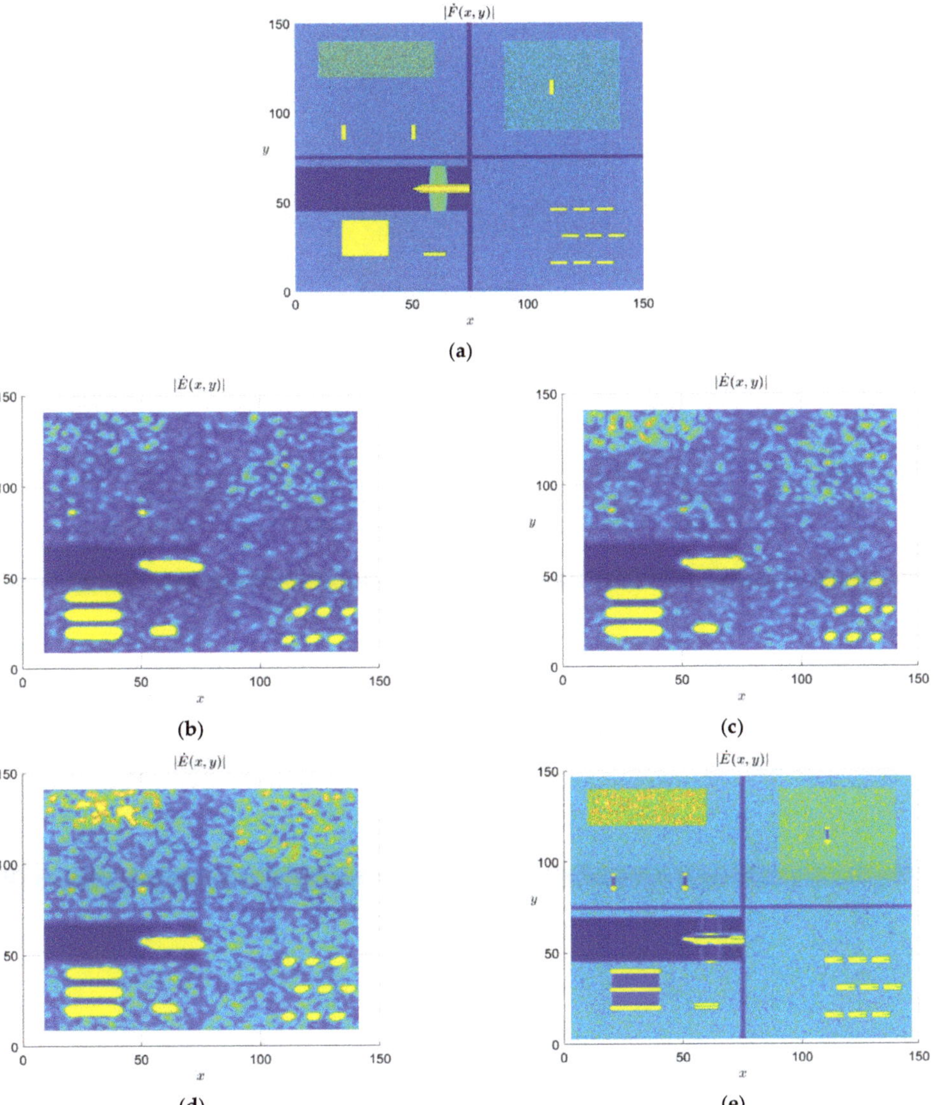

Figure 6. Optimal output effects of the radar system $\left|\dot{Y}(\vec{r})\right|$: (**a**) true coherent image, (**b**) radar system with one beam, (**c**) incoherent addition of processing results from two beams, (**d**) incoherent addition of processing results from three beams, (**d**) optimal method of processing received space–time signals.

Radar images in Figure 6 increase information content with an increase in the number of observation beams. At the same time, at high resolution in Figure 6e, resonant scattering of waves is observed only at observation angles. Some surfaces scatter electromagnetic waves and are not visible for these viewing angles.

For the obtained radar images in Figure 6, their entropy, accuracy, and degree of difference from the test image were calculated according to the following quality metrics (completely referenced and without reference): root mean square error (MSE), peak signal-to-noise ratio (PSNR), structural similarity index (SSIM), blind/referenceless image spatial quality evaluator (BRISQUE), natural image quality evaluator (NIQE), and perception-based image quality evaluator (PIQE). All results are shown in Table 1.

Table 1. The results of the accuracy evaluation of coherent imaging.

Metrics	Figure 6a	Figure 6b	Figure 6c	Figure 6d	Figure 6e
MSE	0	0.0384	0.0349	0.0544	0.0650
PSNR	Inf	14.1592	14.5726	12.6468	11.8705
SSIM	1	0.1939	0.2046	0.1925	0.2116
BRISQUE	43.0494	57.4215	56.2384	56.6353	46.8832
NIQE	8.5282	9.2167	8.9705	8.7231	5.9386
PIQE	69.4806	84.3591	82.8793	83.2645	46.6039

The MSE metric measures the average sum of the squared difference between the elements of the true image y_i and the result of the formation of the radar image $\hat{y}_i$ for all its N points

$$\text{MSE} = \frac{1}{N}\sum_{i=1}^{N}(y_i - \hat{y}_i)^2. \tag{37}$$

Raising to the second power is performed so that negative values are not offset by positive ones. Also, due to the properties of this metric, the influence of errors statistically increases by quadrature from the original value. The smaller the MSE, the more accurate our prediction. The optimum is reached at point 0, that is, we predict perfectly.

The PSNR metric is presented in decibels and is equal to the ratio of the maximum possible image amplitude $\max[y_i]$ to the root mean square error

$$\text{pSNR} = 20\log_{10}\frac{\max[y_i]}{\sqrt{\text{MSE}}}. \tag{38}$$

Statistically, the greater the similarity between the images, the lower the MSE value, and, therefore, the greater the PSNR. The PNSR is dimensionless because both the numerator and denominator are measured in pixel values. It is expedient to use the PSNR values in this case to compare the quality of different methods of radar imaging and to study the effect of different parameters on the performance of a particular algorithm.

SSIM (Structural Similarity Index) is one method to measure the similarity between two images. The SSIM index is a full matching method. In other words, it measures the quality based on the true image according to the following algorithm:

$$\text{SSIM}(x,y) = \frac{(2\mu_x\mu_y + c_1)(2\sigma_{xy} + c_2)}{\left(\mu_x^2 + \mu_y^2 + c_1\right)\left(\sigma_x^2 + \sigma_y^2 + c_2\right)}. \tag{39}$$

where μ_x is the mean of the x image, μ_y is the mean of the y image, σ_{xy} is the covariance of the x and y images, σ_x^2 is the variance of the x image, σ_y^2 is the variance of the y image, and c_1 and c_2 are the constants that depend on the images' dynamic range.

The SSIM index is an evolution of traditional methods such as PSNR and MSE, which have proven to be incompatible with the physiology of human perception.

For a more accurate statistical assessment of the quality of the obtained algorithm, the referenceless metrics BRISQUE, NIQE, and PIQE were also considered. These metrics are less often used in such tasks, but they should show the general statistics and confirm or refute the results.

According to the metrics (MSE, PSNR, SSIM) that use the reference image, the optimal output effect is not always the best because the dynamic range of the images was adjusted manually. In practice, to see the structure of the image in the presence of powerful reflectors, the dynamic range is also adjusted during the secondary processing of radar images. At the same time, metrics (BRISQUE, NIQE, and PIQE) that do not require information about the reference image show that the quality of radar images for optimal output is even better than the reference image. This is due to the fact that the modulus of the ideal complex scattering coefficient has a stochastic nature, and the reconstructed radar images are smoothed by the uncertainty function.

From the obtained results it follows that the proposed optimal method of image construction in an aerospace radar system with planar antenna array is more informative, allows us to form radar images without gaps, and has an improved spatial azimuth resolution of 35% compared to single-beam viewing mode, at 33% incoherent processing of signals from two beams of antenna pattern and by 31% in incoherent processing of signals from three beams.

The main idea of this article is the development of the statistical theory of the optimization of signal processing methods in airborne radars. The optimization of the theory should lead to specific practical results. This is exactly what happened in the example of signal processing methods optimization in the radar with a multi-channel two-dimensional antenna array, which was placed on platform moving at a constant speed. It is necessary to compare the obtained practical results with the existing ones.

The first method of radar imaging, which must be compared with the received one, is called Stripmap. This is the first imaging technique in SAR history that has medium detail and medium swath. It is being carefully studied now, particularly in [44,45]. The work [44] analyzes the capabilities of SAR at the Sentinel 1 satellite, which has 5 m by 5 m spatial resolution and acquires data with an 80 km swath. In [45], the TerraSAR-X satellite is considered, which can achieve a spatial resolution of up to 3 m for a standard scene size of 30 km x 50 m (width x length) in this mode. The method proposed in this article shows that such a mode in the antenna array is not enough, and this observation mode does not use the potential of the antenna array. It is necessary to form multiple rays in space to increase the swath width.

The designers of Sentinel 1 and TerraSAR-X also understand the need for the spatial distribution of beams. For this, as shown in [46,47], TerraSAR-X implements the ScanSAR (a swath width of 100 km, resolution of up to 18.5 m) and WideScanSAR (a swath width of up to 270 km, a spatial resolution of 40 m) methods). At the same time, these methods lose both the method proposed in the article and the StripMap method in terms of resolution due to the small interval of signal accumulation and spatial scanning. The optimal synthesized method says that it is necessary to form many beams, not just move one beam.

Observing the experiences of other developers and realizing the low accuracy of scanning methods, Sentinel 1 released new methods of viewing with many beams and fast movement in space, which are called Interferometric Wide swath and Extra Wide swath [48,49]. Interferometric Wide swath acquires data with a 250 km swath at 5 m by 20 m spatial resolution. Extra Wide swath acquires data over a 400 km swath at 20 m by 40 m spatial resolution. This method is very close to the one proposed by the authors of this article, but does not implement the beam focusing procedure.

Beam focusing is implemented by the SpotLight method [50,51], which now makes it possible to form radar images with a resolution of up to 0.25 m from a territory of 4 km $\times$ 4 km. It is clearly seen that the construction of high-precision images of large areas will require a lot of time. To overcome this contradiction, the method in this article implements MultiBeam SpotLight mode with the optimal displacement of the focus spot beam so that there are no gaps. To date, this method has not been implemented, but it can potentially be implemented in new satellites, since we have proposed combining the two observation methods [52].

5. Conclusions

The statistical theory of the synthesis of optimal methods of processing space–time signals in aerospace radars with antenna arrays with the restoration of a coherent image of the observation area was further developed. In contrast to the known simplified problem statements with a given space survey mode and field processing in the plane of the antenna array, the synthesized optimal methods allowed us to overcome the contradictions in aerospace radars between high spatial resolution and a wide field of view. Modifications of optimal methods for their technical implementation in aerospace-based radars are proposed and analytical expressions for the marginal errors of coherent image reconstruction are obtained.

Using the developed theory, the problem of synthesis and analysis of the method of coherent imaging of aerospace radars with a planar antenna array is solved. The peculiarity of the obtained results is the generalized statement of the problem of registration of scattered electromagnetic fields without concretization of spatial processing in antenna arrays, which allowed us to synthesize the optimal method of signal processing and the optimal mode of space survey. The proposed underlying surface scanning mode combines the advantages of two existing modes—multi-beam and Spot-light, which allowed us to obtain both the highest resolution and the widest field of view.

The obtained results allow us to make a significant contribution to the implementation of the concept of cognitive radars with adaptive antenna pattern formation depending on the tasks. It should be noted that the obtained methods and structures of aerospace radars are common and can be used for both pulsed and continuous operation.

Further development of the theory should be carried out in the direction of optimizing the processing of ultra-wideband functionally deterministic signals processing in multi-channel aerospace imaging radar systems. Ultra-wideband creates prospects for another source of increasing the accuracy of radar imaging and will allow the creation of lightweight sparse antenna arrays for satellites and aircraft. Valery Volosyuk has already developed such a theory for passive multi-positional radiometric systems.

Author Contributions: Conceptualization, V.V. and methodology, S.Z. All authors have read and agreed to the published version of the manuscript.

Funding: The work was funded by the Ministry of Education and Science of Ukraine, the state registration numbers of the projects are 0120U102082, 0119U100968, 0121U109600, and 0121U109598.

Institutional Review Board Statement: Not applicable.

Informed Consent Statement: Not applicable.

Data Availability Statement: Not applicable.

Conflicts of Interest: The authors declare no conflict of interest.

References

1. Falkovich, S.E.; Ponomarev, V.I.; Shkvarko, I.V. *Optimal Reception of Space-Time Signals in Radio Channels with Scattering*; Radio i Svyaz: Moscow, Russia, 1989. (In Russian)
2. Falkovich, S.E.; Kostenko, P.Y. *Fundamentals of the Statistical Theory of Radio Engineering Systems: A Tutorial*; Kharkiv Aviation Institute: Kharkov, Russia, 2005. (In Russian)
3. Falkovich, S.E. *Signal Parameters Estimation*; Sovetskoye Radio: Moscow, Russia, 1970. (In Russian)
4. Falkovich, S.E.; Khomyakov, E.N. *Statistical Theory of Radio Measuring Systems*; Radio i Svyaz: Moscow, Russia, 1981. (In Russian)
5. Tikhonov, V.I. *Statistical Radio Engineering*; Radio i Svyaz: Moscow, Russia, 1982. (In Russian)
6. Tikhonov, V.I. *Optimal Signal Reception*; Radio i Svyaz: Moscow, Russia, 1983. (In Russian)
7. Bakut, P.A.; Tartakovsky, G.P. *Questions of the Statistical Theory of Radar*; Sovetskoye Radio: Moscow, Russia, 1963. (In Russian)
8. Amiantov, I.N. *Selected Problems of the Statistical Communication Theory*; Sovetskoye Radio: Moscow, Russia, 1971. (In Russian)

9.   Kotel'nikov, V.A. *The Theory of Optimum Noise Immunity*; McGraw-Hill Book, Co.: New York, NY, USA, 1959.
10.   Gutkin, L.S. *Theory of Optimal Radio Reception Methods with Fluctuation Interference*; Gosenergoizdat: Moscow, Russia, 1961. (In Russian)
11.   Levin, B.R. *Theoretical Foundations of Statistical Radio Engineering*; Sovetskoye Radio: Moscow, Russia, 1969. (In Russian)
12.   Van Tris, G. *Theory of Detection, Estimation and Modulation*; Sovetskoye Radio: Moscow, Russia, 1972–1977; Volumes 1–3. (In Russian)
13.   Middlton, D. *Introduction to Statistical Communication Theory*; Sovetskoye Radio: Moscow, Russia, 1961–1962; Volumes 1–2. (In Russian)
14.   Shirman, Y.D.; Manzhos, V.N. *Theory and Technique of Processing Radar Information against the Background of Interference*; Radio i Svyaz: Moscow, Russia, 1981. (In Russian)
15.   Kraus, J.D. *Radio Astronomy*, 2nd ed.; Cygnus-Quasar Books: Powell, OH, USA, 1986.
16.   Tseitlin, N.M. *Aerial Equipment and Radio Astronomy*; Soviet Radio: Moscow, Russia, 1976; 352p.
17.   Thompson, A.R.; Moran, J.M.; Swenson, G.W. Antennas and arrays. In *Interferometry and Synthesis in Radio Astronomy*; Springer International Publishing: Cham, Switzerland, 2017; pp. 153–206. [CrossRef]
18.   Van Schooneveld, C. Image Formation from Coherence Functions in Astronomy. In Proceedings of the IAU Colloquium No. 49 on the Formation of Images from Spatial Coherence Functions in Astronomy, Groningen, The Netherlands, 10–12 August 1978.
19.   Kondratenkov, G.S.; Potekhin, V.A.; Reutov, A.P.; Feoktistov, Y.A. *Earth Survey Radars*; Radio i Svyaz: Moscow, Russia, 1983. (In Russian)
20.   Reutov, A.P.; Mikhailov, V.A.; Kondratenkov, G.S.; Boyko, B.V. *Side-Looking Radars*; Sovetskoye Radio: Moscow, Russia, 1970. (In Russian)
21.   Karavayev, V.V.; Sazonov, V.V. *Fundamentals of Synthesized Antenna Theory*; Sovetskoye Radio: Moscow, Russia, 1974. (In Russian)
22.   Antipov, V.N.; Goryainov, V.T.; Kulin, A.V.; Mansurov, V.V.; Okhonskiy, A.G.; Sazonov, N.A.; Titov, M.P.; Tolstov, E.F.; Shapovalov, A.V. *Digital Synthesized Aperture Radars*; Radio i Svyaz: Moscow, Russia, 1988. (In Russian)
23.   Volosyuk, V.K.; Kravchenko, V.F. *Statistical Theory of Radio Engineering Systems for Remote Sensing and Radar*; Fizmatlit: Moscow, Russia, 1988. (In Russian)
24.   Moreira, A.; Prats-Iraola, P.; Younis, M.; Krieger, G.; Hajnsek, I.; Papathanassiou, K.P. A tutorial on synthetic aperture radar. *IEEE Geosci. Remote Sens. Mag.* **2013**, *1*, 6–43. [CrossRef]
25.   Krieger, G.; Moreira, A.; Fiedler, H.; Hajnsek, I.; Werner, M.; Younis, M.; Zink, M. TanDEM-X: A Satellite Formation for High-Resolution SAR Interferometry. *IEEE Trans. Geosci. Remote Sens.* **2007**, *45*, 3317–3341. [CrossRef]
26.   Reigber, A.; Moreira, A. First demonstration of airborne SAR tomography using multibaseline L-band data. *IEEE Trans. Geosci. Remote. Sens.* **2000**, *38*, 2142–2152. [CrossRef]
27.   Charvat, G.L.; Kempel, L.C.; Rothwell, E.J.; Coleman, C.M.; Mokole, E.L. A Through-Dielectric Radar Imaging System. *IEEE Trans. Antennas Propag.* **2010**, *58*, 2594–2603. [CrossRef]
28.   Goodman, J.W. *Introduction to the Fourier Optics*; Mir: Moscow, Russia, 1970. (In Russian)
29.   Born, M. *Fundamentals of Optics*; Nauka: Moscow, Russia, 1973. (In Russian)
30.   Zommerfel'd, A. *Optics, Izdat*; Inostrannoy Literatury: Moscow, Russia, 1953. (In Russian)
31.   Zubkovich, S.G. *Statistical Characteristics of Radio Signals Reflected from the Earth's Surface*; Sovetskoye Radio: Moscow, Russia, 1968. (In Russian)
32.   Fradin, A.Z. *Ultrahigh Frequency Antennas*; Sovetskoye Radio: Moscow, Russia, 1957. (In Russian)
33.   Rytov, S.M.; Kravtsov, Y.A.; Tatarskiy, V.I. *Introduction to Statistical Radiophysics*; Nauka: Moscow, Russia, 1978. (In Russian)
34.   Kravchenko, V.F.; Kutuza, B.G.; Volosyuk, V.K.; Pavlikov, V.V.; Zhyla, S.S. Super-resolution SAR imaging: Optimal algorithm synthesis and simulation results. In Proceedings of the 2017 Progress in Electromagnetics Research Symposium—Spring (PIERS), St. Petersburg, Russia, 22–25 May 2017; pp. 419–425. [CrossRef]
35.   Pavlikov, V.V.; Zhyla, S.S.; Odokienko, O.V. Structural optimization of Dicke-type radiometer. In Proceedings of the 2016 II International Young Scientists Forum on Applied Physics and Engineering (YSF), Kharkiv, Ukraine, 10–14 October 2016; pp. 171–174. [CrossRef]
36.   Ostroumov, I.; Kuzmenko, N.; Sushchenko, O.; Pavlikov, V.; Zhyla, S.; Solomentsev, O.; Zaliskyi, M.; Averyanova, Y.; Tserne, E.; Popov, A.; et al. Modelling and Simulation of DME Navigation Global Service Volume. *Adv. Space Res.* **2021**, *68*, 3495–3507. [CrossRef]
37.   Volosyuk, V.K.; Pavlikov, V.V.; Zhyla, S.S. Algorithms Synthesis and Potentiality Analysis of Optimum Ultrawideband Signal Processing in the Radiometric System with Modulation. In Proceedings of the 2011 VIII International Conference on Antenna Theory and Techniques, Kyiv, Ukraine, 20–23 September 2011. [CrossRef]
38.   Pavlikov, V.V.; Volosyuk, V.K.; Zhyla, S.S.; Van Huu, N. Active Aperture Synthesis Radar for High Spatial Resolution Imaging. In Proceedings of the 2018 9th International Conference on Ultrawideband and Ultrashort Impulse Signals (UWBUSIS), Odessa, Ukraine, 4–7 September 2018. [CrossRef]
39.   Pavlikov, V.; Volosyuk, V.; Zhyla, S.; Van, H.N.; Van, K.N. A New Method of Multi-Frequency Active Aperture Synthesis for Imaging of SAR Blind Zone Under Aerospace Vehicle. In Proceedings of the 2017 14th International Conference The Experience of Designing and Application of CAD Systems in Microelectronics (CADSM), Lviv, Ukraine, 21–25 February 2017. [CrossRef]

40. Gorovyi, I.; Bezvesilniy, O.; Vavriv, D.; Ievgen, G. Multi-look SAR processing with road location and moving target parameters estimation. In Proceedings of the 2015 16th International Radar Symposium (IRS), Dresden, Germany, 24–26 June 2015; pp. 581–586. [CrossRef]

41. Bezvesilniy, O.O.; Kochetov, B.A.; Vavriv, D.M. Moving target detection with multi-look SAR. In Proceedings of the 2014 20th International Conference on Microwaves, Radar and Wireless Communications (MIKON), Gdansk, Poland, 16–18 June 2014; pp. 1–4. [CrossRef]

42. Volosyuk, V.K.; Pavlikov, V.V.; Zhyla, S.S. Phenomenological Description of the Electromagnetic Field and Coherent Images in Radio Engineering and Optical Systems. In Proceedings of the 2018 IEEE 17th International Conference on Mathematical Methods in Electromagnetic Theory (MMET), Kiev, Ukraine, 2–5 July 2018; pp. 302–305.

43. Volosyuk, V.K. Phenomenological description of coherent radar images based on the concepts of the measure of set and stochastic integral. In *Telecommunication and Radio Engineering*; Volosyuk, V.K., Zhyla, S.S., Kolesnikov, D.V., Eds.; Begell House: Danbury, CT, USA, 2019; Volume 78, pp. 19–30.

44. Isar, A.; Nafornita, C. Sentinel 1 Stripmap GRDH image despeckling using two stages algorithms. In Proceedings of the 2016 12th IEEE International Symposium on Electronics and Telecommunications (ISETC), Timisoara, Romania, 27–28 October 2016; pp. 343–348. [CrossRef]

45. Gisinger, C.; Schubert, A.; Breit, H.; Garthwaite, M.; Balss, U.; Willberg, M.; Small, D.; Eineder, M.; Miranda, N. In-Depth Verification of Sentinel-1 and TerraSAR-X Geolocation Accuracy Using the Australian Corner Reflector Array. *IEEE Trans. Geosci. Remote. Sens.* **2020**, *59*, 1154–1181. [CrossRef]

46. Zhang, L.; Liu, H.; Gu, X.; Guo, H.; Chen, J.; Liu, G. Sea Ice Classification Using TerraSAR-X ScanSAR Data With Removal of Scalloping and Interscan Banding. *IEEE J. Sel. Top. Appl. Earth Obs. Remote Sens.* **2019**, *12*, 589–598. [CrossRef]

47. Steinbrecher, U.; Kraus, T.; Castellanos Alfonzo, G.; Grigorov, C.; Schulze, D.; Braeutigam, B. TerraSAR-X: Design of the new operational WideScanSAR mode. In Proceedings of the EUSAR 2014 10th European Conference on Synthetic Aperture Radar, Berlin, Germany, 3–5 June 2014; pp. 1–4.

48. Korosov, A.; Demchev, D.; Miranda, N.; Franceschi, N.; Park, J.-W. Thermal Denoising of Cross-Polarized Sentinel-1 Data in Interferometric and Extra Wide Swath Modes. *IEEE Trans. Geosci. Remote Sens.* **2021**, *60*, 5218411. [CrossRef]

49. Wegmuller, U.; Werner, C.; Wiesmann, A.; Strozzi, T.; Kourkouli, P.; Frey, O. Time-series analysis of Sentinel-1 interferometric wide swath data: Techniques and challenges. In Proceedings of the 2016 IEEE International Geoscience and Remote Sensing Symposium (IGARSS), Beijing, China, 10–15 July 2016; pp. 3898–3901. [CrossRef]

50. Ignatenko, V.; Nottingham, M.; Radius, A.; Lamentowski, L.; Muff, D. ICEYE Microsatellite SAR Constellation Status Update: Long Dwell Spotlight and Wide Swath Imaging Modes. In Proceedings of the 2021 IEEE International Geoscience and Remote Sensing Symposium IGARSS, Brussels, Belgium, 11–16 July 2021; pp. 1493–1496. [CrossRef]

51. Ignatenko, V.; Dogan, O.; Muff, D.; Lamentowski, L.; Radius, A.; Nottingham, M.; Leprovost, P.; Seilonen, T. ICEYE Microsatellite SAR Constellation Status Update: Spotlight Extended Area Mode, Daily Coherent Ground Tracks and Waveform Diversity. In Proceedings of the IGARSS 2022—2022 IEEE International Geoscience and Remote Sensing Symposium, Lumpur, Malaysia, 17–22 July 2022; pp. 4145–4148. [CrossRef]

52. Kraus, T.; Ribeiro, J.P.T.; Bachmann, M.; Steinbrecher, U.; Grigorov, C. Concurrent Imaging for TerraSAR-X: Wide-Area Imaging Paired With High-Resolution Capabilities. *IEEE Trans. Geosci. Remote Sens.* **2022**, *60*, 5220314. [CrossRef]

 *computation*

Article

Detection of Shoplifting on Video Using a Hybrid Network

Lyudmyla Kirichenko [1,2], Tamara Radivilova [3], Bohdan Sydorenko [1] and Sergiy Yakovlev [4,5,*]

1   Department of Applied Mathematics, Kharkiv National University of Radio Electronics,
    61166 Kharkiv, Ukraine
2   Applied Mathematics Department, Wroclaw University of Science and Technology, 50-370 Wroclaw, Poland
3   Department of Infocommunication Engineering, Kharkiv National University of Radio Electronics,
    61166 Kharkiv, Ukraine
4   Mathematical Modelling and Artificial Intelligence Department, National Aerospace University "Kharkiv
    Aviation Institute", 61072 Kharkiv, Ukraine
5   Institute of Information Technology, Lodz University of Technology, 90-924 Lodz, Poland
*   Correspondence: s.yakovlev@khai.edu

Abstract: Shoplifting is a major problem for shop owners and many other parties, including the police. Video surveillance generates huge amounts of information that staff cannot process in real time. In this article, the problem of detecting shoplifting in video records was solved using a classifier, which was a hybrid neural network. The hybrid neural network included convolutional and recurrent ones. The convolutional network was used to extract features from the video frames. The recurrent network processed the time sequence of the video frames features and classified the video fragments. In this work, gated recurrent units were selected as the recurrent network. The well-known UCF-Crime dataset was used to form the training and test datasets. The classification results showed a high accuracy of 93%, which was higher than the accuracy of the classifiers considered in the review. Further research will focus on the practical implementation of the proposed hybrid neural network.

Keywords: human behavior; shoplifting; video surveillance; classification; features; neural network; gated recurrent units

Citation: Kirichenko, L.; Radivilova, T.; Sydorenko, B.; Yakovlev, S. Detection of Shoplifting on Video Using a Hybrid Network. *Computation* **2022**, *10*, 199. https://doi.org/10.3390/computation10110199

Academic Editors: Mykola Nechyporuk, Vladimir Pavlikov and Dmitriy Kritskiy

Received: 11 October 2022
Accepted: 1 November 2022
Published: 6 November 2022

Publisher's Note: MDPI stays neutral with regard to jurisdictional claims in published maps and institutional affiliations.

1. Introduction

As one of the many crimes committed in stores, shoplifting attracts a lot of attention. Shoplifting is the theft of goods from an open retail establishment, usually by concealing items from the store in one's pockets, under clothing, or in a bag and leaving the store without paying. Shoplifting is a big problem for store owners and many other parties, including the police, government, and courts. According to a study by the National Association for Shoplifting Prevention, 1 in 11 people is a shoplifter. Moreover, it has been reported that thieves of this kind are arrested only once in every 48 thefts [1].

There are dozens of implementation methods for shoplifting; most of them usually involve hiding things by a person or an accomplice and leaving the store without paying. The key word is "hiding", accompanied by a characteristic sequence of actions. In terms of the behavior of the average shopper, it is abnormal behavior. Of course, when a person reviews a recorded incident of shoplifting, it is not difficult to trace this behavior and discover if the theft actually occurred. However, as the number of shoppers increases, so does the number of shoplifters, which creates the problem of not being able to monitor every one of them.

Every year, the number of video surveillance cameras in public places such as streets, banks, shopping malls, and retail stores grows to improve public safety. Video camera networks generate huge amounts of data and security personnel cannot process all the information in real time. The more recording devices become available, the more difficult the task of monitoring becomes [2].

Thus, there is a need for automatic video surveillance that detects shoplifting events [3]. In recent years, the solution to the problem of real-time monitoring and information processing has been the application of artificial intelligence methods; in particular, algorithms for anomaly detection and event classification [1,4–7].

2. Review of the Literature

The most common approaches to applying artificial intelligence to video monitoring data primarily include motion detection, face recognition, surveillance, inactivity detection, and anomaly behavior detection [8].

In [9], the authors conducted research on shoplifting classifications based on the Jubatus plug-in to extract the feature values from images to assess anomalous customer behavior. In the proposed application, the surveillance video data were classified using a linear classifier and a kNN classifier and the probability of shoplifting was determined.

Tsushita and Zin [10] presented an algorithm for video surveillance detection, violence and theft. Their approach divided the frame into eight areas and looked for speed changes in the person being monitored.

In [11], the authors proposed a model of a genetic algorithm generating neural networks to classify human behavior in videos. The authors developed shallow and deep neural networks that used the posture changes of people in video sequences as the input.

To detect violent theft in video sequences, [12] proposed using a deep learning sequence model in which a feature extractor was trained. The features were then processed with two layers of long-term convolutional memory and passed through fully connected layers for the classification.

To accurately recognize human actions, the authors in [13] proposed to obtain features based on MobileNetV2 and Darknet53 deep learning models. The selected features, based on an improved particle swarm optimization algorithm, were used to classify actions in the video sequences using different classifiers. The effectiveness of the proposed approach was tested by the authors in experiments on six publicly available datasets.

One obvious and popular approach to detect theft is to classify the actions presented in the video using convolutional neural networks. Convolutional neural networks have shown superior performance in computer vision in recent years. In particular, 3D CNN networks, an extension of the convolutional neural networks of CNN, focus on extracting spatial and temporal features from videos. Programs that have been implemented with a 3D CNN include object recognition, human action recognition, and gesture recognition.

In [14], the authors presented an approach to anomaly detection using a pretrained C3D model for the feature extraction and a full-link neural network to perform the regression. Using the UCF-Crime dataset [15], the authors trained their model on 11 classes and tested its results on videos of theft, fights, and traffic incidents.

The authors of [16] presented an approach to detect anomalies in real time; the algorithm was taught to classify 13 anomalous behaviors such as theft and burglary, fighting, shooting, and vandalism. They used a 3D CNN network to extract the features and label the samples into two categories: normal and abnormal. Their model included a rating loss function and trained a fully connected neural network to make decisions.

The authors of [17] used 3D convolutional neural networks to recognize suspicious actions. The 3D cuboid of the motion based on the frame difference method was used to detect and recognize real-time actions in video sequences. The effectiveness of the proposed methods was tested with the implementation of two sets of videos.

In [18], the authors analyzed the presence of violence in surveillance videos. For this purpose, the authors proposed the use of a deep learning model based on 3D convolutional neural networks without using manual functions or RNN architecture exclusively for encoding temporal information. To evaluate the effectiveness and efficiency of the model, the authors experimentally tested it on three benchmark datasets.

In [19], the authors investigated existing video classification procedures in order to recommend the most efficient and productive process. The authors showed that the combined use of a CNN (convolutional neural network) and an RNN (recurrent neural network) performed better than CNN-dependent methods.

In the manuscript of [20], it was proposed to combine a 3D CNN and LSTM to predict human actions. In the approach, the 3D CNN was used for the feature extraction and LSTM for the classification. The result of the model depended on the pose, illumination, and environment. The reliability of these features allowed the prediction of human actions.

In [21], the authors also used a 3D CNN for the feature extraction and classification. They analyzed the effectiveness of this neural network model on a dataset of real-time shoplifting videos.

In [22], the authors presented an expert system for recognizing the actions of thieves. The system used a convolutional neural network to analyze the typical features of the motion of the thief and a deep learning module based on long-term memory was used to train the extracted features. This system alerted shoplifting based on an analysis of the appearance and motion features in a video sequence.

3. Problem Statement

Summarizing the research review, we proposed to solve the problem of shoplifting detection using a binary classification of customer behavior based on a video fragment in two classes: "shoplifting" and "non-shoplifting". For this purpose, it was desirable to develop a classifier that was a symbiosis of two neural networks: convolutional and recurrent. The convolutional neural network removed the features from each frame of the video and the recurrent network processed the time sequence of the processed frames and the further classification.

The input data were video sequences with the same duration and number of frames for which it was known whether a theft had occurred or not. A sequence of frames was formed from each video sequence. Each such video sequence was an object, which was labeled with one of two classes: 0—not shoplifting or 1—shoplifting. The labeled set of frame sequences was the training set. During the classifier training, each object passed the stage of feature acquisition through the hybrid neural network. The training resulted in a classifier, which was further used to classify new objects.

The aim of this work was to develop, train, and test a classifier based on a hybrid neural network to detect shoplifting from video monitoring data.

4. Materials and Methods

4.1. Forming the Input Dataset

The UCF-Crime dataset was used to form the training test dataset [15,16,23]. UCF-Crime is a dataset of 128 h of video. It consists of 1900 long and uncut videos of real-life criminal events, including violent incidents, arrests, arson, assault, traffic accidents, burglaries, explosions, fights, robberies, shootings, thefts, shoplifting, and vandalism.

The dataset with shoplifting had 28 videos recorded from surveillance cameras in retail stores, which had recorded incidents of shoplifting. Due to the fact that the dataset had a small number of videos, the dataset was artificially enlarged by splitting each video into 32 video episodes. Thus, 896 samples with a duration of 3 s each were obtained. The dataset was divided into 2 classes: "non-shoplifting" (Class 0) and "shoplifting" (Class 1). As a result, we obtained two classes of events with a division of the number of videos in each class: 741 videos of "not shoplifting" and 155 of "shoplifting".

Figure 1a shows a frame that corresponded with a video where shoppers simply chose products. The frame in Figure 1b corresponded with a video with the presence of "shoplifting".

Figure 1. Video frames: (**a**) not shoplifting; (**b**) shoplifting.

4.2. Choice of Neural Networks as a Classifier

The two most widely used deep learning architectures for video classifications are convolutional and recurrent neural networks. CNNs are mainly used to learn spatial information from a video whereas RNNs are used to learn temporal information from a video. These network architectures are used for very different purposes. However, the nature of video data with both spatial and temporal information requires the use of both of these network architectures [24,25].

Convolutional neural networks are a variation of neural networks, which are aimed at solving problems of image recognition and the detection of objects in them with the help of computer vision. The property that allows the solving of the above-mentioned problems is that convolutional neural networks form new information segments with reduced dimensionality but preserved features and combine them using several neural layers of different types [26]. Usually CNNs include convolutional layers, aggregated layers, fully connected layers, and normalization layers. A convolutional layer simulates the response of an individual neuron to a visual stimulus. The convolution layers apply a convolution operation to the input, passing the result to the next layer [27].

Recurrent neural networks, or RNNs, are a type of neural network specializing in value sequence processing [28]. Convolutional networks scale to images with a large width and height; recurrent networks can scale to much larger sequences than would be practical for networks without a sequence-based specialization. Most recurrent networks can also handle sequences of a variable length [29].

As the output of a recurrent neuron at a time step is a function of all inputs at previous time steps, we can state that the neuron has a form of memory. The part of a neural network that stores the state through the time steps is called a memory cell.

The most popular long-term memory cell is the LSTM (long short-term memory) cell [30]. LSTM was created to avoid the problem of long-term dependency. An LSTM cell is a set of layers that interact with each other according to certain rules. Layers are connected using an element-by-element addition and multiplication operations.

Gated recurrent units are a type of recurrent neural network [31]. Similar to LSTM, they have been proposed to solve problems such as computing gradients in long-term memory. The two blocks work the same in many cases, but a GRU is trained faster. The block itself has a much lower number of parameters, which makes it much more efficient. Gated recurrence units are a new block and only started to be used in 2014, which explains its lower popularity [32].

In this paper, we used a video classification method that extracted the features from each frame by a CNN and passed the sequence to a separate GRU convolutional neural network. The convolutional neural network was used as a feature extractor so we obtained a sequence of feature vectors.

Feature extraction consists of determining the most relevant characteristics of images and assigning labels to them. In image classification, the decisive step is to analyze the properties of the image features and numerical features into classes. In other words, the image is classified according to its content. The efficiency of a classification model and the degree of classification accuracy mainly depend on the numerical properties of the different image features that represent these classification models. In recent years, many feature extraction methods have been developed; each method has advantages and disadvantages [33].

Transfer learning was used to extract the features especially pretrained from the ImageNet dataset CNN MobileNetV3Large from Keras [23,34]. As this neural network has been designed for low-resource use cases—namely, for phone central processors—it has a satisfactory execution speed, which guarantees its well-functioning performance in a real-time system.

As we used a pretrained model, its weights were loaded along with the model architecture; in this way, we used uploaded parameters. The feature extraction used a neural network without a fully connected layer at the top to obtain a set of features before they went into a prediction. The neural network model preprepared on the ImageNet dataset together with the weights could be loaded with Keras [34,35].

ImageNet is a project that aims to label and classify images into nearly 22,000 categories based on a specific set of words and phrases. At the time of writing, there were over 14 million images in the ImageNet project [34]. MobileNetV3Large received an image interpreted as an array with a shape of $224 \times 224 \times 3$ as the input and at the output, we got a feature vector with size of 960.

4.3. Assessment of the Classification Accuracy

To evaluate the obtained results during a binary classification, an error matrix with true positive (TP), false positive (FP), false negative (FN), and true negative (TN) values was most often obtained.

The following characteristics were chosen as the classification results in such a case.

Accuracy: the fraction of correct answers of the algorithm was found using the fractional expression:

$$\frac{TP + TN}{TP + FP + FN + TN}. \tag{1}$$

However, we should keep in mind that this metric was not very informative in tasks with unbalanced classes.

Precision: the fraction of objects that the classifier called positive and were, in fact positive was defined using the expression:

$$\frac{TP}{TP + FP}. \tag{2}$$

Recall: The proportion of objects of a positive class from all objects of a positive class was found using the expression:

$$\frac{TP}{TP + FN}. \tag{3}$$

This indicator demonstrated the ability of the algorithm to detect this class as a whole the accuracy indicator helped to distinguish this class from other classes.

As with the precision measure, the recall measure was independent of the class corre lation and, therefore, was applicable in unbalanced samples, unlike the accuracy measure

F-measure (F-score): One of several ways to combine precision and recall into one aggregate criterion is to compute an F-measure. In this case, it was the harmonic mean of accuracy and recall.

The error curve or AUC–ROC curve (area under curve–receiver operating characteris tic curve) is a graphical characteristic of the binary classifier quality. The dependence of the percentage of true positive classifications was the true positive rate (TPR):

$$TPR = \frac{TP}{TP + FN} \tag{4}$$

from the false positive rate (FPR) of the false classifications (FPR):

$$FPR = \frac{FP}{FP + TN}. \tag{5}$$

In an ideal case, when the classifier does not make mistakes (FPR = 0, TPR = 1), the square under the curve is equal to one; if the classifier outputs the same number of TP and FP results, the AUC–ROC will approach 0.5. The area under the curve shows the quality of the algorithm so the bigger the square, the better.

5. Experiments

We built an algorithm to solve the shoplifting recognition problem as a classification problem. It is possible to point out the four main stages of the algorithm.

1: At the first stage, the data collection, a set of samples was collected for further use in the training of the model and its testing. In our study, it was a set of videos with recorded cases of shoplifting from surveillance cameras in retail stores. As the dataset was unbalanced, the predominant class was under sampled, i.e., a number of instances with the label "not shoplifting" were removed so that the number of video fragments in the two classes was equal (155).

As the dataset was somewhat small for such a non-trivial task as the classification of human actions (only 310 instances), it was decided to artificially enlarge it. Each video fragment was horizontally mirrored, after which we obtained 620 instances. From each of the 620 video fragments, 2 more copies were formed that were rotated 5 degrees to the left and 5 degrees to the right. Thus, we obtained a dataset of 1860 video fragments.

2: At the second stage of data preprocessing, work was performed on the data. The video fragments were labelled into classes (Class 1: cases with shoplifting and Class 0: normal customer behavior). Processing was also performed before the feature extraction by resizing the image to a size of 224 × 224 pixels and dividing the video into frames. As each video fragment was 3 s long by 10 frames/second, this provided sequences of 30 frames. The dataset was divided into training and test sets (1302 training and 558 test sets).

3: During the third stage, the feature extraction was performed using the convolutional neural network MobileNetV3Large preprepared on the ImageNet-1k dataset.

4: We created, trained, and tested a recurrent neural network with layers of gated recurrent nodes. The features extracted by the convolutional network MobileNetV3Large from each image of the labeled sequences of frames of video fragments were delivered for training in a recurrent network with gated nodes.

Figure 2 shows the main stages of the classification algorithm under consideration.

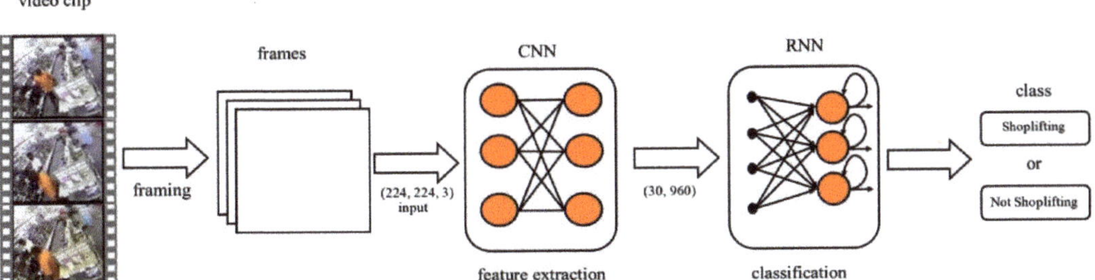

Figure 2. The main stages of the classification algorithm.

The program to solve the problem of video surveillance shoplifting recognition was written in the Python programming language. The web interactive computing environment Jupyter Notebook was used to process the dataset and the Colaboratory environment from Google Research was used to build the neural network architecture and train and test the model. This resource allowed the program to run in a cloud environment with Google Tensor Processors (TPU), which satisfied the need for a high speed when working with machine learning packages; namely, the TensorFlow software library that used the tensor calculations.

6. Results and Discussion

To achieve a sufficiently high accuracy of the classifier, a lot of research and experimentation was conducted in practice; in particular, the choice of video classification method, the search for a dataset, the search for optimal data processing and neural networks, and their configurations with parameters.

Data processing is an important factor in preparing model training. As noted above, we performed an artificial increase in the dataset as training on 310 samples achieved an accuracy of about 77% whereas in the case of a dataset increased to 1860 samples, the accuracy increased by about 5–7%.

The choice and configuration of the neural networks were also very important. The first experiments were performed on a combination of a convolutional InceptionV3 network [35] and a recurrent neural network with gated nodes. For such a set of neural networks, the speed of the program in real time was low because the extraction of the features took a long time, so we had to modify the convolutional neural network. Our task was to find a network that had fewer features in the output and would find the important information in the frame. After a number of experiments, we found an optimum neural network: MobileNetV3Large, which had a size of (1, 960) on the output in contrast to two times larger (1, 2048) in InceptionV3. As a result, the accuracy increased by approximately 10%.

One of the few hyperparameters of the convolutional neural network available for selection was the dimension of the video fragment. This was chosen based on the parameters of the input layer. We reduced all instances to $224 \times 224 \times 3$, meaning that the resolution was 224×224 and used the three-color model (RGB). We also removed the top layer in order to perform the feature extraction. At the output, we had a vector with a dimension of 1×960 (one frame)—that is, 960 features—which was a constant value for this neural network. By trial and error, the optimal number of frames per sequence was obtained, which was 30.

An important step in the optimization was the selection of the batch size parameter, which meant the number of features per training iteration. Table 1 shows the accuracy and loss rates for each parameter value from 4 to 64 for the training and validation samples. The training sample, which occupied 70% of the dataset, was divided for the model training into a training sample and a validation sample at a ratio of 7:3. The validation sample was an intermediate training phase, which was used to select the best model and optimize it.

Table 1. Indicators for each investigated batch size value.

Batch Size	Iterations	Training Accuracy	Training Loss	Validation Accuracy	Validation Loss	Test Accuracy
4	228	99.34	0.026	90.28	0.407	90.14
8	114	98.02	0.066	92.84	0.247	91.27
16	57	99.34	0.027	90.79	0.326	92.11
32	29	98.57	0.051	90.79	0.302	92.83
64	15	99.45	0.015	93.09	0.338	93.19

Such hyperparameters were selected by a large number of experiments: batch size: 64; epochs: 60; the ratio of the training and test data: 70:30; and the quality metric in terms of training: accuracy. As we only had two classes, binary cross-entropy loss was chosen to perform in the recurrent neural network.

We concluded that this configuration (including all the factors listed above) was optimal. Below, we present the results of this particular case.

An important and necessary part of the process of selecting the hyperparameters was the construction of the architecture of the sequence model (recurrent classifier). The set of layers was as follows: (1) at the input, we had a GRU layer with 32 units; (2) the GRU layer was next, with 16 units; (3) a dropout layer was used with a rate of 0.4 to reduce overfitting; (4) a fully connected layer was used with 8 neurons with a Relu activation function; and (5) the output was a fully connected layer with 1 neuron and a sigmoid activation function, which yielded the probability of a theft.

Using the Adam optimizer (Adaptive Momentum) from Keras, we obtained a leaning rate of 0.001 by default and the descent stochastic gradient momentum was adaptively adjusted as determined by the Adam optimizer.

The graph of changes in the values of the training and validation accuracy depending on the epoch of the training is shown in Figure 3.

Figure 3. Training and validation accuracy depending on the epoch.

The graph of changes in the training and validation loss values is shown in Figure 4.

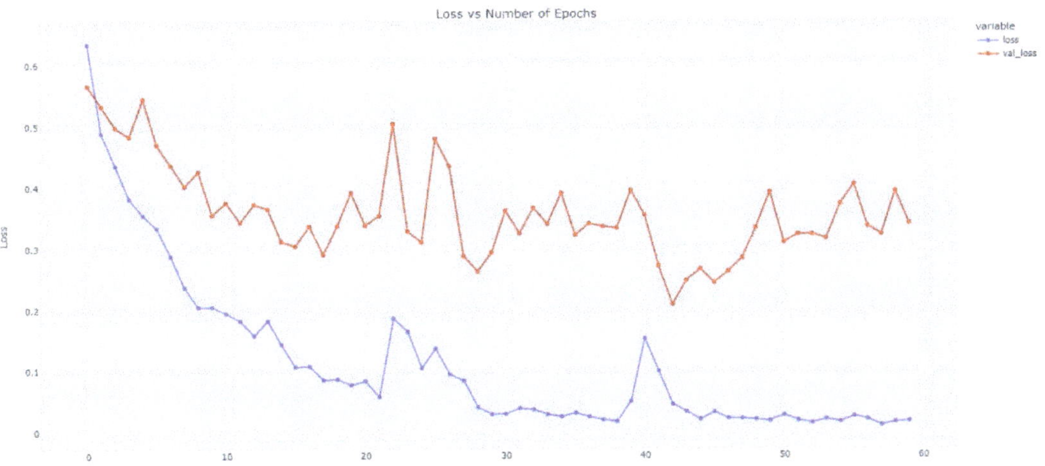

Figure 4. Training and validation loss values depending on the epoch.

The classification results for the test sample are displayed in the confusion matrix in Table 2; the rows show the true positive (TP) and false negative (FN) classification results and the columns show the false positive (FP) and true negative (TN) values.

Table 2. Confusion matrix.

	Normal Behavior (Class 0)	**Shoplifting (Class 1)**
Predicted Class 0	261	19
Predicted Class 1	22	256

Table 3 shows the values of the calculated metrics of the precision, recall, and F1-score in relation to each of the classes.

Table 3. Values of precision, recall, and F1-score.

	Precision	**Recall**	**F1-Score**	**Support**
Normal Behavior (Class 0)	0.92	0.93	0.93	280
Shoplifting (Class 1)	0.93	0.92	0.93	278
Accuracy	–	–	0.93	558

Thus, the classification accuracy was 0.93. As the sample was balanced and given the presented values of the precision, recall, and F1-score, this value fully characterized the classification result.

The quality of the prompted classifier was also evaluated using an additional ROC curve (Figure 5). The area under it (AUC value) was 0.97.

It should be noted that the classification accuracy of 93% was several percent higher than the accuracy obtained in similar studies. In [21], the classification accuracy was 75%. The researchers in [22] obtained an accuracy of 90.26%, but did not use real video surveillance data, only a synthesized set of videos.

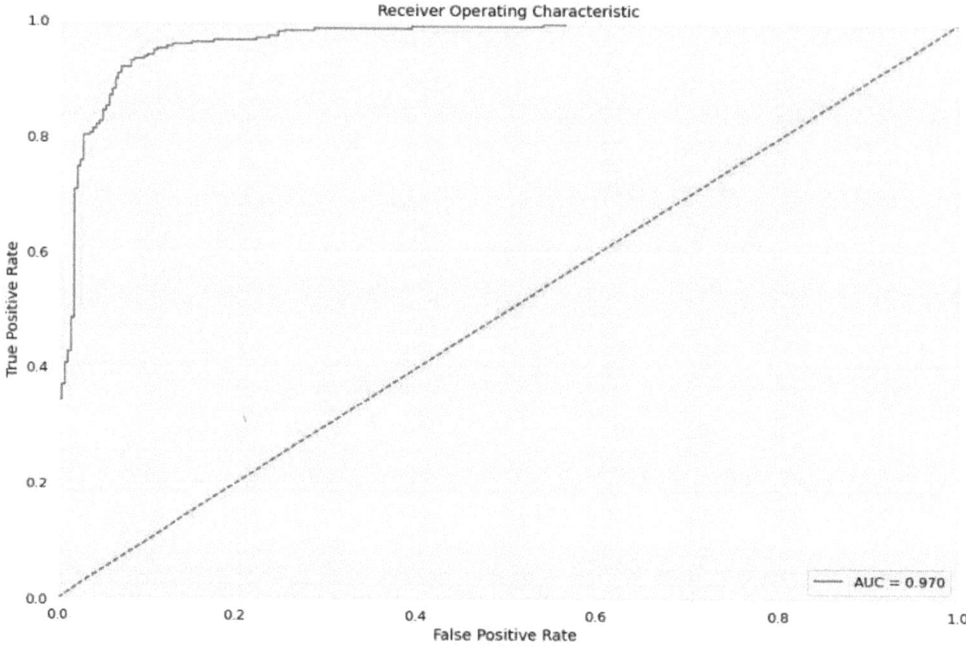

Figure 5. ROC curve and AUC value.

7. Conclusions

In this article, a classifier of video data from surveillance cameras was proposed to identify fragments with cases of shoplifting. The classifier represented a symbiosis of convolutional and recurrent neural networks. With this approach, a convolutional neural network was used to extract the features from each frame of a video recording and a recurrent network was applied to process the time sequence of the video frames and classify them. As a recurrent network, a variety of gated recurrent units was selected.

To teach the hybrid neural network, the popular UCF-Crime dataset was used. This contained video recordings of consumer behavior, including shoplifting. The original dataset contained data that were class-unbalanced; therefore, the sample of the predominant class was reduced. The resulting balanced sample of video data was artificially increased, which made it possible to improve the network training. The experiments were carried out with a pretrained convolutional neural network.

A neural network with ventilated recurrent nodes was used to classify the sequence of the video fragments. The classification results showed a high accuracy of 93%, which was several percent higher than the accuracy of the classifiers considered in the review of similar studies. The trained classifier had a high performance sufficient for a real-time operation. Further research will focus on the practical implementation of the proposed hybrid neural network in shopping malls.

Author Contributions: Conceptualization, L.K. and S.Y.; methodology, L.K. and T.R.; software, B.S.; validation, T.R. and B.S.; formal analysis, L.K. and S.Y.; investigation, L.K. and S.Y.; resources, T.R.; writing—original draft preparation, L.K. and S.Y.; writing—review and editing, L.K. and T.R.; visualization, B.S.; supervision, L.K. and S.Y.; project administration, L.K. and T.R. All authors have read and agreed to the published version of the manuscript.

Funding: This research was funded by Beethoven Grant No. DFG-NCN 2016/23/G/ST1/04083 and by a Grant of the Ministry of Education and Science of Ukraine "Technologies, tools for mathematical modeling, optimization and system analysis of coverage problems in space monitoring systems".

Institutional Review Board Statement: Not applicable.

Informed Consent Statement: Not applicable.

Data Availability Statement: Generated data and test tasks were used.

Conflicts of Interest: The authors declare no conflict of interest.

References

1. Chemere, D.S. Real-time Shoplifting Detection from Surveillance Video. Master Thesis, Addis Ababa University, Addis Ababa, Ethiopia, 2018; p. 94.
2. Kirichenko, L.; Radivilova, T. Analyzes of the distributed system load with multifractal input data flows. In Proceedings of the 2017 14th International Conference The Experience of Designing and Application of CAD Systems in Microelectronics, CADSM 2017, Lviv, Ukraine, 21–25 February 2017; pp. 260–264.
3. Gim, U.J.; Lee, J.J.; Kim, J.H.; Park, Y.H.; Nasridinov, A. An Automatic Shoplifting Detection from Surveillance Videos. In Proceedings of the AAAI Conference on Artificial Intelligence, New York, NY USA, 7–12 2020; Apress: Berkeley, CA, USA, 2020; Volume 34, pp. 13795–13796.
4. Ivanisenko, I.; Kirichenko, L.; Radivilova, T. Investigation of multifractal properties of additive data stream. In Proceedings of the 2016 IEEE 1st International Conference on Data Stream Mining and Processing, Lviv, Ukraine, 23–27 August 2016; pp. 305–308.
5. Kirichenko, L.; Radivilova, T.; Bulakh, V. Machine learning in classification time series with fractal properties. *Data* **2019**, *4*, 5. [CrossRef]
6. Pang, G.; Shen, C.; Cao, L.; van den Hengel, A. Deep Learning for Anomaly Detection: A Review. *ACM Comput. Surv.* **2020**, *1*, 36. [CrossRef]
7. Radivilova, T.; Kirichenko, L.; Ageiev, D.; Bulakh, V. Classification methods of machine learning to detect DDoS attacks. In Proceedings of the 2019 10th IEEE International Conference on Intelligent Data Acquisition and Advanced Computing Systems: Technology and Applications, IDAACS, Metz, France, 18–21 September 2019; pp. 207–210.
8. Rehman, A.; Belhaouari, S.B. Deep Learning for Video Classification: A Review. *TechRxiv* **2021**, *preprint*.
9. Yamato, Y.; Fukumoto, Y.; Kumazaki, H. Security camera movie and ERP data matching system to prevent theft. In Proceedings of the 2017 14th IEEE Annual Consumer Communications & Networking Conference (CCNC), Las Vegas, NV, USA, 8–11 January 2017; pp. 1014–1015.
10. Tsushita, H.; Zin, T.T. A Study on Detection of Abnormal Behavior by a Surveillance Camera Image. In *Big Data Analysis and Deep Learning Applications*; Zin, T.T., Lin, J.C.W., Eds.; Springer: Singapore, 2019; pp. 284–291.
11. Flores-Munguia, C.; Ortiz-Bayliss, J.C.; Terashima-Marin, H. Leveraging a Neuroevolutionary Approach for Classifying Violent Behavior in Video. *Comput. Intell. Neurosci.* **2022**, *2022*, 1279945. [CrossRef] [PubMed]
12. Morales, G.; Salazar-Reque, I.; Telles, J.; Diaz, D. Detecting violent robberies in cctv videos using deep learning, IFIP advances in information and communication technology. In *Artificial Intelligence Applications and Innovations*; Springer International Publishing: Cham, Switzerland, 2019; pp. 282–291.
13. Akbar, M.N.; Riaz, F.; Awan, A.B.; Khan, M.A.; Tariq, U.; Rehman, S. A Hybrid Duo-Deep Learning and Best Features Based Framework for Action Recognition. *Comput. Mater. Contin.* **2022**, *73*, 2555–2576.
14. Nasaruddin, N.; Muchtar, K.; Afdhal, A.; Dwiyantoro, A.P.J. Deep anomaly detection through visual attention in surveillance videos. *Big Data* **2020**, *7*, 87. [CrossRef]
15. University of Central Florida. UCF-Crime Dataset. Available online: https://www.v7labs.com/open-datasets/ucf-crime-dataset (accessed on 17 September 2022).
16. Sultani, W.; Chen, C.; Shah, M. Real-world anomaly detection in surveillance videos. In Proceedings of the IEEE Conference on Computer Vision and Pattern Recognition, Salt Lake City, UT, USA, 18–23 June 2018; pp. 6479–6488.
17. Arunnehru, J.; Chamundeeswari, G.; Bharathi, S.P. Human action recognition using 3D convolutional neural networks with 3D motion cuboids in surveillance videos. *Procedia Comput. Sci.* **2018**, *133*, 471–477. [CrossRef]
18. Li, J.; Jiang, X.; Sun, T.; Xu, K. Efficient Violence Detection Using 3D Convolutional Neural Networks. In Proceedings of the 2019 16th IEEE International Conference on Advanced Video and Signal Based Surveillance (AVSS), Taipei, Taiwan, 18–21 September 2019; pp. 1–8.
19. Islam, M.S.; Sultana, S.; Kumar Roy, U.; Al Mahmud, J. A review on video classification with methods, findings, performance, challenges, limitations and future work. *J. Ilm. Tek. Elektro Komput. Dan Inform. (JITEKI)* **2020**, *6*, 47–57. [CrossRef]
20. Alfaifi, R.; Artoli, A.M. Human action prediction with 3D-CNN. *SN Comput. Sci.* **2020**, *1*, 286. [CrossRef]
21. Martinez-Mascorro, G.A.; Abreu-Pederzini, J.R.; Ortiz-Bayliss, J.C.; Garcia-Collantes, A.; Terashima-Marin, H. Criminal Intention Detection at Early Stages of Shoplifting Cases by Using 3D Convolutional Neural Networks. *Computation* **2021**, *9*, 24. [CrossRef]
22. Ansari, M.A.; Singh, D.K. ESAR, An Expert Shoplifting Activity Recognition System. *Cybern. Inf. Technol.* **2022**, *22*, 190–200. [CrossRef]
23. Harvey, M. Five Video Classification Methods Implemented in Keras and TensorFlow: Exploring the UCF101 Video Action Dataset. 2017. Available online: https://blog.coast.ai/five-video-classification-methods-implemented-in-keras-and-tensorflow-99cad29cc0b5. (accessed on 14 September 2022).

24. Kirichenko, L.; Alghawli, A.S.A.; Radivilova, T. Generalized approach to analysis of multifractal properties from short time series. *Int. J. Adv. Comput. Sci. Appl.* **2020**, *11*, 183–198. [CrossRef]
25. Wang, J.; Yang, Y.; Mao, J.; Huang, Z.; Huang, C.; Xu, W. Cnn-rnn: A unified framework for multi-label image classification. In Proceedings of the IEEE Conference on Computer Vision and Pattern Recognition, Las Vegas, NV, USA, 27–30 June 2016; pp. 2285–2294.
26. Gollapudi, S. *Learn Computer Vision Using OpenCV: With Deep Learning CNNs and RNNs*, 1st ed.; Apress: Berkeley, CA, USA, 2019; p. 171.
27. Nebauer, C. Evaluation of convolutional neural networks for visual recognition. *Neural Netw. IEEE Trans.* **1998**, *9*, 685. [CrossRef] [PubMed]
28. Medsker, L.; Jain, L.C. (Eds.) *Recurrent Neural Networks: Design and Applications (International Series on Computational Intelligence)*, 1st ed.; CRC Press: Boca Raton, FL, USA, 1999; p. 416.
29. Time Series Classification. Welcome to the UEA & UCR Time Series Classification Repository. Available online: http://www.timeseriesclassification.com (accessed on 9 May 2022).
30. Segall, R.S.; Niu, G. *Biomedical and Business Applications Using Artificial Neural Networks and Machine Learning*; IGI Global: Hershey, PA, USA, 2022; p. 394.
31. Shah, S. *Implementation and Evaluation of Gated Recurrent Unit for Speech Separation and Speech Enhancement*; Northern Illinois University: DeKalb, IL, USA, 2019; p. 91.
32. LazyProgrammer. *Deep Learning: Recurrent Neural Networks in Python: LSTM, GRU, and More RNN Machine Learning Architectures in Python and Theano*; Machine Learning in Python; LazyProgrammer: Apress Berkeley, CA, USA, 2021; p. 93.
33. Medjahed, S.A. A Comparative Study of Feature Extraction Methods in Images Classification. *Int. J. Image Graph. Signal Process.* **2015**, *7*, 16. [CrossRef]
34. ImageNet Database. Available online: https://image-net.org/index.php (accessed on 8 July 2020).
35. Keras API Reference/Keras Applications/MobileNet, MobileNetV2, and MobileNetV3. Available online: https://keras.io/api/applications/ (accessed on 14 September 2022).

 *computation*

Article

Modeling of the Stress–Strain of the Suspensions of the Stators of High-Power Turbogenerators

Oleksii Tretiak [1,*], Dmitriy Kritskiy [2,*], Igor Kobzar [3,*], Victoria Sokolova [4,5,*], Mariia Arefieva [5,6,*], Iryna Tretiak [7,*], Hromenko Denys [1,*] and Viacheslav Nazarenko [1,*]

[1] Faculty of Aircraft Engineering, Department of Aerohydrodynamics, National Aerospace University «Kharkiv Aviation Institute», 61070 Kharkiv, Ukraine

[2] Department of Information Technology Design, National Aerospace University «Kharkiv Aviation Institute», 61070 Kharkiv, Ukraine

[3] Special Design Office for Turbogenerators and Hydrogenerators, Joint Stock Company "Ukrainian Energy Machines", 61037 Kharkiv, Ukraine

[4] NNI «Institute of Public Administration», V. N. Karazin Kharkiv National University 61070 Kharkiv, Ukraine

[5] Department of Aerohydrodynamics, National Aerospace University «Kharkiv Aviation Institute», 61070 Kharkiv, Ukraine

[6] Kharkiv Lyceum «IT Step School Kharkiv», 61010 Kharkiv, Ukraine

[7] Aerospace Thermal Engineering Department, National Aerospace University «Kharkiv Aviation Institute», 61070 Kharkiv, Ukraine

[*] Correspondence: alex3tretjak@ukr.net (O.T.); d.krickiy@khai.edu (D.K.); ivkobzar@ukr.net (I.K.); sokolova.v.mk@gmail.com (V.S.); marii.arefieva@gmail.com (M.A.); irina.ii3t@gmail.com (I.T.); dvgromenko@gmail.com (H.D.); my_registrator@ukr.net (V.N.)

Abstract: In the submitted scientific work, the existing types of stator fastening design of turbogenerators and the main causes of the stressed state of the stator suspensions are considered. A detailed calculation of the complex stressed state of the turbogenerator stator suspension was carried out for a number of electrical sheet steels, taking into consideration the unevenness of the heat distribution along the horizontal axis of the unit. It is proposed that the calculation of the mechanical stress is carried out by means of the mechanical and thermal calculation, coordinated with the electrical one. The possibility of replacing steel 38X2H2BA with steel 34CrNiMo6 and 40NiCrMo7 is indicated, subject to compliance with GOST 8479-70 for the same strength group.

Keywords: turbogenerator; spring suspension; limiting conditions; active steel; the stator casing

Citation: Tretiak, O.; Kritskiy, D.; Kobzar, I.; Sokolova, V.; Arefieva, M.; Tretiak, I.; Denys, H.; Nazarenko, V. Modeling of the Stress–Strain of the Suspensions of the Stators of High-Power Turbogenerators. *Computation* 2022, 10, 191. https://doi.org/10.3390/computation10110191

Academic Editor: Dimitris Drikakis

Received: 11 September 2022
Accepted: 25 October 2022
Published: 28 October 2022

Publisher's Note: MDPI stays neutral with regard to jurisdictional claims in published maps and institutional affiliations.

1. Introduction

In recent decades, there has been a tendency not only to increase the efficiency of electric machines (EM), but also to decrease their mass-dimensional indices per unit of power. As a rule, this is carried out by the optimization of calculations and the use of three-dimensional modeling of physical processes (electromagnetic, mechanical, temperature, ventilation, etc.), as well as the emergence of new materials with improved parameters, which allows reduction of the weight of turbogenerator units.

Today, in Ukraine, almost 74% of electrical energy is produced at thermal and nuclear power plants, where turbogenerators are in operation. The service lives of most turbogenerators have already expired, and others are on the verge of expiration; this is due to long-term insufficient financing of the energy industry. At the same time, EM operating modes are complicated by uneven loads in the electrical network, which cause both generator overload (transition to emergency modes of operation, due to malfunction of generators at stations or an increase in the amount of consumed energy), and their shutdown (decrease in the amount of consumed energy). The solution to this problem is the partial modernization of already existing units with an increase in their capacity and the parallel, step-by-step replacement of the rest of the outdated machines with more powerful and lighter ones.

The development of a single methodology for calculating the stress–strain state (STS) for high-power turbogenerator units, based on a combination of analytical and three-dimensional calculations that allows increasing the accuracy of the calculation problem, is of great scientific and practical interest.

Let us consider the design of a two-pole turbogenerator manufactured by the EM WEG Group (USA) (see Figure 1) [1]. Application of the designs by putting into practice additional flexible elements, as a rule, is the most suitable method for turbogenerators rated over 100 MW.

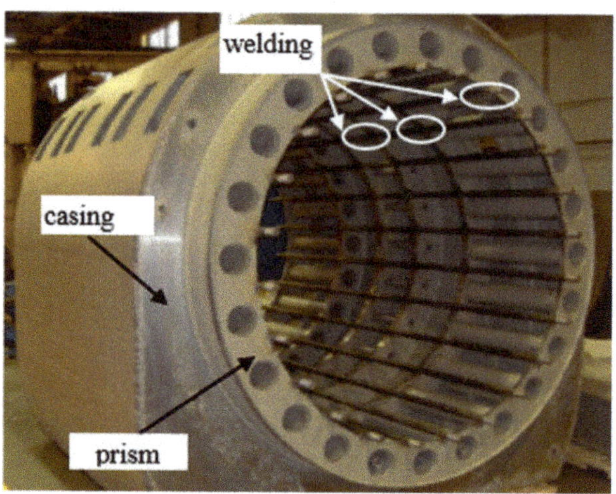

Figure 1. Stator Casing of EM WEG Group.

The core fastening units include a system of prisms, which, with the help of pressing down flanges, keep the assembled core in a monolithic state. However, there are no springs in these suspensions that dampen the vibrational components of the harmonics of electric forces caused by gravity.

Classic designs of turbogenerators include two types of suspension, namely internal and external. One of the most striking and reliable representatives of turbogenerators with internal suspension are Turbogenerators rated 200 MW and 300 MW of the TGV series produced by the JSC "Ukrainian Energy Machines". The design of the stator housing of an electric machine has been submitted in the scientific paper [2]. In Refs. [3,4] the results of calculations of thermal fields in the end parts of the turbogenerator are presented. However, as operating experience shows, active cooling inside the rods and the high speed of the flowing refrigerant solve the problem of lowering temperatures in the end parts and their fastenings for the winding rods. Based on operating experience, the numerical solution for the frontal parts does not give an answer regarding the state of the grooves. Figure 2 shows a serial sample and a longitudinal section.

Vibration is one of the important factors leading to the damage and destruction of turbogenerator housing parts. In Ref. [5], the effect of vibration on turbogenerator assemblies is considered in detail. In turbogenerator design, the suspension (plate springs) takes the load caused by vibration forces and must ensure the safety of the structure when the unit enters the short-circuit mode. Other methods require optimization of the design to increase power while maintaining weight and dimensions [1,2,6]. When solving these problems, it is necessary to pay attention to the level of vibration when carrying out tests [7]. To ensure reliable operation, it is necessary to calculate the design of the suspension, which reduces vibration while maintaining the basic power.

(a)   (b)

Figure 2. Turbogenerator type TGV-300: (**a**) at the base of operations, TPP; (**b**) cross section: 1—stator casing; 2—external shield; 3—brush-holders device; 4—stator (active steel and bars); 5—vertical row of springs; 6—horizontal row of springs.

2. Turbogenerator Suspension Design by JSC "Ukrainian Energy Machines"

In the given scientific paper, the strength of the structures of the internal suspensions of the stators of hydrogen-cooled turbogenerators rated 200 MW, 250 MW, 300 MW, and 325 MW, produced by JSC "Ukrainian Energy Machines" is studied. All the researched machines have a similar design of internal suspensions; the difference is in the number and geometric parameters of the suspension used.

In Figure 3, the general view of the turbogenerator is submitted.

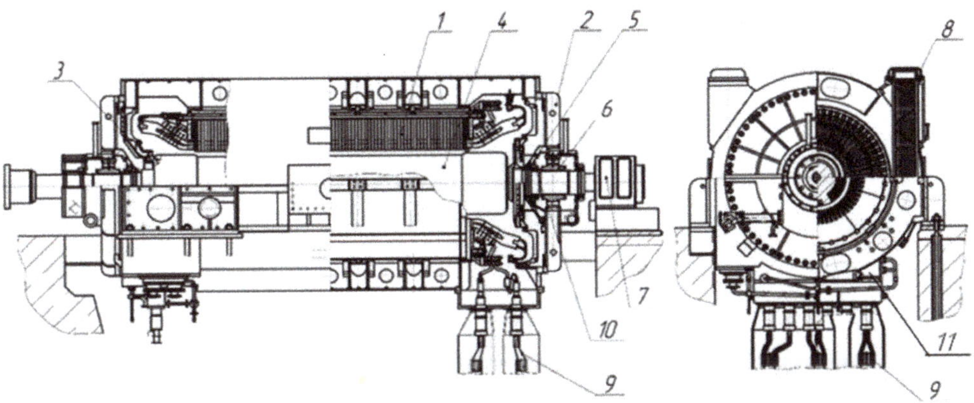

Figure 3. Turbogenerator TGV-200 design includes the following, namely: 1—stator; 2—external shield; 3—internal fairing; 4—rotor; 5—rotor shaft sealing; 6—bearing; 7—brush-holder device; 8—gas-cooler; 9—end terminals; 10—oil trap; 11—stator suspension.

An analysis of the stress–strain state of the suspension unit, which includes a spring, a support plate, a strap, and a system of pins and bolted connections, was carried out (see Figure 4).

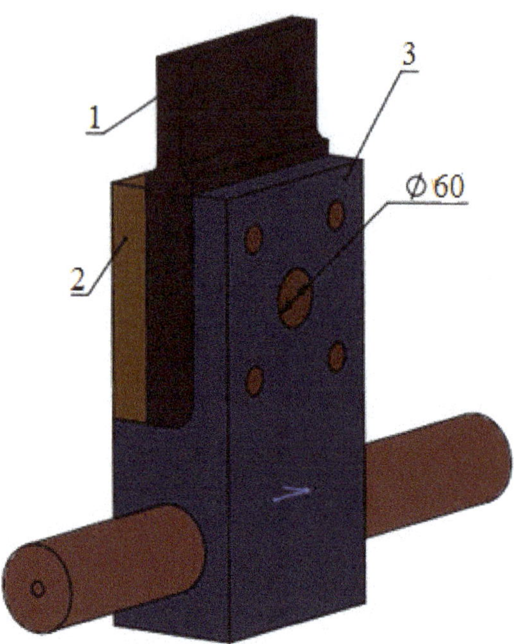

Figure 4. Three-dimensional image of the assembly of the spring to the frame: 1—spring; 2—support plate; 3—system of pins.

The strength of the suspension unit was studied at the moment of a two-phase short-circuit, which corresponds to the maximum load on the suspension system. The occurrence of a short-circuit is characterized by the emergence of a moment of short-circuit of the MRC, which leads to compressive/stretching forces of the RKZ acting on the springs. The magnitude of these forces is determined for each machine by classical analytical methods of calculating the suspension of the stator core during a two-phase short-circuit, in accordance with the technical conditions for Turbogenerators TGV-200, TGV-300 series, manufactured by JSC "Ukrainian Energy Machines". The calculation is carried out for the case of static loading of the suspension system by the compression/tension force of the RKZ, while a dynamic factor is chosen equal to 2.

The rated torque acting on the stator of the generator is determined by the following formula:

$$M_{\text{H}} - 9560\frac{N}{n},\tag{1}$$

where N is the power of the generator and n is the rotational speed of the generator per minute (rpm).

Forces on the vertical spring at rated mode are calculated as per the following formula:

$$P_{\text{H}} = \frac{Gg}{z_{\text{B}}} + \frac{M_{\text{H}} \times cos\varphi}{z} \times \frac{1}{R},\tag{2}$$

where G is the stator mass, z_{B} is quantity of vertical springs, $cos\varphi$ is the power factor, z is the general quantity of the springs, and R is the radius of the spring arrangement.

At short-circuit mode, this force is determined as:

$$P_{\text{K3}} = \frac{Gg}{z_{\text{B}}} + P_{max},\tag{3}$$

where P_{max} is force, which acts on the spring at a sudden two-phase short-circuit on the terminals.

In addition, the influence of temperature loads on the suspension is taken into account, which vary along the length of the stator core and are determined to take into consideration the internal heating of the stator with the solution of the gas-dynamic problem. According to the experience of operating turbogenerators rated 325 MW with hydrogen cooling, the temperature difference between the core and the stator casing can be 60 °C which is confirmed by calculation data. According to the calculation data for a turbogenerator rated 250 MW with hydrogen–water cooling, the temperature of the "active steel" of the stator on the side of the slip rings is 36 °C, in the middle of the machine is 39 °C, and on the side of the turbine is 41 °C. The gas temperature in the radial channel is in the range of 25 °C to 47 °C. The hottest point of the gas on the stator is 47 °C and is located in the second compartment in the area of the backrest. The gas temperature increases from the tooth to the back by ~10 °C. The turbogenerator cooling system is shown in Figure 5.

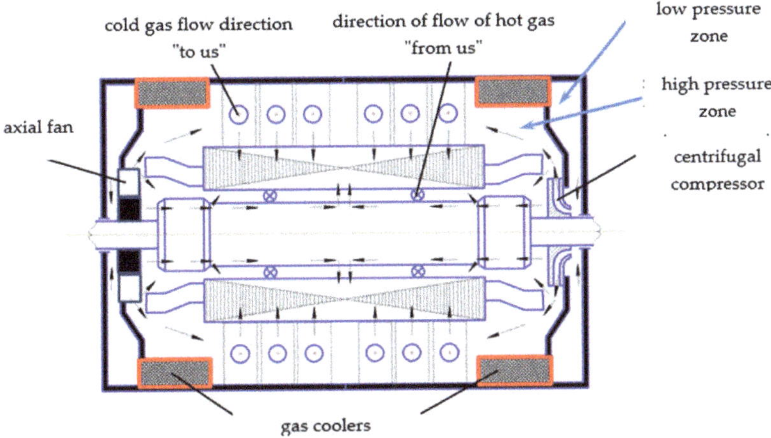

Figure 5. Turbogenerator cooling system.

The longitudinal ventilation channels are shown in Figure 6. The springs are inside the stator.

Therefore, for each row of springs, it is necessary to determine the mechanical stresses, taking into account the change in their thermal state. Determination of the temperature field in parts of the suspension unit is performed by solving the unsolved thermal problem, applying boundary conditions of the first kind.

Thus, the calculation of the suspension is carried out for the axial tension/compression forces acting on the spring at short-circuit and are determined according to the classical engineering methods used for the calculation of the suspension of turbogenerators. With that, the temperature loads on the suspension assembly unit mentioned above are taken into account. This allows for a more accurate description of the real stress–strain state (STS) in the suspension assembly unit.

The general view of the suspension of the considered turbogenerators is shown above in Figure 4. The suspension consists of vertical and horizontal flat springs, one end of which is fixed to the stator casing, and the other is fastened to the frame. The number of stator spring suspensions for generators, even those of the same power, may vary. At the same time, depending on the power of the generator, the geometric parameters of the suspensions and the load acting on them change. Further, a study of the strength of the suspension unit for generators rated 200 MW, 250 MW, and 325 MW is performed.

The purpose of the calculation is to define the stresses in the suspension spring and the details of its fastening to the stator casing and to the frame in case of a short-circuit in the stator winding, taking into account the unevenness of temperature loads and possible assembly inaccuracies.

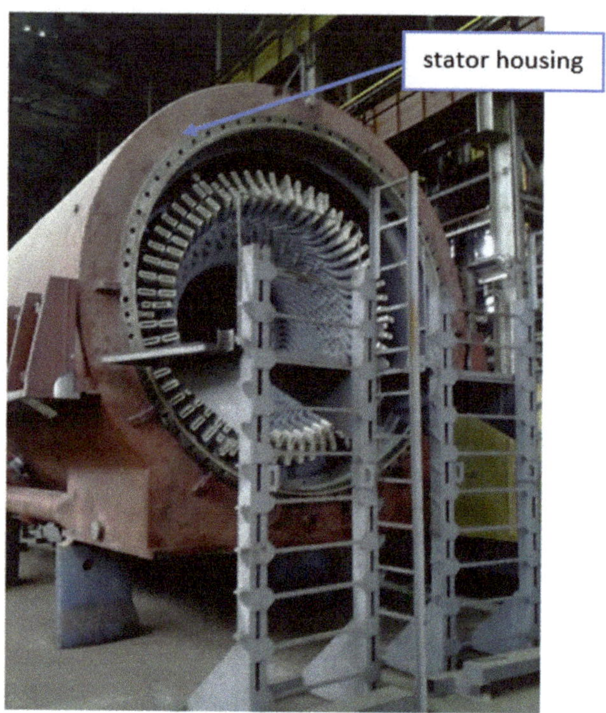

Figure 6. The Stator of Turbogenerator TGV-300-2U3.

3. Study of the Strength of the Stator Suspension of Generators Rated 325 MW and 250 MW

The main geometric and physical characteristics of the suspension of the turbogenerator rated 325 MW, 3000 rpm are as follows:

- weight of the stator core with the winding $G = 185{,}000$ kg;
- quantity of springs $Z = 20$ pcs.;
- thickness of a spring $h = 1.8$ cm;
- widths of a spring $b = 20$ cm;
- calculation length of a spring $l = 65$ cm;
- cross-sectional area of a spring $F = b \times h = 20 \times 1.8 = 36$ cm^2;
- the distance between the eye bolts of a spring $L = 85$ cm;
- the radius of springs arrangement $R = 147.4$ cm.

The amplitude value of the moment during a short-circuit is equal to Msh.c. (MKZ) = 2.62×108 kg cm, and the maximum force on one spring from the torque during a short-circuit is $P2 = 89{,}000$ kg.

The stator core of the generator rated 325 MW is attached to the housing with 10 horizontal and 10 vertical springs. Fastening with the help of 16 springs is also used. Figure 7 shows the fastening unit for attaching the spring to the frame.

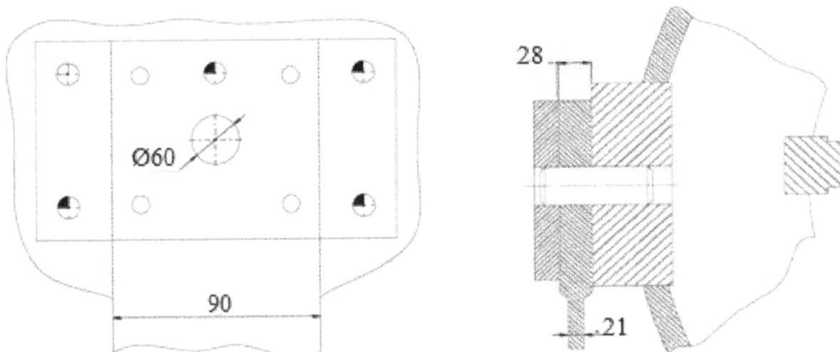

Figure 7. Drawing of the fastening unit for attaching the spring to the frame.

Similar images of the method of the fastening of the spring to the stator casing are shown in Figures 8 and 9.

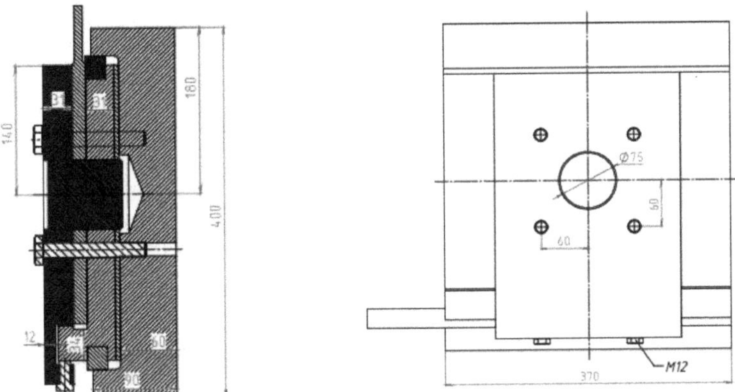

Figure 8. Drawing of the fastening unit of a spring to the stator casing.

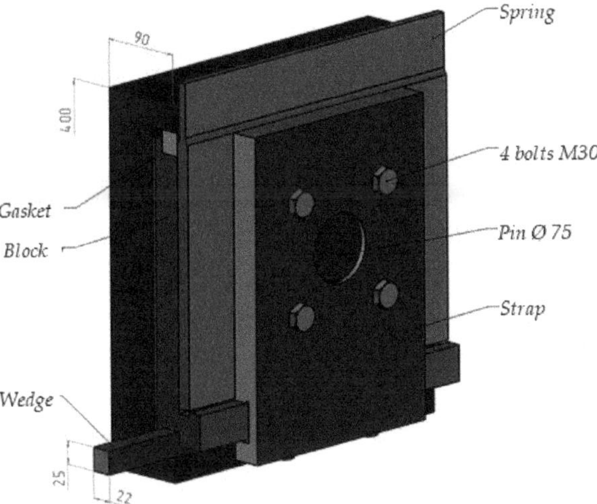

Figure 9. Three-dimensional image of the assembly of the spring attachment to the stator casing.

A plate with mounting holes is welded to the frame, to which a spring is attached and fixed with a strap. In the generator rated 250 MW and 320 MW, the spring is attached to the frame with a conical pin with a diameter of 60 mm, 5 conical pins of 30 mm, and 4 bolts, M30.

The spring is attached to the stator casing with a pin and a strap. In the generator rated 250 MW, the spring is attached for support with a conical pin with a diameter of 60 mm and three bolts, M30. A sleeve welded to the rings of the stator casing is installed on the cylindrical ends of the supports with tension. The conical pin is held on one end by a support, and on the other by a cheek, which is attached to the support with two bolts, M36, working in tension. Thus, the pin has two planes of cutting. On a generator rated 320 MW, fastening is carried out with one pin with a diameter of 60 mm and 4 bolts, M36.

The material of the suspension and pins is alloyed steel.

In Figure 10, a calculation diagram with the main loads acting on the support elements of the suspension and the temperature of its elements is shown, where the numbers 1, 2, and 3 indicate the support pins, and 4 indicates the main body of the spring, the directions of forces are indicated by arrows. T1, T2, T3, and T4 are the calculated temperatures of the suspension elements.

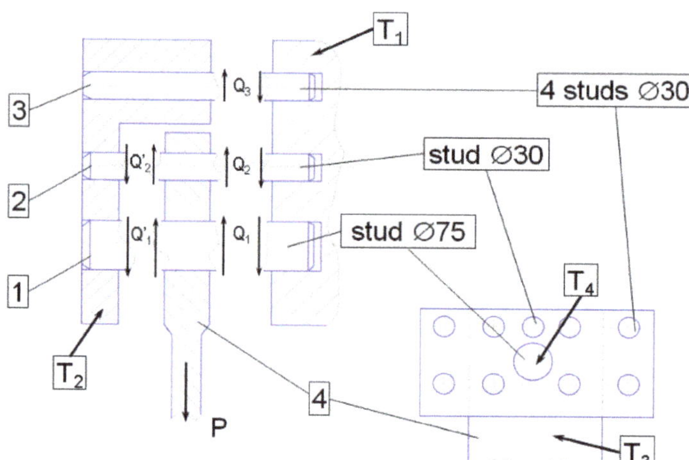

Figure 10. Calculation Diagram of the Suspension: 1, 2, 3—support pins; 4—spring.

One of the defining moments in solving MSE problems is the choice of a finite element. Two types of tetrahedrons (Figure 11) are used as basic finite elements with different approximations of movements inside the element. The first tetrahedron has units at the tops (Figure 11a) and is based on a linear approximation of movements inside the element, and the second is an oblique tetrahedron that has units at the tops of the element and in the middle of its edges; it is based on a quadratic approximation of movements inside the element (Figure 11b).

An oblique tetrahedron allows more accurate description of the geometry and deformation process of the research object; however, it has 10 internal units and contains 30 unknown values, which is almost three times the number of unknowns for an ordinary tetrahedron with four units and, accordingly, with 12 unknowns values sought for.

Therefore, a classical tetrahedron will be used for a qualitative study of VAT parameters, and an oblique one will be used for more accurate, final calculations. In SolidsWork, which is used to solve these problems, these are TETRA4 and TETRA10 finite elements, respectively [8].

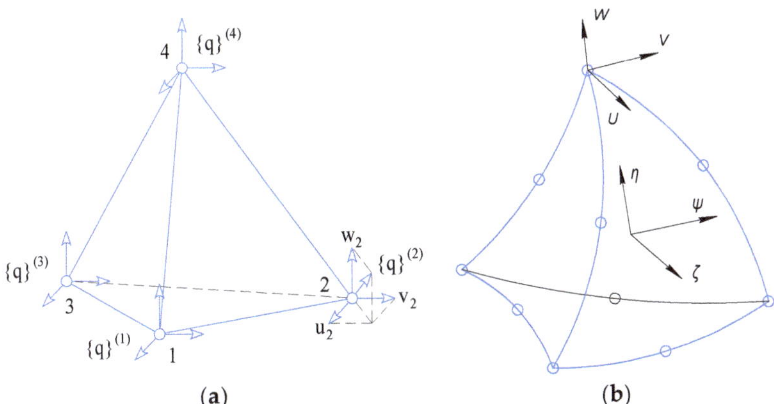

Figure 11. A finite element in the form of a tetrahedron: (**a**) TETRA4 and (**b**) TETRA10.

The influence of temperature fields leads to the emergence of additional temperature deformations [9–11]

$$\varepsilon_x^T = \alpha(T - T_0), \varepsilon_y^T = \alpha(T - T_0), \varepsilon_z^T = \alpha(T - T_0), \quad \gamma_{xy}^T = \gamma_{xz}^T = \gamma_{yz}^T = 0, \tag{4}$$

where α is the coefficient of linear temperature expansion of the material, $T = T(x, y, z)$—is the temperature distribution obtained from solving the thermal conductivity problem, and T_0 is the temperature at which there are no thermal stresses in the material.

The general deformation of the body consists of elastic deformations and temperature deformations [10]:

$$\begin{aligned}
\varepsilon_x &= \varepsilon_x^y + \varepsilon_x^T, \ldots \varepsilon_y = \varepsilon_y^y + \varepsilon_y^T, \ \varepsilon_z = \varepsilon_z^y + \varepsilon_z^T \\
\gamma_{xy} &= \gamma_{xy}^y + \gamma_{xy}^T, \ \gamma_{yz} = \gamma_{yz}^y + \gamma_{yz}^T, \ \gamma_{xz} = \gamma_{xz}^y + \gamma_{xz}^T
\end{aligned} \tag{5}$$

where ε_x^y, ε_y^y, ε_z^y, γ_{xy}^y, γ_{yz}^y, γ_{xz}^y elastic deformations.

Taking into account temperature deformations, the relationship between stresses and deformations takes on the form:

$$\begin{aligned}
\sigma_x &= \frac{E}{1+v}\left(\frac{v}{1-2v}\varepsilon + \varepsilon_x\right) - \frac{E\alpha}{1-v}(T - T_0), \\
\sigma_y &= \frac{E}{1+v}\left(\frac{v}{1-2v}\varepsilon + \varepsilon_y\right) - \frac{E\alpha}{1-v}(T - T_0), \\
\sigma_z &= \frac{E}{1+v}\left(\frac{v}{1-2v}\varepsilon + \varepsilon_z\right) - \frac{E\alpha}{1-v}(T - T_0), \\
\tau_{xy} &= G\gamma_{xy}, \tau_{yz} = G\gamma_{yz}, \quad \tau_{xz} = G\gamma_{xz}.
\end{aligned} \tag{6}$$

Let us imagine (6) in matrix form:

$$\{\sigma\} = [B]\{\varepsilon\} - [T], \tag{7}$$

where $\{\sigma\} = \{\sigma_x \sigma_y \sigma_z \tau_{xz} \tau_{yz} \tau_{zy}\}^T$.

In Ref. [12] the methods for mesh decomposition have been submitted; however, this method cannot be fully applied due to the significant difference between fasteners and main structural elements. A significant contribution to mesh adaptation can be made by improved shape optimization based on mesh [13]. When constructing the mesh, we relied on the results given in Ref. [14]. It should be noted that, when using second order equations in numerical methods, the mesh was smoothed at the contact points of the elements. Taking into account the results of [14–18], the mesh was adapted, while at the attachment points and small elements, the mesh was thickened from the condition that, for conical surfaces, the stress during mesh discrediting should not differ by more than 0.1%.

In Figure 12 show basic grid for calculation.

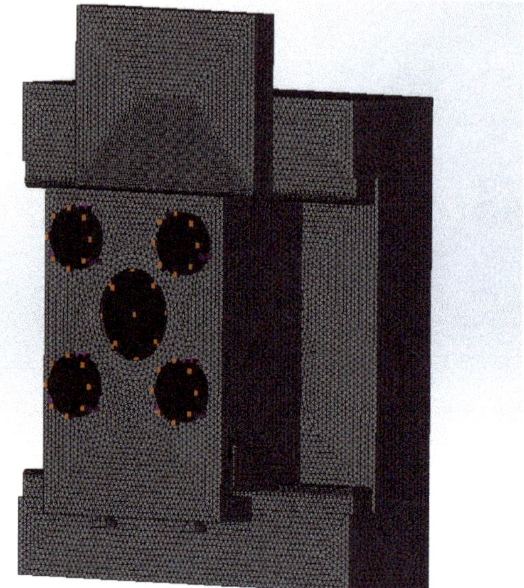

Figure 12. Calculation grid.

The stress–strain state of the assembled unit at compression is shown in Figure 13a, and when stretched is shown in Figure 13b.

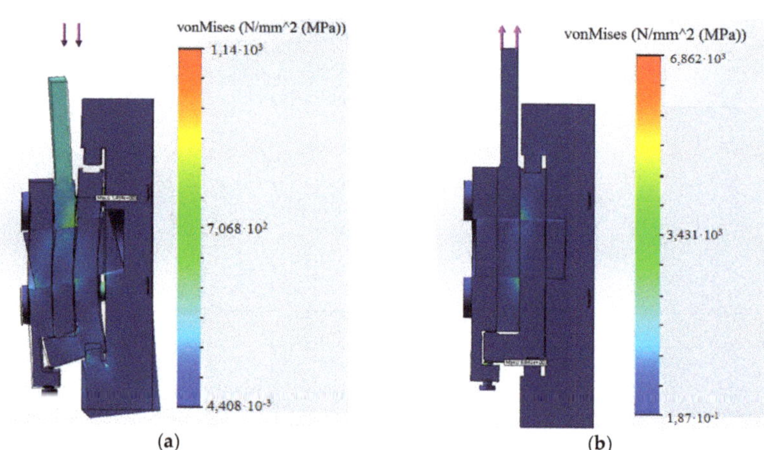

Figure 13. The Stress–Strain State of the Suspension: at Compression (**a**) and Stretching (**b**).

In Figures 14 and 15, the stress field on the spring surface and the diagram of stress changes along the curve, which is marked with numbers from 1 to 8 in the Figure, during compression is shown. As expected, there is a significant stress concentration near the holes.

The total stress values were analyzed for von Mises stresses. With that, the experience presented in Refs. [19–22] was taken into account.

In the process of correlation of the calculations obtained by the FEM, the values calculated using engineering methods were compared, as well as the actual stresses obtained by measuring with strain gauges.

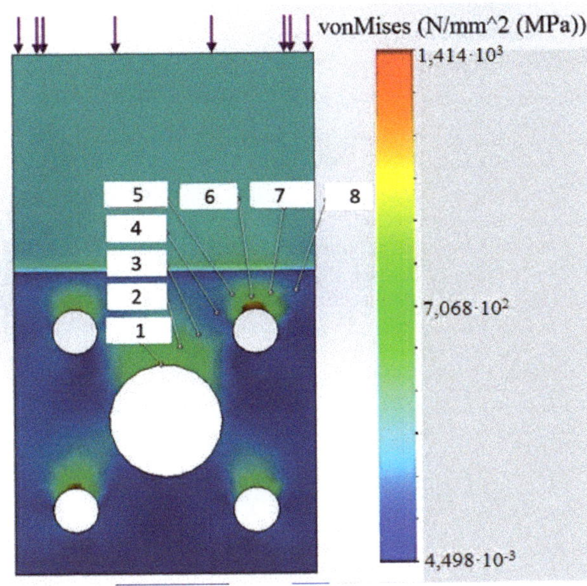

Figure 14. Stresses Field on the Spring Surface.

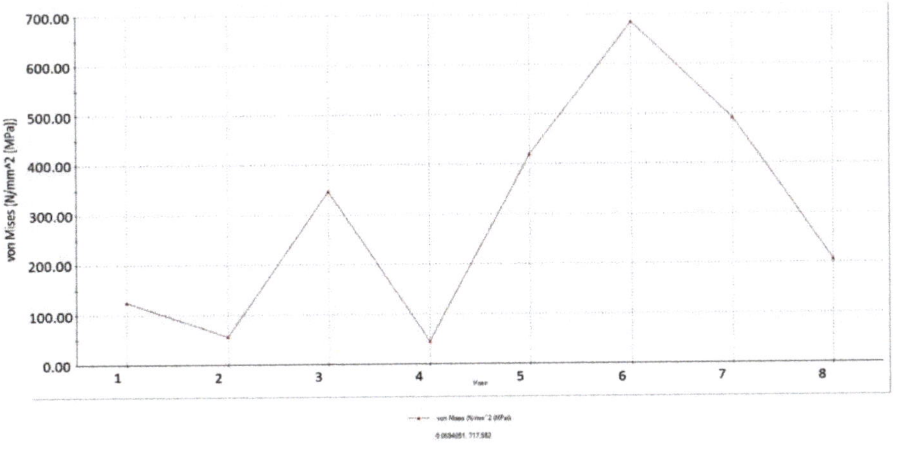

Figure 15. Stresses Change Diagram.

4. Discussion

Among the submitted scientific papers on the design of turbogenerators, a great number of works have been devoted to optimization of the design. Moreover, a significant amount of them are aimed at reducing the mass of the structure. In recent decades, new designs from Chinese and European schools have appeared, which have the most affordable prices with a moderate weight, but further optimization leads to a decrease in the projected resource. It is possible that in ten or twenty years we will face fatigue phenomena caused by vibration loads. As evidenced by the works of American and Canadian scientists, the new designs will have a reduced weight of the body parts, but the issue of increasing the reliability of the suspension of turbogenerators is widespread. The replacement of these units is not possible during operation, and if the springs fail, it will lead to the destruction of the entire structure, including the turbines. Therefore, it is necessary to further determine the permissible limits of yield and quality of steel and declare them at the global worldwide level.

5. Conclusions

Table 1 shows the calculation data of the suspension assembly unit obtained by the engineering method and proposed method based on three-dimensional modeling. The stresses in the spring are caused only by compressive forces. It can be seen that the maximum deviation of calculation results according to the proposed method and according to the engineering method does not exceed 15%. This, on the one hand, confirms the reliability of the obtained results, and on the other hand, indicates the need to carry out final calculations on the strength of the suspension unit, using three-dimensional modeling to clarify the obtained stress values.

Table 1. Stress in the suspension of a generator rated 325 MW.

Parameters	Calculation Method	
	Engineering Calculation	**3D Calculation (Proposed Method)**
maximum values of stresses in the spring, MPa	40	44.6
stress in pins Ø60: from crumpling (spring-pin), MPa	85.6	84.4
crumpling stress between pin Ø60 and support (strap), MPa	22.7	25

Author Contributions: Writing—review and editing O.T., D.K., I.K., V.S., M.A., I.T., H.D. and V.N. All authors have read and agreed to the published version of the manuscript.

Funding: This research received no external funding.

Conflicts of Interest: The authors declare no conflict of interest.

References

1. Shumilov, Y.A.; Demidyuk, B.M.; Shtorgin, A.V. Vibrodiagnostics as a component of monitoring the technical condition of power units. *Pract. Inst. Electrodyn. Natl. Acad. Sci. Ukr.* **2008**, *19*, 76–80.
2. Undrill, J.M.; Casazza, J.A.; Gulachenski, E.M.; Kirchmayer, L.K. Electromechanical equivalents for use in power system stability studies. *IEEE Trans. PAS* **1971**, *90*, 2060–2071. [CrossRef]
3. Jichao, H.; Ping, Z.; Yutian, S.; Yanling, L.; Dajun, T.; Baojun, G.; Weili, L. Numerical analysis of end part temperature in the turbogenerator end region with magnetic shield structure under the different operation conditions. *Int. J. Therm. Sci.* **2018**, *132*, 267–274.
4. Jichao, H.; Baojun, G. Influence of different underexcited operations on thermal field in the turbogenerator end region with magnetic shield structure. *Int. J. Therm. Sci.* **2019**, *138*, 534–544. [CrossRef]
5. Finley, W.R.; Hodowanec, M.M.; Holter, W.G. An Analytical Approach to Solving Motor Vibration Problems. In Proceedings of the Industry Applications Society 46th Annual Petroleum and Chemical Technical Conference (Cat No. 99CH37000), San Diego, CA, USA, 13–15 September 1999; Paper No. PCIC-99-20; p. 16.
6. Shevchenko, V.V. Forecasting the operational state of turbogenerators. *Elektrika* **2015**, *1*, 3–7.
7. Detlev, W. Gross Partial Discharge "Measurement and Monitoring on Rotating Machines". In Proceedings of the 2002 IEEE International Symposium on Electrical Insulation (Cat. No.02CH37316), Boston, MA, USA, 7–10 April 2002.
8. Akin, J. *Finite Element Analysis Concepts via SolidWorks*; World Scientific: New Jersey, NJ, USA; London, UK; Singapore; Beijing, China; Shanghai, China; Hong Kong, China; Taipei, Taiwan; Chennai, India, 2009; p. 303.
9. Gatewood, B.E. *Thermal Stresses in Relation to Aircraft, Projectiles, Turbines and Nuclear Reactors/Translated from English*; Grigorovsky, N.I., Ed.; Publishing House of Foreign Literature: Moscow, Rusia, 1959; p. 350.
10. Kvitka, A.L.; Voroshko, P.P.; Bobrovitskaya, S.D. *Stress-Strain State of Bodies of Revolution*; Naukova Dumka: Kyiv, Ukraine, 1977; p. 209.
11. Leibenzon, L.S. *Course of the Theory of Elasticity: Textbook*, 2nd ed.; Leningrad: Gostekhizdat: Moscow/Saint Petersburg, Russia, 1947; p. 465.
12. Diekmann, R.; Preis, R.; Schlimbach, F.; Walshaw, C. Shape-optimized mash partitioning and load balancing for parallel adaptive FEM. *Parallel Comput.* **2000**, *26*, 1555–1581. [CrossRef]

3. Upadhyay, B.; Sonigra, S.; Daxini, S. Numerical analysis perspective in structural shape optimization: A review post 2000. *Adv. Eng. Softw.* **2021**, *155*, 102992. [CrossRef]
4. Karban, P.; Panek, D.; Orosz, T.; Petrasova, I.; Dolezel, I. FEM based robust design optimization with Agros and Artap. *Comput. Math. Appl.* **2021**, *81*, 618–633. [CrossRef]
5. Li, H.; Yamada, T.; Jolivet, P.; Furuta, K.; Kondoh, T.; Izui, K.; Nishiwaki, S. Full-scale 3D structural topology optimization using adaptive mesh refinement based on the level-set method. *Finite Elem. Anal. Des.* **2021**, *194*, 103561. [CrossRef]
6. Lopez, J.; Anitescu, C.; Rabczuk, T. CAD-compatible structural shape optimization with a movable Bézier tetrahedral mesh. *Comput. Methods Appl. Mech. Eng.* **2020**, *367*, 113066. [CrossRef]
7. Carson, H.; Huang, A.; Galbraith, M.; Allmaras, S.; Darmofal, D. Anisotropic mesh adaptation for continuous finite element discretization through mesh optimization via error sampling and synthesis. *J. Comput. Phys.* **2020**, *420*, 109620. [CrossRef]
8. Hasani, M.; Rahaghi, M. The optimization of an electromagnetic vibration energy harvester based on developed electromagnetic damping models. *Energy Convers. Manag.* **2022**, *254*, 115271. [CrossRef]
9. Vinuela, Z.; Perez-Castellanos, J. The anisotropic criterion of von Mises (1928) as a yield condition for PMMCs. A calibration procedure based on numerical cell-analysis. *Compos. Struct.* **2015**, *134*, 613–632. [CrossRef]
20. Waqar, S.; Guo, K.; Sun, J. FEM analysis of thermal and residual stress profile in selective laser melting of 316L stainless steel. *J. Manuf. Process.* **2021**, *66*, 81–100. [CrossRef]
21. Dafeng, W.; Xizhong, A.; Peng, H.; Qian, J.; Hatao, F.; Hao, Z.; Xiaohong, Y.; Qingchuan, Z. Multi-particle FEM modelling on hot pressing of TiC-316L composite powders. *Powder Technol.* **2020**, *361*, 389–399.
22. Sheikh, U.A.; Jameel, A. Elasto-plastic large deformation analysis of bi-material components by FEM. *Mater. Today Proc.* **2020**, *26*, 1795–1802. [CrossRef]

MDPI
St. Alban-Anlage 66
4052 Basel
Switzerland
www.mdpi.com

Computation Editorial Office
E-mail: computation@mdpi.com
www.mdpi.com/journal/computation